Logic Colloquium '90

Since their inception, the Perspectives in Logic and Lecture Notes in Logic series have published seminal works by leading logicians. Many of the original books in the series have been unavailable for years, but they are now in print once again.

This volume, the 2nd publication in the Lecture Notes in Logic series, is the proceedings of the Association for Symbolic Logic meeting held in Helsinki, Finland, in July 1990. It contains 18 papers by leading researchers. These papers cover all fields of mathematical logic from the philosophy of mathematics, through model theory, proof theory, recursion theory, and set theory, to the connections of logic to computer science. The articles published here are still widely cited and continue to provide ideas for ongoing research projects.

JUHA OIKKONEN works in the Department of Mathematics at the University of Helsinki.

JOUKO VÄÄNÄNEN works in the Department of Mathematics at the University of Helsinki.

LECTURE NOTES IN LOGIC

A Publication of The Association for Symbolic Logic

This series serves researchers, teachers, and students in the field of symbolic logic, broadly interpreted. The aim of the series is to bring publications to the logic community with the least possible delay and to provide rapid dissemination of the latest research. Scientific quality is the overriding criterion by which submissions are evaluated.

More information, including a list of the books in the series, can be found at http://www.aslonline.org/books-lnl.html

LECTURE NOTES IN LOGIC 2

Logic Colloquium '90

ASL Summer Meeting in Helsinki

Edited by

JUHA OIKKONEN

University of Helsinki

JOUKO VÄÄNÄNEN

University of Helsinki

CAMBRIDGE
UNIVERSITY PRESS

University Printing House, Cambridge CB2 8BS, United Kingdom

One Liberty Plaza, 20th Floor, New York, NY 10006, USA

477 Williamstown Road, Port Melbourne, VIC 3207, Australia

4843/24, 2nd Floor, Ansari Road, Daryaganj, Delhi – 110002, India

79 Anson Road, #06–04/06, Singapore 079906

Cambridge University Press is part of the University of Cambridge.

It furthers the University's mission by disseminating knowledge in the pursuit of education, learning, and research at the highest international levels of excellence.

www.cambridge.org
Information on this title: www.cambridge.org/9781107169029
10.1017/9781316718254

Association for Symbolic Logic
Richard A. Shore, Publisher
Department of Mathematics, Cornell University, Ithaca, NY 14853
http://www.aslonline.org

A catalogue record for this publication is available from the British Library.

ISBN 978-1-107-16902-9 Hardback

FOREWORD

The 1990 European Summer Meeting of the Association for Symbolic Logic was held in Finland from July 15 to July 22, 1990. The meeting was called *Logic Colloquium '90* and it took place in the Porthania building of the University of Helsinki as part of the program of the 350th anniversary of the university.

The meeting was attended by 140 registered participants and 31 accompanying persons, from 23 different countries. The organizing bodies were the Department of Mathematics of the University of Helsinki, The Philosophical Society of Finland, and The Finnish Mathematical Society. Financial support was received from the Ministry of Education of Finland, The Academy of Finland, Suomen Kulttuurirahasto Foundation, UNESCO, Rolf Nevanlinna Institute, and IUHPS.

The Organizing Committee of the meeting consisted of Aapo Halko, Heikki Heikkilä, Lauri Hella, Taneli Huuskonen, Tapani Hyttinen, Kerkko Luosto, Ilkka Niiniluoto (vice chairman), Juha Oikkonen, and Jouko Väänänen (chairman), all from the University of Helsinki.

The Program Committee consisted of Peter Aczel (Manchester), Max Dickmann (Paris), Heinz-Dieter Ebbinghaus (Freiburg), Jens Fenstad (Oslo), Jaakko Hintikka (chairman, Boston), Wilfrid Hodges (London), Alistair Lachlan (Vancouver), Azriel Levy (Jerusalem), Heikki Mannila (Helsinki), Ilkka Niiniluoto (Helsinki), Juha Oikkonen (Helsinki), and Jouko Väänänen (secretary, Helsinki).

The program of the meeting is listed on the following pages. Warren Goldfarb, Ronald Jensen, Phokion Kolaitis, Per Martin-Löf, Alan Mekler, and Hugh Woodin did not contribute a paper to the proceedings. As comparison between the contents of this book and the actual program reveals, some authors made an agreement with the editors to contribute a slightly different paper from the one read in the meeting. Also Joan Moschovakis and Alan Silver were approached by the editors and they submitted the paper they read in a contributed papers session of the meeting.

The editors are indebted to Heikki Heikkilä, Yiannis Moschovakis, Martti Nikunen, and Hannele Salminen for their help during the preparation of this volume. We owe special thanks to Herbert Enderton for the substantial work he has done in putting together the final manuscript.

Juha Oikkonen
Jouko Väänänen

Invited talks of Logic Colloquium '90

WILFRIED BUCHHOLZ (München)
Cut-elimination in uncountable logic and collapsing functions

BARRY COOPER (Leeds)
Definability and global degree theory

PATRICK DEHORNOY (Caen)
About the word problem for free left distributive groupoids

HANS-DIETER DONDER (München)
On ω_1-complete filters

DOV GABBAY (London)
Labelled deductive systems

WARREN GOLDFARB (Harvard)
On Gödel's philosophy

JAAKKO HINTIKKA (Boston)
Is there completeness in mathematics after Gödel?

IAN HODKINSON (London)
An axiomatisation of the temporal logic with until and since over real numbers

RONALD JENSEN (Oxford)
Remarks on the core model

HAIM JUDAH (Bar-Ilan)
Δ^1_3-sets of reals

PHOKION KOLAITIS (Santa Cruz)
1. *Logical definability and complexity classes*
2. *Model theory of finite structures*
3. *0-1 laws*

RICHARD LAVER (Boulder)
Elementary embeddings of a rank into itself

PER MARTIN-LÖF (Stockholm)
Logic and metaphysics

ALAN MEKLER (Vancouver)
Almost free algebras: 20 years of progress

GRIGORI MINTS (Stanford)
Gentzen-type systems and resolution rule for modal predicate logic

YIANNIS MOSCHOVAKIS (Los Angeles)
Sense and denotation as algorithm and value

TULENDE MUSTAFIN (Karaganda)
On similarities of complete theories

LUDOMIR NEWELSKI (Wrocław)
Geometry of finite rank types

FRANÇOISE POINT (Paris)
Decidability problems for theories of modules

JEAN-PIERRE RESSAYRE (Paris)
Discrete subrings of real closed fields and applications to polynomial time computability

SAHARON SHELAH (Jerusalem)
Indiscernibility

HUGH WOODIN (Berkeley)
Large cardinals and descriptive set theory

CONTENTS

A NOTE ON THE ORDINAL ANALYSIS OF KPM

W. BUCHHOLZ [1]

This note extends our method from (Buchholz [2]) in such a way that it applies also to the rather strong theory KPM. This theory was introduced and analyzed proof-theoretically in (Rathjen [6]), where Rathjen establishes an upper bound for its proof theoretic ordinal |KPM|. The bound was given in terms of a primitive recursive system $\mathcal{T}(\mathrm{M})$ of ordinal notations based on certain ordinal functions χ , ψ_κ ($\omega < \kappa < \mathrm{M}$, κ regular) [2] that had been introduced and studied in (Rathjen [5]). [3] In section 1 of this note we define and study a slightly different system of functions ψ_κ ($\kappa \leq \mathrm{M}$)—where ψ_M plays the rôle of Rathjen's χ—that is particularly well suited for our purpose of extending [2]. In section 2 we describe how one obtains, by a suitable modification of [2], an upper bound for |KPM| in terms of the ψ_κ's from section 1. We conjecture that this bound is best possible and coincides with the bound given in [6]. In section 3 we prove some additional properties of the functions ψ_κ which are needed to set up a primitive recursive ordinal notation system of ordertype $> \vartheta^\star$, where $\vartheta^\star := \psi_{\Omega_1}\varepsilon_{\mathrm{M}+1}$ is the upper bound for |KPM| determined in section 2.

Remark: Another ordinal analysis of KPM has been obtained independently by T. Arai in *Proof theory for reflecting ordinals II: recursively Mahlo ordinals* (handwritten notes, 1989).

§1. Basic properties of the functions ψ_κ ($\kappa \leq \mathrm{M}$). *Preliminaries.* The letters $\alpha, \beta, \gamma, \delta, \mu, \sigma, \xi, \eta, \zeta$ always denote ordinals. On denotes the class of all ordinals, and Lim the class of all limit numbers. Every ordinal α is identified with the set $\{\xi \in \mathrm{On} : \xi < \alpha\}$ of its predecessors. For $\alpha \leq \beta$ we set $[\alpha, \beta[:= \{\xi : \alpha \leq \xi < \beta\}$. By $+$ we denote ordinary (noncommutative) ordinal *addition.* An ordinal $\alpha > 0$ which is closed under $+$ is called an *additive principal number.* The class of all additive principal numbers is denoted by AP. The *Veblen function* φ is defined by $\varphi\alpha\beta := \varphi_\alpha(\beta)$, where φ_α is the ordering function of the class $\{\beta \in \mathrm{AP} : \forall\xi < \alpha(\varphi_\xi(\beta) = \beta)\}$. An ordinal $\gamma > 0$ which is closed under φ (and thus also under $+$) is said to be *strongly critical.* The class of all strongly critical ordinals is denoted by SC.

---

[1]The final version of this paper was written while the author was visiting Carnegie Mellon University during the academic year 1990/91. I would like to express my sincere thanks to Wilfried Sieg (who invited me) and all members of the Philosophy Department of CMU for their generous hospitality.

[2]M denotes the first weakly Mahlo cardinal.

[3]The essential new feature of [5] is the function χ, while the ψ_κ's ($\kappa < \mathrm{M}$) are obtained by a straightforward generalization of previous constructions in [1], [3], [4].

Some basic facts:

1. $AP = \{\omega^\alpha : \alpha \in On\}$
2. $\varphi 0\beta = \omega^\beta$, $\varphi 1\beta = \varepsilon_\beta$
3. For each $\gamma > 0$ there are uniquely determined $n \in \mathbb{N}$ and additive principal numbers $\gamma_0 \geq \cdots \geq \gamma_n$ such that $\gamma = \gamma_0 + \cdots + \gamma_n$.
4. For each $\gamma \in AP \setminus SC$ there are uniquely determined $\xi, \eta < \gamma$ such that $\gamma = \varphi\xi\eta$.
5. Every uncountable cardinal is strongly critical.

Definition of $SC(\gamma)$:

1. $SC(0) := \emptyset$
2. $SC(\gamma) := \{\gamma\}$, if $\gamma \in SC$
3. $SC(\gamma_0 + \cdots + \gamma_n) := SC(\gamma_0) \cup \cdots \cup SC(\gamma_n)$, if $n \geq 1$ and $\gamma_0 \geq \cdots \geq \gamma_n$ are additive principal numbers.
4. $SC(\varphi\xi\eta) := SC(\xi) \cup SC(\eta)$, if $\xi, \eta < \varphi\xi\eta$.

We assume the existence of a weakly Mahlo cardinal M. So every closed unbounded (club) set $X \subseteq M$ contains at least one regular cardinal, and M itself is a regular cardinal.

DEFINITION 1.1.

$R := \{\alpha : \omega < \alpha \leq M \ \& \ \alpha \text{ regular}\}$
$M^\Gamma := \min\{\gamma \in SC : M < \gamma\} =$ closure of $M \cup \{M\}$ under $+, \varphi$
$SC_M(\gamma) := SC(\gamma) \cap M$
$\Omega_0 := 0$, $\Omega_\sigma := \aleph_\sigma$ for $\sigma > 0$.
$\Omega :=$ the function $\sigma \mapsto \Omega_\sigma$ restricted to $\sigma < M$

Remark: $\forall\kappa \in R(\ \kappa = \Omega_\kappa$ or $\kappa \in \{\Omega_{\sigma+1} : \sigma < M\}\ )$

Convention. *In the following the letters* κ, π, τ *always denote elements of* R.

DEFINITION 1.2 (The collapsing functions ψ_κ).
By transfinite recursion on α we define ordinals $\psi_\kappa\alpha$ and sets $C(\alpha, \beta) \subseteq On$ as follows. Under the induction hypothesis that $\psi_\pi\xi$ and $C(\xi, \eta)$ are already defined for all $\xi < \alpha$, $\pi \in R$, $\eta \in On$ we set

1. $C(\alpha, \beta) :=$ closure of $\beta \cup \{0, M\}$ under $+, \varphi, \Omega, \psi|\alpha$,
   where $\psi|\alpha$ denotes the binary function given by
   $$\mathrm{dom}(\psi|\alpha) := \{(\pi, \xi) : \xi < \alpha \ \& \ \pi \in R \ \& \ \pi, \xi \in C(\xi, \psi_\pi\xi)\}$$
   $$(\psi|\alpha)(\pi, \xi) := \psi_\pi\xi.$$
2. $\psi_\kappa\alpha := \min\{\beta \in \mathcal{D}_\kappa(\alpha) : C(\alpha, \beta) \cap \kappa \subseteq \beta\}$
   $$\text{with } \ \mathcal{D}_\kappa(\alpha) := \begin{cases} \{\beta \in R : \alpha \in C(\alpha, M) \Rightarrow \alpha \in C(\alpha, \beta)\} & \text{if } \kappa = M \\ \{\beta : \kappa \in C(\alpha, \kappa) \Rightarrow \kappa \in C(\alpha, \beta)\} & \text{if } \kappa < M \end{cases}$$

Abbreviation: $C_\kappa(\alpha) := C(\alpha, \psi_\kappa\alpha)$

The first two lemmata are immediate consequences of Definition 1.2.

LEMMA 1.1.

a) $\alpha_0 \leq \alpha \ \& \ \beta_0 \leq \beta \Longrightarrow C(\alpha_0, \beta_0) \subseteq C(\alpha, \beta)$
b) $\emptyset \neq X \subseteq On \ \& \ \beta = \sup(X) \Longrightarrow C(\alpha, \beta) = \bigcup_{\eta \in X} C(\alpha, \eta)$
c) $\beta < \kappa \Longrightarrow \mathsf{card}(C(\alpha, \beta)) < \kappa$

LEMMA 1.2.
$C(\alpha,\beta) = \bigcup_{n<\omega} C^n(\alpha,\beta)$, where $C^n(\alpha,\beta)$ is defined by
(i) $C^0(\alpha,\beta) := \beta \cup \{0, \mathrm{M}\}$,
(ii) $C^{n+1}(\alpha,\beta) := \{\gamma : SC(\gamma) \subseteq C^n(\alpha,\beta)\} \cup \{\Omega_\sigma : \sigma \in C^n(\alpha,\beta)\} \cup$
$\cup \{\psi_\pi\xi : \xi < \alpha \ \& \ \pi, \xi \in C^n(\alpha,\beta) \cap C_\pi(\xi)\}$

LEMMA 1.3.
a) $C_\kappa(\alpha) \cap \kappa = \psi_\kappa\alpha < \kappa$
b) $\kappa < \mathrm{M} \Longrightarrow \psi_\kappa\alpha \notin \mathrm{R}$
c) $\psi_\kappa\alpha \in \mathrm{SC} \setminus \{\Omega_\sigma : \sigma < \Omega_\sigma\}$
d) $\kappa \in C(\alpha,\kappa) \Longleftrightarrow \kappa \in C_\kappa(\alpha)$
e) $C(\alpha,\mathrm{M}) = \mathrm{M}^\Gamma = \{\xi : \xi \in C_\mathrm{M}(\xi)\}$
f) $\gamma \in C_\kappa(\alpha) \implies \gamma \in C_\mathrm{M}(\gamma) \ \& \ SC_\mathrm{M}(\gamma) = SC(\gamma) \setminus \{\mathrm{M}\}$
g) $\gamma < \alpha \ \& \ \gamma \in C(\alpha,\beta) \implies \psi_\mathrm{M}\gamma \in C(\alpha,\beta)$

Proof.
a),b) 1. $C_\kappa(\alpha) \cap \kappa = \psi_\kappa\alpha$ is a trivial consequence of the definition of $\psi_\kappa\alpha$. 2. Let $\kappa = \mathrm{M}$. Obviously there exists a $\delta < \kappa$ such that $\mathrm{R} \cap [\delta, \kappa[\subseteq \mathcal{D}_\kappa(\alpha)$. Therefore in order to get $\psi_\kappa\alpha < \kappa$ it suffices to prove that the set $U := \{\beta \in \kappa : C(\alpha,\beta) \cap \kappa \subseteq \beta\}$ is closed unbounded (club) in κ.
i) *closed*: Let $\emptyset \neq X \subseteq U$ and $\beta := \sup(X) < \kappa$. Then $C(\alpha,\beta) \cap \kappa = \bigcup_{\xi \in X}(C(\alpha,\xi) \cap \kappa) \subseteq \bigcup_{\xi \in X} \xi = \beta$, i.e. $\beta \in U$.
ii) *unbounded*: Let $\beta_0 < \kappa$. We define $\beta_{n+1} := \min\{\eta : C(\alpha,\beta_n) \cap \kappa \subseteq \eta\}$ and $\beta := \sup_{n<\omega} \beta_n$. Using L.1.1c we obtain $\beta_n \leq \beta_{n+1} < \kappa$. Hence $\beta_0 \leq \beta < \kappa$ and $C(\alpha,\beta) \cap \kappa = \bigcup_{n<\omega}(C(\alpha,\beta_n) \cap \kappa) \subseteq \bigcup_{n<\omega} \beta_{n+1} = \beta$, i.e. $\beta_0 \leq \beta \in U$. 3. Let $\kappa < \mathrm{M}$. Starting with $\beta_0 := \min(\mathcal{D}_\kappa(\alpha))$ we define the ordinals β_n and β as in 2.(ii). Then we have $\beta \in \mathcal{D}_\kappa(\alpha) \cap U$ and therefore $\psi_\kappa\alpha \leq \beta < \kappa$. — Now assume that $\psi_\kappa\alpha \in \mathrm{R}$. We prove $\beta_n < \psi_\kappa\alpha$ $(\forall n)$. By definition of β_0 and by L.1.1a we have $\beta_0 \leq \psi_\kappa\alpha$ & $\beta_0 \notin \mathrm{Lim}$. Hence $\beta_0 < \psi_\kappa\alpha$. From $\beta_n < \psi_\kappa\alpha \in \mathrm{R}$ it follows that $C(\alpha,\beta_n) \cap \kappa \subseteq \psi_\kappa\alpha$ and $\mathrm{card}(C(\alpha,\beta_n) \cap \kappa) < \psi_\kappa\alpha$, and therefore $\beta_{n+1} < \psi_\kappa\alpha$. From $\forall n(\beta_n < \psi_\kappa\alpha \in \mathrm{R})$ we get $\beta < \psi_\kappa\alpha$. *Contradiction.*
c) 1. Obviously $C_\kappa(\alpha) \cap \kappa$ is closed under φ. Together with a) this implies $\psi_\kappa\alpha \in \mathrm{SC}$. — 2. We have $(\psi_\kappa\alpha = \Omega_\sigma > \sigma \Rightarrow \psi_\kappa\alpha \in C_\kappa(\alpha))$ and (by a)) $\psi_\kappa\alpha \notin C_\kappa(\alpha)$. Hence $\psi_\kappa\alpha \notin \{\Omega_\sigma : \sigma < \Omega_\sigma\}$.
d) follows from L.1.1a, L.1.3a and the definition of $\psi_\kappa\alpha$.
e) By L.1.3a $\forall\pi\in\mathrm{R}(\psi_\pi\xi < \mathrm{M})$ and therefore $C(\alpha,\mathrm{M}) = \mathrm{M}^\Gamma$. As in d) one obtains $(\alpha \in C(\alpha,\mathrm{M}) \Leftrightarrow \alpha \in C_\mathrm{M}(\alpha))$.
f) and g) follow from e).

LEMMA 1.4.
a) $\gamma \in C(\alpha,\beta) \Longleftrightarrow SC(\gamma) \subseteq C(\alpha,\beta)$
b) $\Omega_\sigma \in C(\alpha,\beta) \Longleftrightarrow \sigma \in C(\alpha,\beta)$
c) $\kappa = \Omega_{\sigma+1} \implies \Omega_\sigma < \psi_\kappa\alpha < \Omega_{\sigma+1}$
d) $\Omega_\kappa = \kappa \implies \Omega_{\psi_\kappa\alpha} = \psi_\kappa\alpha$
e) $\Omega_{\psi_\mathrm{M}\alpha} = \psi_\mathrm{M}\alpha$
f) $\Omega_\sigma \leq \gamma \leq \Omega_{\sigma+1} \ \& \ \gamma \in C(\alpha,\beta) \implies \sigma \in C(\alpha,\beta)$

Proof. a) and b) follow from L.1.2 and L.1.3c. — e) follows from d), since

$M \in R$ and $\Omega_M = M$. — f) follows from a),b),c),d) and L.1.2.
c) Let $\kappa = \Omega_{\sigma+1}$. Then $\kappa \in C(\alpha, \kappa)$ and thus $\kappa \in C_\kappa(\alpha)$. By a) and b) from $\kappa = \Omega_{\sigma+1} \in C_\kappa(\alpha)$ we get $\Omega_\sigma \in C_\kappa(\alpha) \cap \kappa = \psi_\kappa\alpha$.
d) Take $\sigma \in On$ such that $\Omega_\sigma \le \psi_\kappa\alpha < \Omega_{\sigma+1}$. Then we have $\sigma + 1 < \kappa$ and thus $C_\kappa(\alpha) \cap \kappa = \psi_\kappa\alpha < \Omega_{\sigma+1} < \Omega_\kappa = \kappa$. This implies $\Omega_{\sigma+1} \notin C_\kappa(\alpha)$ and then (by a),b)) $\sigma \notin C_\kappa(\alpha)$. Hence $\psi_\kappa\alpha \le \sigma \le \Omega_\sigma \le \psi_\kappa\alpha$.

LEMMA 1.5.
a) $\alpha_0 < \alpha \;\&\; \alpha_0 \in C_M(\alpha) \implies \psi_M\alpha_0 < \psi_M\alpha$
b) $\psi_M\alpha_0 = \psi_M\alpha_1 \;\&\; \alpha_0, \alpha_1 < M^\Gamma \implies \alpha_0 = \alpha_1$

Proof.
a) From the premise we get $\psi_M\alpha_0 \in C_M(\alpha) \cap M = \psi_M\alpha$ by L.1.3a,g.
b) Assume $\psi_M\alpha_0 = \psi_M\alpha_1 \;\&\; \alpha_0 < \alpha_1 < M^\Gamma$. Then $\alpha_0 \in C_M(\alpha_0) \subseteq C_M(\alpha_1)$ and therefore by a) $\psi_M\alpha_0 < \psi_M\alpha_1$. *Contradiction.*

LEMMA 1.6.
For $\kappa < M$ the following holds
a) $\alpha_0 < \alpha \implies \psi_\kappa\alpha_0 \le \psi_\kappa\alpha$
b) $\alpha_0 < \alpha \;\&\; \kappa, \alpha_0 \in C_\kappa(\alpha_0) \implies \psi_\kappa\alpha_0 < \psi_\kappa\alpha$

Proof.
a) From $\alpha_0 < \alpha$ it follows that $C(\alpha_0, \psi_\kappa\alpha) \cap \kappa \subseteq \psi_\kappa\alpha$. By definition of $\psi_\kappa\alpha_0$ it therefore suffices to prove $\psi_\kappa\alpha \in \{\beta : \kappa \in C(\alpha_0, \kappa) \Rightarrow \kappa \in C(\alpha_0, \beta)\}$. So let $\kappa \in C(\alpha_0, \kappa)$. — We have to prove $\kappa \in C(\alpha_0, \psi_\kappa\alpha)$.
CASE 1: $\kappa = \Omega_{\sigma+1}$. By Lemma 1.4c we have $\Omega_\sigma < \psi_\kappa\alpha$ and therefore $\sigma + 1 \in C(\alpha_0, \psi_\kappa\alpha)$ which implies $\kappa \in C(\alpha_0, \psi_\kappa\alpha)$.
CASE 2: $\kappa = \Omega_\kappa$. From $\kappa \in C(\alpha_0, \kappa) \subseteq C(\alpha, \kappa)$ we obtain $\kappa \in C_\kappa(\alpha_0) \cap C_\kappa(\alpha)$. From this by L.1.2, L.1.3b, L.1.5b it follows that $\kappa = \psi_M\xi$ with $\xi < \alpha_0$ and $\xi \in C_\kappa(\alpha)$. Now by L.1.4a, L.1.3a,e we get $SC_M(\xi) \subseteq C_\kappa(\alpha) \cap C_M(\xi) \cap M = C_\kappa(\alpha) \cap \kappa = \psi_\kappa\alpha$, and then $\xi \in C(\alpha_0, \psi_\kappa\alpha)$ (by L.1.3f). From this together with $\xi < \alpha_0$ we obtain $\kappa = \psi_M\xi \in C(\alpha_0, \psi_\kappa\alpha)$ (by L.1.3g).
b) The premise together with a) implies $\alpha_0 < \alpha \;\&\; \kappa, \alpha_0 \in C_\kappa(\alpha) \cap C_\kappa(\alpha_0)$ which gives us $\psi_\kappa\alpha_0 \in C_\kappa(\alpha) \cap \kappa = \psi_\kappa\alpha$.

DEFINITION 1.3.
For each set $X \subseteq On$ we set $\mathcal{H}_\gamma(X) := \bigcap\{C(\alpha, \beta) : X \subseteq C(\alpha, \beta) \;\&\; \gamma < \alpha\}$.

§2. Ordinal analysis of KPM.

In this section we show how one has to modify (and extend) [2] in order to establish that the ordinal $\psi_{\Omega_1}\varepsilon_{M+1}$ is an upper bound for |KPM|. Of course we now assume that the reader is familiar with [2].

The theory KPM is obtained from KPi by adding the following axiom scheme:

(Mahlo) $\forall x \exists y \phi(x, y, \vec{z}) \rightarrow \exists w[Ad(w) \wedge \forall x {\in} w \exists y {\in} w \phi(x, y, \vec{z})]$ $\qquad (\phi \in \Delta_0)$

We extend the infinitary system RS^∞ introduced in Section 3 of [2] by adding the following inference rule:

$$\text{(Mah)} \quad \frac{\Gamma, B(\mathsf{L}_{\mathrm{M}}) : \alpha_0}{\Gamma, \exists w{\in}\mathsf{L}_{\mathrm{M}}(Ad(w) \wedge B(w)) : \alpha} \quad (\alpha_0 + \mathrm{M} < \alpha)$$

where $B(w)$ is of the form $\forall x{\in}w\exists y{\in}wA(x,y)$ with $\mathrm{k}(A) \subseteq \mathrm{M}$.

We set $\mathrm{R} := \{\alpha : \omega < \alpha \le \mathrm{M} \ \& \ \alpha \text{ regular}\}$.

Then all lemmata and theorems of Section **3** [4] are also true for the extended system RS^∞(with almost literally the same proofs)[5], and as an easy consequence from Theorem **3.12** one obtains the

EMBEDDING THEOREM for KPM.
If $\mathrm{M} \in \mathcal{H}$ and if $\mathcal{H}$ is closed under $\xi \mapsto \xi^{\mathrm{R}}$ then for each theorem ϕ of KPM *there is an $n \in \mathbb{N}$ such that $\mathcal{H}\vert\frac{\omega^{\mathrm{M}+n}}{\mathrm{M}+n}\ \phi^{\mathrm{M}}$.*

Some more severe modifications have to be carried out on Section **4**. The first part of this section (down to Lemma **4.5**) has to be replaced by Section 1 of the present paper. Then the sets $C(\alpha, \beta)$ are no longer closed under $(\pi, \xi) \mapsto \psi_\pi\xi$ $(\xi < \alpha)$, but only under $(\psi|\alpha)$ as defined in Definition 1.2 above. Therefore we have to add "$\pi, \xi \in C_\pi(\xi)$" to the premise of Lemma **4.6c**, and accordingly a minor modification as to be made in the proof of Lemma **4.7**($\mathcal{A}1$). But this causes no problems. A little bit problematic is the fact that the function ψ_{M} is not weakly increasing. In order to overcome this difficulty we prove the following lemma.

DEFINITION 2.1.
For $\gamma = \omega^{\gamma_0} + \cdots + \omega^{\gamma_n}$ with $\gamma_0 \ge \cdots \ge \gamma_n$ we set $\mathsf{e}(\gamma) := \omega^{\gamma_n + 1}$.
Further we set $\mathsf{e}(0) := \mathrm{On}$.

LEMMA 2.1.
For $\gamma \in C_{\mathrm{M}}(\gamma+1)$ and $0 < \alpha < \mathsf{e}(\gamma)$ the following holds
a) $\psi_{\mathrm{M}}(\gamma+1) \le \psi_{\mathrm{M}}(\gamma+\alpha)$ & $C_{\mathrm{M}}(\gamma+1) \subseteq C_{\mathrm{M}}(\gamma+\alpha)$
b) $0 < \alpha_0 < \alpha$ & $\alpha_0 \in C_{\mathrm{M}}(\gamma+1) \implies \psi_{\mathrm{M}}(\gamma+\alpha_0) < \psi_{\mathrm{M}}(\gamma+\alpha)$

Proof:
a) follows from b).
b) We will prove $(*)$ $\psi_{\mathrm{M}}(\gamma+1) \le \psi_{\mathrm{M}}(\gamma+\alpha)$. From this we get $\gamma+\alpha_0 \in C_{\mathrm{M}}(\gamma+1) \subseteq C_{\mathrm{M}}(\gamma+\alpha)$ and then by L.1.5a the assertion.
For $\gamma = 0$ $(*)$ is trivial. If $\gamma \ne 0$ then $\gamma + \alpha < \mathrm{M}^\Gamma$ and therefore $\gamma+\alpha \in C_{\mathrm{M}}(\gamma+\alpha)$ which (together with $\alpha < \mathsf{e}(\gamma)$) implies $\gamma + 1 \in C_{\mathrm{M}}(\gamma+\alpha)$. Hence $\psi_{\mathrm{M}}(\gamma+1) \le \psi_{\mathrm{M}}(\gamma+\alpha)$ by L.1.5a.

Now we give a complete list of all modifications which have to be carried out in [2] subsequent to Lemma **4.6** .

[4]We use boldface numerals to indicate reference to [2]

[5]In Theorem **3.8** one has to add the clause which corresponds to the new inference rule (Mah). The last line in the proof of Lemma **3.14** has to be modified to "*... cannot be the main part of a* (Ref)- *or* (Mah)-*inference.*". At the end of the proof of Lemma **3.17** one may add the remark "*Due to the premise $\alpha \le \beta < \kappa$ we have $\alpha <$* M*, and therefore the given derivation of Γ, C does not contain any applications of* (Mah)."

(1) Replace I by M in the definition of $\bar{K}$.
(2) Add "$\eta < \gamma + e(\gamma)$" to the premise of Lemma **4.7**($\mathcal{A}2$).
(3) Add "$\omega^{\mu+\alpha} < e(\gamma)$" to the premise of Theorem **4.8**.
(4) Add "$\pi \le e(\gamma')$" to the premise of ($\Box$) in the proof of Theorem **4.8**.
(5) Insert the following proof of "$\psi_\kappa\alpha^* \le \psi_\kappa\widehat{\alpha}$" at the end of the proof of ($\Box$):
"*From* $\gamma', \mu', \alpha' \in \mathcal{H}_{\gamma'}[\Theta]$ *we get* $\alpha^* \in \mathcal{H}_{\gamma'}[\Theta]$. *From* $k(\Theta) \subseteq C_\kappa(\gamma+1) \subseteq C_\kappa(\widehat{\alpha})$ & $\gamma' < \widehat{\alpha}$ *it follows that* $\mathcal{H}_{\gamma'}[\Theta] \subseteq C_\kappa(\widehat{\alpha})$. *Hence* $\alpha^* \in C_\kappa(\widehat{\alpha})$ *and thus* $\psi_\kappa\alpha^* \le \psi_\kappa\widehat{\alpha}$, *since* $\alpha^* < \widehat{\alpha}$."
(6) Extend the proof of Theorem **4.8** by the following treatment of the case where the last inference in the given derivation of Γ is an application of (Mah):
"*5. Suppose that* $\exists w{\in}L_M(Ad(w) \wedge B(w)) \in \Gamma$ *and* $\mathcal{H}_\gamma[\Theta] \mid\!\frac{\alpha_0}{\mu}\, \Gamma, B(L_M)$ *with* $B(w) \equiv \forall x{\in}w\exists y{\in}wA(x,y)$ & $\alpha_0 + M < \alpha$ & $k(A) \subseteq M$.
Then $\kappa = M$ *(since* $\Gamma \subseteq \Sigma(\kappa)$ *and* $\kappa \le M$*).*
For $\iota \in \mathcal{T}_M$ *we set* $\gamma_\iota := \gamma + \omega^{\mu+\alpha_0+|\iota|}$. *Then* $C_M(\gamma+1) \subseteq C_M(\gamma_\iota)$, *and since* $SC(|\iota|) \subseteq SC_M(\gamma_\iota) \subseteq \psi_M\gamma_\iota$, *we have* $|\iota| < \psi_M\gamma_\iota$ *and thus* $k(\Theta,\iota) \subseteq C_M(\gamma_\iota)$.
From $\gamma, \mu, \alpha_0 \in \mathcal{H}_\gamma[\Theta]$ *we get* $\gamma_\iota \in \mathcal{H}_\gamma[\Theta,\iota]$. *Consequently* $\mathcal{A}(\Theta,\iota;\gamma_\iota,M,\mu)$, *and the Inversion-Lemma gives us* $\mathcal{H}_\gamma[\Theta][\iota] \mid\!\frac{\alpha_0}{\mu}\, \Gamma, \iota \notin L_0 \to \exists y \in L_M A(\iota,y)$.
Now we apply the I.H. and obtain $\mathcal{H}_{\alpha_\iota^*}[\Theta][\iota] \mid\!\frac{\psi_M\alpha_\iota^*}{\cdot}\, \Gamma, \iota \notin L_0 \to \exists y \in L_M A(\iota,y)$ *with* $\alpha_\iota^* := \gamma_\iota + \omega^{\mu+\alpha_0} < \gamma + \omega^{\mu+\alpha_0+M} =: \alpha^* < \widehat{\alpha}$.
Let $\pi := \psi_M\alpha^*$ & $\beta_\iota := \psi_M\alpha_\iota^*$. *Then by L.4.7* $\pi \in \mathcal{H}_{\widehat{\alpha}}[\Theta]$ & $\pi < \psi_M\widehat{\alpha}$.
We also have $\forall\iota{\in}\mathcal{T}_\pi(\alpha_\iota^* \in C_M(\alpha^*))$ *and thus* $\forall\iota{\in}\mathcal{T}_\pi(\beta_\iota < \pi)$.
The Boundedness-Lemma gives us now
$\forall\iota \in \mathcal{T}_\pi(\, \mathcal{H}_{\widehat{\alpha}}[\Theta][\iota] \mid\!\frac{\beta_\iota}{\pi}\, \Gamma, \iota \notin L_0 \to \exists y{\in}L_\pi A(\iota,y)\,)$.
From this by an application of ($\bigwedge$) *we obtain* $\mathcal{H}_{\widehat{\alpha}}[\Theta] \mid\!\frac{\pi}{\cdot}\, \Gamma, B(L_\pi)$.
*From L.***2.5h** *and L.***3.10** *we get* $\mathcal{H}_{\widehat{\alpha}}[\Theta] \mid\!\frac{\delta}{0}\, \Gamma, Ad(L_\pi)$ *with* $\delta := \omega^{\pi+5}$. *We also have* $\mathcal{H}_{\widehat{\alpha}}[\Theta] \mid\!\frac{0}{\cdot}\, \Gamma, L_\pi \notin L_0$. *Hence* $\mathcal{H}_{\widehat{\alpha}}[\Theta] \mid\!\frac{\delta+2}{\cdot}\, \Gamma, L_\pi \notin L_0 \wedge Ad(L_\pi) \wedge B(L_\pi)$. *Now we apply* ($\bigvee$) *and obtain* $\mathcal{H}_{\widehat{\alpha}}[\Theta] \mid\!\frac{\psi_M\widehat{\alpha}}{\cdot}\, \Gamma$."
(7) Replace I by M in the Corollary to Theorem **4.8** and in Theorem **4.9**.

This yields the following Theorem.

THEOREM.
Let $\vartheta^* := \psi_{\Omega_1}(\varepsilon_{M+1})$. *Then for each* Σ_1*-sentence* ϕ *of* $\mathcal{L}$ *we have:*
$\mathrm{KPM} \vdash \forall x(Ad(x) \to \phi^x) \implies L_{\vartheta^*} \models \phi$.

COROLLARY. $|\mathrm{KPM}| \le \psi_{\Omega_1}(\varepsilon_{M+1})$.

§3. Further properties of the functions ψ_κ.

We prove four theorems which together with L.1.3a,b,c and L.1.4a-e provide a complete basis for the definition of a primitive recursive well-ordering (OT,$\prec$) which is isomorphic to $(C(M^\Gamma,0),<)$. (The set OT consists of terms built up from the constants $\underline{0}$, $\underline{M}$ by the function symbols $\underline{+}$, $\underline{\varphi}$, $\underline{\Omega}$, $\underline{\psi}$, such that for each $\gamma \in C(M^\Gamma,0)$ there is a unique term $t \in$ OT with $|t| = \gamma$, and for all $s,t \in$ OT one has $(s \prec t \Leftrightarrow |s| < |t|)$. Here $|t|$ denotes the canonical value of t. For details see [1], [4], [5].)

Now the letters $\alpha, \beta, \gamma, \delta, \mu, \sigma, \xi, \eta, \zeta$ always denote ordinals less than M^Γ. So, for all α we have $\alpha \in C_{\mathrm{M}}(\alpha)$ and $SC(\alpha) \setminus \{\mathrm{M}\} = SC_{\mathrm{M}}(\alpha) \subseteq \psi_{\mathrm{M}}\alpha$.

DEFINITION 3.1.

$$\mathsf{sc}_\kappa(\alpha) := \begin{cases} \max SC_{\mathrm{M}}(\alpha) & \text{if } \kappa = \mathrm{M} \ \& \ SC_{\mathrm{M}}(\alpha) \neq \emptyset \\ 0 & \text{otherwise} \end{cases}$$

LEMMA 3.1.

a) $\mathsf{sc}_\kappa(\alpha) < \psi_\kappa\alpha$

b) $\pi = \mathrm{M} \ \& \ \mathsf{sc}_\pi(\beta) < \psi_\kappa\alpha \implies \beta \in C_\kappa(\alpha)$

Proof. Trivial (cf. L.1.a,e,f and L.1.4a).

LEMMA 3.2.

Let $\kappa \in C_\kappa(\alpha) \ \& \ \pi \in C_\pi(\beta)$. *Then*

$\psi_\pi\beta < \kappa < \pi \ \& \ \mathsf{sc}_\pi(\beta) < \psi_\kappa\alpha \implies \psi_\pi\beta < \psi_\kappa\alpha.$

Proof. By L.1.4c,d it follows that $\Omega_\pi = \pi$ and $\Omega_{\psi_\pi\beta} = \psi_\pi\beta$. Therefore if $\kappa = \Omega_{\sigma+1}$ then $\psi_\pi\beta \leq \Omega_\sigma < \psi_\kappa\alpha$, and we may now assume that $\Omega_\kappa = \kappa$. Then by L.1.2 and L.1.3b we obtain $\kappa = \psi_{\mathrm{M}}\gamma$ with $\gamma < \alpha \ \& \ \gamma \in C_\kappa(\alpha) \cap C_{\mathrm{M}}(\gamma)$. By L.1.4a and L1.3a we get $SC_{\mathrm{M}}(\gamma) \subseteq C_\kappa(\alpha) \cap C_{\mathrm{M}}(\gamma) \cap \mathrm{M} = C_\kappa(\alpha) \cap \kappa = \psi_\kappa\alpha$. From $\psi_\pi\beta < \kappa = \psi_{\mathrm{M}}\gamma < \pi$ it follows that $\psi_{\mathrm{M}}\gamma \notin C_\pi(\beta)$ and thus $\beta \leq \gamma$ or $\psi_\pi\beta \leq \mathsf{sc}_{\mathrm{M}}(\gamma)$. — If $\psi_\pi\beta \leq \mathsf{sc}_{\mathrm{M}}(\gamma)$ then $\psi_\pi\beta < \psi_\kappa\alpha$, since $SC_{\mathrm{M}}(\gamma) \subseteq \psi_\kappa\alpha$. If $\mathsf{sc}_{\mathrm{M}}(\gamma) < \psi_\pi\beta \ \& \ \pi = \mathrm{M}$ then we have $\beta \leq \gamma < \alpha$ and $\beta \in C_\kappa(\alpha)$ (since $\mathsf{sc}_\pi(\beta) < \psi_\kappa\alpha$), from which we get $\psi_\pi\beta \in C_\kappa(\alpha) \cap \kappa = \psi_\kappa\alpha$. — For $\pi = \mathrm{M}$ the proof is now finished. — If $\mathsf{sc}_{\mathrm{M}}(\gamma) < \psi_\pi\beta \ \& \ \pi < \mathrm{M}$ then $\psi_{\mathrm{M}}\gamma < \pi < \mathrm{M} \ \& \ \mathsf{sc}_{\mathrm{M}}(\gamma) < \psi_\pi\beta$ which (according to what we already proved for $\pi = \mathrm{M}$) implies $\kappa = \psi_{\mathrm{M}}\gamma < \psi_\kappa\alpha$. *Contradiction.*

DEFINITION 3.2.

$\mathcal{K}(\pi, \beta, \kappa, \alpha)$ abbreviates the disjunction of $(\mathcal{K}1), \ldots, (\mathcal{K}4)$ below:

$(\mathcal{K}1)$ $\pi \leq \psi_\kappa\alpha$

$(\mathcal{K}2)$ $\psi_\pi\beta \leq \mathsf{sc}_\kappa(\alpha)$

$(\mathcal{K}3)$ $\pi = \kappa \ \& \ \beta < \alpha \ \& \ \mathsf{sc}_\pi(\beta) < \psi_\kappa\alpha$

$(\mathcal{K}4)$ $\psi_\pi\beta < \kappa < \pi \ \& \ \mathsf{sc}_\pi(\beta) < \psi_\kappa\alpha$

LEMMA 3.3.

Let $\kappa \in C_\kappa(\alpha) \ \& \ \pi \in C_\pi(\beta)$.

a) $\neg\mathcal{K}(\pi, \beta, \kappa, \alpha) \ \& \ \neg\mathcal{K}(\kappa, \alpha, \pi, \beta) \implies \kappa = \pi \ \& \ \alpha = \beta$

b) $\mathcal{K}(\pi, \beta, \kappa, \alpha) \implies \psi_\pi\beta \leq \psi_\kappa\alpha$

c) $\mathcal{K}(\pi, \beta, \kappa, \alpha) \ \& \ \beta \in C_\pi(\beta) \implies \psi_\pi\beta < \psi_\kappa\alpha$

Proof. a) is a logical consequence of the linearity of $<$. b) and c) follow immediately from L.1.3a, L.1.5a, L.1.6, L.3.1, L.3.2.

As an immediate consequence from lemma 3.3 we get

THEOREM 3.1.

$\kappa, \alpha \in C_\kappa(\alpha) \ \& \ \pi, \beta \in C_\pi(\beta) \ \& \ \psi_\kappa\alpha = \psi_\pi\beta \implies \kappa = \pi \ \& \ \alpha = \beta.$

THEOREM 3.2.

Let $\kappa \in C_\kappa(\alpha) \ \& \ \pi, \beta \in C_\pi(\beta)$.

a) $\psi_\pi\beta < \psi_\kappa\alpha \iff \mathcal{K}(\pi, \beta, \kappa, \alpha)$

b) $\psi_\pi\beta \in C_\kappa(\alpha) \iff (\psi_\pi\beta < \psi_\kappa\alpha \text{ or } [\beta < \alpha \ \& \ \pi, \beta \in C_\kappa(\alpha)])$

Proof. a) "$\Leftarrow$" follows from L.3.3b. "$\Rightarrow$" follows from L.3.3a,c.
b) The "$\Leftarrow$" part is trivial. So let us assume that $\psi_\kappa\alpha \le \psi_\pi\beta \in C_\kappa(\alpha)$. By L.1.2 and L.1.3c this implies the existence of $\tau, \xi \in C_\kappa(\alpha) \cap C_\tau(\xi)$ with $\xi < \alpha$ and $\psi_\pi\beta = \psi_\tau\xi$. From this by Theorem 3.1 we obtain $\pi = \tau \in C_\kappa(\alpha)$ and $\beta = \xi \in C_\kappa(\alpha) \cap \alpha$.

THEOREM 3.3.
$\kappa \in C_\kappa(\alpha) \iff \kappa \in \{\Omega_{\sigma+1} : \sigma < \mathrm{M}\} \cup \{\psi_\mathrm{M}\xi : \xi < \alpha\} \cup \{\mathrm{M}\}$

Proof. 1. "$\Rightarrow$" follows from L.1.2 and L.1.3b. — 2. By L.1.3d we have $(\kappa \in C_\kappa(\alpha) \Leftrightarrow \kappa \in C(\alpha, \kappa))$. — 3. If $\kappa = \Omega_{\sigma+1}$ then $\sigma + 1 < \kappa$ and thus $\kappa \in C(\alpha, \kappa)$. — 4. If $\kappa = \psi_\mathrm{M}\xi$ with $\xi < \alpha$ then $\xi \in C_\mathrm{M}(\xi) = C(\xi, \kappa) \subseteq C(\alpha, \kappa)$ and thus $\kappa \in C(\alpha, \kappa)$.

THEOREM 3.4.
$\kappa = \Omega_{\sigma+1} \implies C_\kappa(\alpha) = C(\alpha, \Omega_\sigma + 1)$

Proof by induction on α. So let us assume that $C_\kappa(\xi) = C(\xi, \Omega_\sigma + 1)$, for all $\xi < \alpha$. — We have to prove $\psi_\kappa\alpha \subseteq C(\alpha, \Omega_\sigma + 1)$. As we will show below the I.H. implies that $\beta := C(\alpha, \Omega_\sigma + 1) \cap \kappa$ is in fact an ordinal. Obviously $\kappa \in C(\alpha, \beta)$ and $C(\alpha, \beta) \cap \kappa \subseteq C(\alpha, \Omega_\sigma + 1) \cap \kappa = \beta$ and thus $\psi_\kappa\alpha \le \beta$, i.e. $\psi_\kappa\alpha \subseteq C(\alpha, \Omega_\sigma + 1)$.
—CLAIM: $\gamma \in C(\alpha, \Omega_\sigma + 1) \cap \kappa \implies \gamma \subseteq C(\alpha, \Omega_\sigma + 1)$.
Proof. 1. $\Omega_\sigma < \gamma \in \mathrm{SC}$. Then $\gamma = \psi_\pi\xi$ with $\xi < \alpha$ & $\xi \in C_\pi(\xi)$. Since $\Omega_\sigma < \gamma < \kappa = \Omega_{\sigma+1}$, we have $\pi = \kappa$ and therefore by the above I.H. $C_\kappa(\xi) = C(\xi, \Omega_\sigma + 1)$. Hence $\gamma = \psi_\kappa\xi \subseteq C(\xi, \Omega_\sigma + 1) \subseteq C(\alpha, \Omega_\sigma + 1)$.
2. Let γ be arbitrary and $\gamma_0 := \max(\{0\} \cup SC(\gamma))$. Then (by 1. above) $\gamma_0 \cup \{\gamma_0\} \subseteq C(\alpha, \Omega_\sigma + 1)$. From this we get $\gamma \subseteq \gamma^* \subseteq C(\alpha, \Omega_\sigma + 1)$, where $\gamma^* := \min\{\eta \in \mathrm{SC} : \gamma_0 < \eta\}$.

COROLLARY. $\psi_{\Omega_1}\alpha = C(\alpha, 0) \cap \Omega_1$

REFERENCES

[1] BUCHHOLZ, W. *A new system of proof-theoretic ordinal functions.* ***Annals of Pure and Applied Logic***, vol. 32 (1986), pp. 195–207.

[2] BUCHHOLZ, W. *A simplified version of local predicativity.* In: P. Aczel, H. Simmons, S. Wainer (eds.), ***Proof Theory***, Cambridge University Press, 1993.

[3] BUCHHOLZ, W., and SCHÜTTE, K. *Ein Ordinalzahlsystem für die beweistheoretische Abgrenzung der Π^1_2-Separation und Bar-Induktion.* ***Sitzungsberichte der Bayerischen Akademie der Wissenschaften***, Mathematisch-Naturwissenschaftliche Klasse (1983).

[4] JÄGER, G. *ρ-inaccessible ordinals, collapsing functions and a recursive notation system.* ***Archiv für mathematische Logik und Grundlagenforschung***, vol. 24 (1984), pp. 49–62.

[5] RATHJEN, M. *Ordinal notations based on a weakly Mahlo cardinal.* ***Archiv für mathematische Logik und Grundlagenforschung***, vol. 29 (1990), pp. 249–263.

[6] RATHJEN, M. *Proof-theoretic analysis of KPM.* ***Archiv für mathematische Logik und Grundlagenforschung***, vol. 30 (1991), pp. 377–403.

Mathematisches Institut der Universität München
Theresienstr. 39, D-80333 München, Germany

ON THE GEOMETRY OF U-RANK 2 TYPES

STEVEN BUECHLER and LUDOMIR NEWELSKI

Abstract. Let T be a countable superstable theory with $< 2^{\aleph_0}$ countable models. We solve the algebraic problem from [Ne4, §4]. In particular, in some cases we complete the countable classification of skeletal p of U-rank 2 (cf. [Bu4]).

§0. Introduction. Throughout the paper we assume that T is a complete countable superstable theory with $< 2^{\aleph_0}$ countable models. For the background from stability theory see [Sh], [Ba], [Bu1], or [P]. The results in [Bu2] suggest that if T has infinite U-rank then every countable model M of T is determined by a subset A of M, called its skeleton (cf. [Bu4]). Hence in the course of proving Vaught's conjecture we have to determine possible isomorphism types of skeletons. The easiest non-trivial case we faced in [Bu4] and [Ne4] was as follows. Assume $p \in S(\emptyset)$ is stationary, non-isolated, has U-rank 2, and if b realizes p then for some $a \in \mathrm{acl}(b)$, $U(a) = 1$ and $\mathrm{tp}(b/a)$ is non-isolated. Let $I(p,\kappa)$ be the number of isomorphism types of sets $p(M)$ of power κ, where M is a model of T. We wanted to prove that $I(T,\aleph_0) < 2^{\aleph_0}$ implies $I(p,\aleph_0) \leq \aleph_0$. Anyway, considering $I(p,\aleph_0)$ seems to be a necessary step on a way to prove Vaught's conjecture for superstable T. Let us recall the main path of reasoning from [Bu4] and [Ne4] thus far.

For a, b as above let $q = \mathrm{tp}(a/\emptyset)$ and $p_a = \mathrm{tp}(b/a)$. We want to count, up to isomorphism of the monster model $\mathfrak{C}$, the number of sets $p(M)$, where M is countable. $p(M)$ is the union of sets $p_{a'}(M)$, where $a' \in M$ realizes q. q has finite multiplicity hence by adding an element of $\mathrm{acl}(\emptyset)$ to the signature we can assume that q is stationary. Throughout we assume $T = T^{\mathrm{eq}}$. Further on in determining the structure of $p(M)$ we can easily dispose with the cases when p_a is strongly minimal or trivial. Hence we can assume that p_a is properly minimal and non-trivial. Then [Ne1] implies that p_a has finite multiplicity, and [Bu1] gives that every stationarization of p_a is locally modular. Similarly we can assume that for b realizing p_a, $\mathrm{stp}(b/a)$ is not modular, non-orthogonal to $\emptyset$ and almost orthogonal to $\emptyset$. In particular, p_a is weakly orthogonal to $q \mid a$. Also, we can assume that all stationarizations of types p_a, $a \in q(\mathfrak{C})$, are non-orthogonal. If $q(M)$ has finite acl-dimension then $p(M)$ can be characterized up to isomorphism just as in [Bu2]. Hence we can assume that for every countable M we consider, $q(M)$ has dimension $\aleph_0$, and $Q = q(M)$ is fixed. As $p(M) = \bigcup\{p_a(M) : a \in Q\}$, classifying the structure of $p(M)$ amounts to describing how the weakly minimal sets $p_a(M)$, $a \in Q$, can be arranged together to form $p(M)$. The types p_a, $a \in Q$, are non-orthogonal, so the main difficulty lies in that we are not free in deciding whether p_a is realized in M or not: if $p_{a_1}, \dots, p_{a_n}$ are realized in M and $a \in Q$ then possibly p_a is realized in $\mathrm{acl}(p_{a_1}(M) \cup \dots \cup p_{a_n}(M))$. This determines a kind of dependence relation on types p_a, $a \in Q$. For various reasons it is easier to work with stationary types rather than with types of finite multiplicity. Thus instead of dependence on $\{p_a : a \in Q\}$ we define a dependence relation on the set of stationarizations of types p_a, $a \in Q$. Also, to make this dependence modular we have to consider some other weakly minimal types as well. The formal definition follows the idea from [Ne2]. Let P^* be the set of strong weakly minimal non-modular types r over Q (in T^{eq}) such that r is non-orthogonal to some (every) p_a, $a \in Q$, and for some finite set $A \subseteq Q$, r does not fork over A and has finitely many conjugates over A. Hence all types in P^* are non-orthogonal.

For $r \in P^*$ and $R \subseteq P^*$, $a \in \mathrm{ACL}(R)$ iff whenever A contains a realization of every type in R then r is realized in $\mathrm{acl}(A \cup Q)$. ACL is a modular dependence relation on P^* ([Bu4, 1.14] or [Ne3, 1.2]). For $A \subseteq Q$ let $P^*_A = \{r \in P^* : r$ is based on $A\}$, $P^0_A = \{\mathrm{stp}(b/a) \mid Q : a \in \mathrm{acl}(A) \cap Q$ and b realizes $p_a\}$, and $P_A = \mathrm{ACL}(P^0_A) \cap P^*_A$. By [Ne4, 1.1], P^*_A is essentially ACL-closed in P^*, meaning that every $r \in \mathrm{ACL}(P^*_A)$ is ACL-interdependent with some $r' \in P^*_A$. Let $P = P_Q$. Let us say that ACL-closed $X, Y \subseteq P$ are isomorphic if there is an automorphism f of $\mathfrak{C}$ with $f[Q] = Q$ and $f[X] = Y$. To compute $I(p, \aleph_0)$ it suffices to determine the isomorphism types of ACL-closed subsets of P.

For X, $Y \subseteq P^*$, $\mathrm{DIM}(X/Y)$ denotes the ACL-dimension of X over Y, and $\mathrm{DIM}(X)$ denotes the ACL-dimension of X. We say that X, $Y \subseteq P^*$ are independent over $Z \subseteq P^*$ ($X \perp Y(Z)$) iff any ACL basis of X over Z remains an ACL-basis of X over $Y \cup Z$. We have that for finite $A \subseteq Q$, $\mathrm{DIM}(P^*_A)$ is finite as well ([Ne4, 1.7], [Bu2, 5.2(a)], or [Bu4, 1.14]). [Bu4, §2] proves that for A, $B \subseteq Q$, $P^*_{A \cup B} \subseteq \mathrm{ACL}(P^*_A \cup P^*_B)$ (we call this a "local character of ACL"), and that q is locally modular. As a consequence we prove in [Ne4, 1.3] that if $C \subseteq A \cap B$, A, $B \subseteq Q$, and $A \perp B(C)$ then $P_A \perp P_B(P_C)$ (and $P^*_A \perp P^*_B(P^*_C)$ as well, also the assumption that $C \neq \emptyset$ is redundant there). We can assume that q is non-trivial. Also, by the local character of ACL, we can assume that $n_0 = \mathrm{DIM}(P_a) > 1$.

Now applying [H] we can assume that q is the generic type of some connected weakly minimal type-definable (in $\mathfrak{C}^{eq}$) group $(G, +)$, in particular that q is modular. By modularity we can associate with q a division ring K such that Q with acl may be regarded as a projective space over K. In fact [H] gives more. acl on $Q \cup \{0\}$ is just a K-vector space dependence (0 is the neutral element of G). Similarly we can associate with any stationarization r of p_a a division ring L. As indicated in [Ne4], P^* with ACL-dependence may be regarded as a projective space over the same L (after identifying ACL-interdependent types). We fix the meaning of K and L for the rest of the paper, unless indicated otherwise. We say that an $A \subseteq Q$ is closed if $\mathrm{acl}(A) \cap Q = A$. As in [Ne4], for $r \in P^*$ we define $A(r)$ as the minimal closed $A \subseteq Q$ such that for some $r_0 \in P^*_A$, $r \in \mathrm{ACL}(r_0)$. By local character of ACL, if for some closed A, $A' \subseteq Q$ and $r_0 \in P^*_A$, $r_1 \in P^*_{A'}$, $r \in \mathrm{ACL}(r_0) \cap \mathrm{ACL}(r_1)$, then for some $r_2 \in P^*_{A \cap A'}$, $r \in \mathrm{ACL}(r_2)$, hence the above definition is correct. Let $n(r) = \dim(A(r))$. In [Ne4, 1.13] and [Bu4] we prove that $n^* = \max\{n(r) : r \in P\}$ is finite (in [Ne4, 1.13] n^* is denoted by n_b). In [Ne4] we reduce the problem of counting isomorphism types of ACL-closed subsets of P to a problem from algebra in the following way. Suppose F_0 is a countable division ring, $n < \omega$, and $F_1 \subseteq M_{n\times n}(F_0)$ is a division subring of the ring of matrices $M_{n\times n}(F_0)$, meaning that addition and multiplication in F_1 are addition and multiplication of matrices, and 1_{F_1} is the identity matrix I. Let $\mathcal{K}(F_0, F_1)$ be the class of pairs (V, W) where V is an F_0-vector space and $W \subseteq V^n$ is an F_1-vector subspace of V^n. V^n is an F_1-space: regard elements of V^n as columns, and F_1 acts on them by matrix multiplication on the left. We say that $(V, W), (V', W') \in \mathcal{K}(F_0, F_1)$ are isomorphic if there is an F_0-linear isomorphism $f : V \to V'$ such that $\hat{f}[W] = W'$ for the induced mapping $\hat{f} : V^n \to (V')^n$. The elements of $\mathcal{K}(F_0, F_1)$ we call (F_0, F_1)-structures.

Assume C is a finite subset of Q, R is a basis of P_C, and E is a selector from $\{r(\mathfrak{C}) : r \in R\}$. In our reduction we need to add $C \cup E$ for some C and E to the signature. Then we replace p and q by $p \mid C \cup E$ and $q \mid C \cup E$, and make other changes accordingly. Notice that in doing so we do not need to change K and L. ACL on the new P corresponds to the old ACL on the old P, localized modulo R. Also, the new n^* equals the old one. Now we prove in [Ne4] that after adding this $C \cup E$ to the set of constants, there is an embedding of L into $M_{n^*\times n^*}(K)$ (so we can assume that $L \subseteq M_{n^*\times n^*}(K)$ is a division subring of $M_{n^*\times n^*}(K)$). Let us work in $T(C \cup E)$. We can regard $Q \cup \{0\}$ as a K-vector space V, acl-dependence in Q corresponding to K-linear dependence in V. We find a correspondence α between types in P and elements of V^{n^*} such that α is onto and translates ACL-dependence into L-linear dependence. We show that all stationarizations of a single p_a are ACL-interdependent, and if $r \in P$ is a stationarization of p_a then $\alpha(r) = (a, 0, 0, \dots) \in V^{n^*}$. This gives a full description of ACL on P. In particular, $p_a \in \mathrm{ACL}(p_{a_1}, \dots p_{a_n})$ iff $(a, 0, 0, \dots) \in$ L-span$((a_1, 0, 0, \dots), \dots, (a_n, 0, 0, \dots))$ and we get a 1-1 correspondence between ACL-closed subsets of P and L-closed $W \subseteq V^{n^*}$ such that non-isomorphic ACL-closed subsets of P correspond to non-isomorphic pairs (V, W) in $\mathcal{K}(K, L)$. Of course this is a translation of a localized version of the original problem. In many cases if there are $2^{\aleph_0}$-many non-isomorphic $(V, W) \in \mathcal{K}(K, L)$, then this still gives $I(T, \aleph_0) = 2^{\aleph_0}$ for the original T. In this paper we exhibit a solution of the problem of counting countable (K, L)-structures, and in some cases show how to apply this to compute $I(p, \aleph_0)$ for the original p.

Many conjectures in stability theory (like these of Zil'ber or Cherlin) indicate that "classifiable" stable structures correspond to a few general patterns, often appearing already in classical mathematics. One of the results in this direction was the work of Hrushovski [H] showing how group structures occur in the stable context. In particular he proved that any modular, stationary regular non-trivial type may be regarded as the generic type of some type-definable group, and forking dependence on it is just a linear dependence over some division ring. We used this result above. But to obtain this he needed some parameters. We may think of these parameters as needed to recover the original pattern in the regular type, which may be distorted due to some special features of the theory.

For example we can construct a stable structure in the following way. We may start with a stable group G, and then forget about a part of its structure, so that G will be stable, but will not be a group anymore. So the original pattern of G is distorted. Hrushovski's theorem says that sometimes we can recover a group structure, possibly in an imaginary extension G^{eq} of G. Returning to our context we think that this may be the role of the added parameters $C \cup E$. The question remains how much distorted the structure of the original p may be with respect to its regularized version.

An example. Now suppose K is any countable division ring, $n < \omega$, and $L \subseteq M_{n\times n}(K)$ is a division subring of $M_{n\times n}(K)$. We do not know too many complicated types of U-rank 2. This example is intended to fill this gap. We shall show that K and L give rise to a stationary type p of U-rank 2 so that ACL on P corresponds to L-dependence. That is, ACL-closed subsets of P correspond to (K, L)-structures.

Let V be a K-space, and V_1 be a subspace of V with $\dim(V_1) = \dim(V/V_1) = \aleph_0$. V^n and $(V/V_1)^n$ are (left) L-spaces. $(V/V_1)^n$ contains $(V/V_1, 0, 0, \dots) = V_2 \cong V/V_1$ as a K-subspace. Define $Q = V_2$. For $a \in Q$ let $P_a = a + V_1^n$. So P_a is an affine L-space (a translation of V_1^n).

If $\vec{\alpha} = (\alpha_1, \dots, \alpha_K) \subseteq L$, $a_1, \dots, a_k \in Q$, and $b_i \in a_i + V_1^n$ then $\sum_i \alpha_i b_i \in V^n$. If $\sum_i \alpha_i a_i = a$ for some $a \in V_2$ then $\sum_i \alpha_i B_i \in A + V_1^n$. Let $f_{\vec{\alpha}}$ be a k-ary partial function acting on $\bigcup_{a\in Q} P_a$ defined as follows. $f_{\vec{\alpha}}(b_1, \dots, b_k)$ is defined if $b_i \in P_{a_i}$ and $\sum_i \alpha_i a_i = a \in V_2$, and then $f_{\vec{\alpha}}(b_1, \dots, b_k) = \sum_i \alpha_i b_i$. Notice that whether $f_{\vec{\alpha}}$ is defined on $(b_1, \dots, b_k)$, with $b_i \in P_{a_i}$, depends only on the linear type of $a_1, \dots, a_k$.

Let $M = (Q \cup \bigcup_{a\in Q} P_a; Q(x), P(x,y), f_{\vec{\alpha}})_{\vec{\alpha}\subseteq L}$, where $Q(M) = Q$, $P(M^2) = \{(a, b) : a \in Q, b \in P_a\}$, equipped with the following additional structure: the structure of K-space on Q, the structure of L-space on P_0, and for every $a \in Q$ the structure of affine L-space on P_a, i.e., the binary subtraction function mapping $P_a \times P_a$ into P_0.

$T = \mathrm{Th}(M)$ is ω-stable; Q and every P_a is strongly minimal. Let p_a be the strongly minimal type isolated by $P_a(x)$ over a. Then p_a is locally modular, and ACL-dependence on $\{p_a : a \in Q\}$ is described by functions $f_{\vec{\alpha}}$, i.e., is just an L-dependence. If $b \in P_a$ for $a \neq 0$ then $p = \mathrm{tp}(b/\emptyset)$ is stationary, has U-rank 2 and is not almost orthogonal to $q = \mathrm{tp}(a/\emptyset)$.

Now we shall modify the construction to get properly weakly minimal p_a and a small superstable T. Then we need of course to assume that L is locally finite. For simplicity we assume that L is finite.

Let $W_0 = W^0 > W^1 > \dots > W^i \dots, i < \omega$, be a sequence of L-spaces such that $[W^i : W^{i+1}]$ is finite and $\bigcap_i W^i$ is $\aleph_0$-dimensional. We identify V_1^n with $\bigcap_i W^i$. Add an independent copy W_a of W_0 over every P_a, $a \neq 0$, i.e., form a formal affine space $a + W_0$ so that $a + V_1^n = P_a$. For $a \neq 0$ extend subtraction from P_a onto W_a so that for $x, y \in W_a$, $x - y \in W_0$ (if $a + x, a + y \in W_a$ then $(a + x) - (a + y) = x - y \in W_0$).

We have to extend also the functions $f_{\vec{\alpha}}$, $\vec{\alpha} \subseteq L$, onto the larger sets W_a, $a \in Q$. For $f_{\vec{\alpha}}$ and $\overline{a} = (a_1, \dots, a_k) \subseteq Q$ of suitable length with $\sum_i \alpha_i a_i = a \in Q$

let us define $f_{\vec{\alpha}}(b_1,\ldots,b_k)$ for $b_i \in W_{a_i}$ as follows: Take some $b'_i \in a_i + V_1^n$. Let $f_{\vec{\alpha}}(b_1,\ldots,b_k) = \sum_i \alpha_i b'_i + \sum_i \alpha_i(b_i - b'_i)$. The first sum in this definition is taken in V^n, the second one in W_0. $\sum_i \alpha_i b'_i + \sum_i \alpha_i(b_i - b'_i)$ is the only element y of W_a such that $y - \sum_i \alpha_i b'_i = \sum_i \alpha_i(b_i - b'_i)$. It is easy to check that this definition does not depend on the choice of b'_i. Now let $M = (Q \cup \bigcup_{a\in Q} W_a; Q(x), W(x,y), W^i(x,y)\ (0 < i < \omega), f_{\vec{\alpha}}\ (\vec{\alpha} \subseteq L)$, the structure of K-space on Q, the structure of L-space on W_0, and the affine L-structure on each W_a given by subtraction), where $Q(M) = Q$, $W(M^2) = \{(a,b) : b \in W_a\}$, $W^i(M^2) = \{(a,b) : b \in a + W^i\}$. Then $T = \mathrm{Th}(M)$ is small and superstable, $p_a =$ the type over a generated by $W^i(a,x)$, $0 < i < \omega$, is properly weakly minimal, locally modular, non-isolated. ACL on $\{p_a : a \in Q\}$ is the L-dependence given by $f_{\vec{\alpha}}$'s. For $0 \neq a \in Q$ and b realizing p_a, $p = \mathrm{tp}(b/\emptyset)$ is stationary, of U-rank 2, not almost orthogonal to $q = \mathrm{tp}(a/\emptyset)$.

One could wonder what description of ACL we obtain here if we apply the analysis from [Ne4] to this case. Notice that the set of first columns of elements of L is a right K-space, a subspace of K^n. It turns out that if K-span(first columns of L) $= K^n$ (equivalently: L-span$(K,0,\ldots,0)^t = K^n$, or there are $\alpha_1,\ldots,\alpha_n \in L$ with first columns K-independent), then through the construction from [Ne4] we recover the original embedding $L \subseteq M_{n\times n}(K)$ (compare [Ne4, 3.11]).

This example shows that the general pattern of a skeletal p of U-rank 2 obtained in [Ne4] occurs in reality.

§1. Counting (K, L)-structures.

In this section we prove that there are either $2^{\aleph_0}$ or countably many countable (K, L)-structures. Also, we show that if K, L are finite (and by [Ne4], K being finite is equivalent to L being finite) and $n^* > 1$ then there are $2^{\aleph_0}$-many $(V,W) \in \mathcal{K}(K,L)$ with $\dim(V) = \aleph_0$. In case when $n^* = 1$ and K is a field, the number of countable (K, L)-structures depends on $[L : K]$. The proofs consist in applying in our context results and methods from algebra and from the "grey zone" between algebra and logic. The detailed analysis of $\mathcal{K}(K, L)$ was carried out in [DR]. Here we adapt their results.

Notice that there is a natural notion of direct sum in $\mathcal{K}(K,L)$: $(V,W) = (V_0,W_0) \oplus (V_1,W_1)$ iff $V = V_0 \oplus V_1$ (which determines V^{n^*}, and embeddings $V_0^{n^*}, V_1^{n^*} \subseteq V^{n^*}$, hence $W_0, W_1 \subseteq V^{n^*}$) and $W = W_0 + W_1$ (in V^{n^*}). It turns out that we can regard every $(K,\ L)$ structure as a left R-module for some matrix ring R. For any ring R, let $\mathcal{M}(R)$ denote the class of left R-modules. Let $R_0 = M^*_{1\times n}(K)$, and let $R = \begin{pmatrix} K & R_0 \\ 0 & L \end{pmatrix}$. For an ideal (possibly one-sided) J of R and an R-module M let $\mathrm{Ker}_M(J)$ denote $\{a \in M : Ja = 0\}$. $\mathrm{Ker}_M(J)$ is a subgroup of M, which is an R-module if J is a right-ideal. Let

$$I_L = \begin{pmatrix} 0 & 0 \\ 0 & L \end{pmatrix} \qquad I_K = \begin{pmatrix} K & 0 \\ 0 & 0 \end{pmatrix} \qquad I_{R_0} = \begin{pmatrix} 0 & R_0 \\ 0 & 0 \end{pmatrix}.$$

Then $(I_{R_0}+I_L)$ and $(I_K+I_{R_0})$ are two-sided ideals of R, I_K and I_{R_0} are left ideals and I_L is a right ideal of R. Now, with any $S = (V,W) \in \mathcal{K}(K,L)$ we associate the R-module $\begin{pmatrix} V \\ W \end{pmatrix}$ and call it $\underline{S}$. Notice that $(I_K + I_{R_0})\begin{pmatrix} V \\ W \end{pmatrix} = \begin{pmatrix} V \\ 0 \end{pmatrix}$. Hence we get

REMARK 1.1. *$S = (V, W) \in \mathcal{K}(K, L)$ is decomposable in $\mathcal{K}(K, L)$ iff $\underline{S}$ is decomposable in $\mathcal{M}(R)$. Also, for S, $S' \in \mathcal{K}(K, L)$, S and S' are isomorphic iff $\underline{S}$ and $\underline{S}'$ are isomorphic as R-modules.*

The model theory of modules is well developed. Unfortunately the mapping $\mathcal{K}(K, L) \ni S \mapsto \underline{S} \in \mathcal{M}(R)$ is not onto. We shall rely on the following results.

THEOREM 1.2. ([PR, 2.1]). *The following conditions on a ring are equivalent.*
(i) Every R-module is a direct sum of indecomposable modules.
(ii) Every R-module is totally transcendental.
If R satisfies these conditions then every indecomposable R-module is finitely generated, and there are at most $|R| + \aleph_0$ indecomposable R-modules.

THEOREM 1.3. ([BK, 8.7] or [Pr, 2.10]). *The following conditions are equivalent.*
(i) Up to isomorphism, there are countably many countable R-modules.
(ii) There are $< 2^{\aleph_0}$ countable R-modules.
(iii) R is of finite representation type (i.e., every R-module is a direct sum of indecomposable modules and there are finitely many indecomposable R-modules).

Let $\mathcal{M}'(R) = \{\underline{S} : S \in \mathcal{K}(K, L)\}$. We say that $\mathcal{K}(K, L)$ is of finite representation type if every (K, L)-structure is a direct sum of indecomposables, and there are finitely many indecomposable (K, L)-structures. Otherwise we say that $\mathcal{K}(K, L)$ is of infinite representation type. Theorems 1.2 and 1.3 deal with $\mathcal{M}(R)$. However, their proofs work as well for the smaller class $\mathcal{M}'(R)$ (this may be checked directly). Hence, modulo Remark 1.1 we get a proof of Theorem 1.4 below. We shall give also another, more direct proof of this theorem based on Theorems 1.2 and 1.3.

THEOREM 1.4. *The following conditions on K, L are equivalent.*
(i) Up to isomorphism, there are countably many countable (K, L)-structures.
(ii) There are $< 2^{\aleph_0}$ countable (K, L)-structures.
(iii) $\mathcal{K}(K, L)$ is of finite representation type.
If K, L satisfy these equivalent conditions, then every indecomposable (K, L)-structure is finitely generated.

Proof. We will show that every R-module N is a direct sum of M_0 and M_1, where $M_0 \in \mathcal{M}'(R)$, $M_1 \in \mathcal{M}(R)$, and $M_1 \subseteq \mathrm{Ker}_N(I_K + I_{R_0})$. Hence R acts on M_1 as I_L and M_1 is essentially an L-space. Modulo Theorems 1.2, 1.3, and Remark 1.1 this will prove Theorem 1.4.

We have $R + I_K + I_{R_0} + I_L$, $I_L I_K I_K I_L = I_{R_0} I_{R_0} = I_L I_{R_0} = 0$. Let $N_1 = \mathrm{Ker}_N(I_K + I_{R_0})$, $N_2 = \mathrm{Ker}_N(I_{R_0} + I_L)$, $N_3 = I_L N$, $N_4 = (I_{R_0}\text{-span})N_3$, where $(I_{R_0}\text{-span})N_3$ is the subgroup of N generated by $I_{R_0} N_3$. Notice that N_3 is a subgroup of N. N_1, N_2, $N_3 + N_4$, and N_4 are submodules of N. $N_1 \cap N_2 = 0$.

The action of R on N_2 is that of I_K, hence N_2 is essentially a K-space. As $(I_{R_0}+I_L)(I_K+I_{R_0})=0$, $(I_K+I_{R_0})N \subseteq N_2$, hence $N=RN=I_1N+(I_K+I_{R_0})N=N_3+N_2=(N_3+N_4)+N_2$. Let $N_2'=(N_3+N_4)\cap N_2$. As N_2 is essentially a K-space, we can find N_2'' such that $N_2=N_2'\oplus N_2''$. Hence $N=(N_3+N_4)\oplus N_2''$. $N_1\cap(N_2''+N_4)=0$, because $N_2''+N_4\subseteq N_2$. Let $N_1'=N_1\cap N_3$. N_1' is a submodule of N_3+N_4, and $(I_K+I_{R_0})N_1'=0$. Also, N_1' is an I_L-space. Choose a subgroup $N_3'<N_3$ so that $N_3=N_1'+N_3'$, $I_LN_3'\subseteq N_3'$ and $N_1'\cap N_3'=0$. Then still $N_4=(I_{R_0}\text{-span})N_3'$, and $N_3'+N_4$ is an R-module. We show that $N_3+N_4=N_1'\oplus(N_3'+N_4)$.

Indeed, if $a\in N_3'$, $b\in N_4$, and $a+b\in N_1'$, then as $I_LN_4=0$, $I_La=I_L(a+b)\subseteq N_1'$, Also, $I_La\subseteq N_3'$, hence $I_La=I_L(a+b)=0$. But N_1' is an I_L-space, so $a+b=0$. As $I_KI_L=0$, $I_KN_3'=0$. $N_4\subseteq \operatorname{Ker}(I_{R_0}+I_L)$, hence $N_3'\cap N_4'=0$. This implies $a=b=0$. Hence we get

$$N=N_1'\oplus N_2''\oplus(N_3'+N_4)\quad\text{and}\quad N_3'\cap N_4=0.$$

Now let $M_1=N_1'$, $M_0=N_2''\oplus(N_3'+N_4)$. It suffices to show $M_0\in\mathcal{M}'(R)$. Obviously, $N_2''\in\mathcal{M}'(R)$ ($N_2''\cong\begin{pmatrix}V\\0\end{pmatrix}$ for some V), so it suffices to show $N_3'+N_4\in\mathcal{M}'(R)$. Consider the mapping

$$\varphi: N_3'+N_4\to\begin{pmatrix}N_4\\(N_4)^{n^*}\end{pmatrix}$$

defined by $\varphi(x)=\begin{pmatrix}x\\0\end{pmatrix}$ for $x\in N_4$, and $\varphi(x)=\begin{pmatrix}0\\(A_kx)_{1\le k\le n^*}\end{pmatrix}$ for $x\in N_3'$, where $A_k=(a_{ij}^k)\in R$ is defined by $a_{ij}^k=1$ if $i=0$, $j=k+1$ and $a_{ij}^k=0$ otherwise. If $x\in N_3'$, $y\in N_4$, let $\varphi(x+y)=\varphi(x)+\varphi(y)$. We check that φ is 1-1. It suffices to see that $\varphi\restriction N_3'$ is 1-1. If $x\in N_3'$ and $\varphi(x)=0$, then $I_{R_0}x=0$. Also, $I_Kx=0$ as $x\in I_LN$ and $I_KI_L=0$, hence $X\in N_1$, contradicting the choice of N_3'. Let $\begin{pmatrix}V\\W\end{pmatrix}$ be the image of φ. V is a K-space, and by direct checking we see that φ translates the action of I_L on N_3' into the action of $L\cong I_L$ on $\begin{pmatrix}0\\W\end{pmatrix}$, hence W is an L-subspace of V^{n^*}, and $\begin{pmatrix}V\\W\end{pmatrix}\in\mathcal{M}'(R)$. It is easy to see that φ is an isomorphism of R-modules.

In Theorem 1.4 we reduced the problem of counting countable (K,L)-structures to determining representation type of $\mathcal{K}(K,L)$. If this representation type is infinite then there are $2^{\aleph_0}$ countable (K,L)-structures, and we get $I(T,\aleph_0)=2^{\aleph_0}$ as well (at least when K, L are finite). If $\mathcal{K}(K,L)$ has finite representation type then there are countably many countable (K,L)-structures, also there are finitely many finite dimensional indecomposable (K,L)-structures and every (K,L)-structure can be presented as a direct sum of indecomposables. As in the proof of [Pr, 2.10], this decomposition is unique up to isomorphism, that is if $\oplus_{i\in I}(K_i, L_i) = \oplus_{j\in J}(K_j, L_j)$ and (K_i, L_i), (K_j, L_j) are indecomposable then there is a bijection $f : I \to J$ such that (K_i, L_i) is isomorphic to $(K_{f(i)}, L_{f(i)})$. Still in the case of finite representation type of $\mathcal{K}(K, L)$ we can not determine $I(p, \aleph_0)$. We need a more detailed information, furnished in [DR]. $\mathcal{K}(K, L)$ is a special case of the structures considered in [DR]. Unfortunately, Dlab and Ringel assume throughout that there is a central field F contained in $K \cap L$ such that $[K : F]$ and $[L : F]$ are finite. In our case, in general probably we can not hope for that much. However, this assumption is obviously true if both K and L are finite, and as we indicate below, is true also when $n^* = 1$ and K is a field.

Our (K, L)-structures correspond to representations of "F-species" $\mathcal{S} = (L, K, {}_KM_L)$, where $M = R_0$ is a K, L-bimodule: K acts on M in the natural way, L acts on M by matrix multiplication on the right. ***Representation*** of $\mathcal{S}$ is a tuple $({}_LW, {}_KV, \varphi)$, where $\varphi : {}_K(M \oplus {}_LW) \to {}_KV$. Let $\mathfrak{R}$ be the category of representations of $\mathcal{S}$. Let $\mathfrak{R}\mathfrak{m}$ be those representations of $\mathcal{S}$ for which the adjoint mapping $\varphi^* : {}_LW \to \mathrm{Hom}_K({}_KM_L, {}_KV)$ of φ is monomorphism. Elements of $\mathfrak{R}\mathfrak{m}$ correspond precisely to R-modules of the form $\begin{pmatrix} V \\ W \end{pmatrix}$. As mentioned before Proposition 5.2 in [DR], $\mathfrak{R}$ is of finite representation type iff $\mathfrak{R}\mathfrak{m}$ is of finite representation type. Our feeling is that $n^* > 1$ should imply $I(T, \aleph_0) = 2^{\aleph_0}$. We were only able to prove this for finite K, L. In fact by [Ne4, 3.11], K is finite iff L is finite (in [Ne4, §3], F_q, F_p stand for K, L respectively).

PROPOSITION 1.5. *If $n^* > 1$ and K, L are finite then $\mathcal{K}(K, L)$ has infinite representation type. In particular, in this case $I(T, \aleph_0) = 2^{\aleph_0}$.*

Proof. Let ${}_LM' = M^*_{n\times 1}(K)$, L acts on M' by matrix multiplication on the left. By [Ne4, 3.11], $\dim({}_LM') \geq 2$. Now, K, L being finite implies that also $\dim(M_L) \geq 2$. Thus $\dim({}_KM) \times \dim(M_L) \geq 4$, hence the assumptions of [DR, 5.2] are satisfied. This proposition (particularly part (ii) of its proof) gives that $\mathfrak{R}\mathfrak{m}$, hence also $\mathcal{K}(K, L)$, has infinite representation type.

Now let us discuss the case $n^* = 1$. Then the situation is much simpler; we have just $L \subseteq K$. [Ne4] gives us in this case that (in $T(C \cup E)$) every $r \in R$ is ACL-interdependent with a stationarization of some p_a and all stationarizations of a single p_a are ACL-interdependent (hence we can consider ACL as a dependence on Q: $a \in \mathrm{ACL}(B)$ iff $p_a \in \mathrm{ACL}(\{p_b : b \in B\})$. $Q \cup \{0\}$ is a K-vector space, hence also an L-space, and ACL-dependence on Q is just L-linear dependence.

By [Ne1] or [Bu3] we know that L is a locally finite field. By [Ne4, 0.3], $\dim({}_LK)$ is finite. By [Bu4], K is also a locally finite field. The elements of $\mathcal{K}(K, L)$ are what Dlab and Ringel call in [DR] representations of L-structures. From [DR, Theorem A] we get the following.

COROLLARY 1.6. *If $n^* = 1$, K is a field, and $[K : L] \geq 4$ then $\mathcal{K}(K, L)$ has infinite representation type. If $[K : L] \leq 3$ then $\mathcal{K}(K, L)$ has finite representation type. Hence if $n^* = 1$, K is a field, and $[K : L] \geq 4$ then $I(T, \aleph_0) = 2^{\aleph_0}$.*

In case $n^* = 1$, that $[K : L] \geq 4$ implies $I(T, \aleph_0) = 2^{\aleph_0}$ was proved also in [Bu4, §4]. In case $n^* = 1$, and $[K : L] \leq 3$ we get that there are countably many countable (K, L)-structures. This still does not automatically yield the value of $I(p.\aleph_0)$, as we have added parameters $C \cup E$ to the signature. In case when $[K : L] = 2$ or 1, we proved in [Ne4] that $I(T, \aleph_0) < 2^{\aleph_0}$ implies $I(p, \aleph_0) = \aleph_0$. In case when $n^* = 1$ and $[K : L] = 3$ we shall prove this in the next section. We shall rely on the special form of decomposition of a (K, L)-structure into indecomposables, implied by the proof of Proposition 4.2 in [DR]. From now on in this section we assume that $n^* = 1$, K is a field, and $[K : L] = 3$. Let $\{1, e, f\}$ be a basis of K as an L-vector space. In [DR, 4.2] it is proved that there are exactly five indecomposable (K, L)-structures: $\alpha^0 = (K, K)$, $\alpha^1 = (K, L + eL)$, $\alpha^2 = (K \times K, (L \times L) + (e, f)L)$, $\alpha^3 = (K, L)$, and $\alpha^4 = (K, 0)$.

For (K, L)-structures $(V, W), (V', W')$ we say that (V', W') is a strong substructure of (V, W) ($(V', W') < (V, W)$) if V' is a K-subspace of V and $W' = W \cap V'$ (that is we regard W in (V, W) as a predicate). For $\alpha \in \mathcal{K}(K, L)$ we stipulate $\alpha = (V_\alpha, W_\alpha)$. Let $S = (V, W)$ be a (K, L)-structure. We will indicate an algorithm of decomposing S. We define by induction sets X_0, X_1, X_2, X_3, X_4. Let $X_i = \{\alpha \in \mathcal{K}(K, L) : \alpha < S, \alpha \cong \alpha^i \text{ and } V_\alpha \cap K\text{-span}(X_{<i}) = 0\}$. Here K-span$(X_{<i})$ is the K-subspace of V generated by $\bigcup\{V_\alpha : \alpha \in X_j, j < i\}$. If $\alpha, \beta \in \mathcal{K}(K, L)$, $\alpha, \beta < S$, then let $\alpha + \beta = (V_\alpha + V_\beta, W_\alpha + W_\beta)$, if $X \subseteq \mathcal{K}(K, L)$ is a family of $\alpha < S$, then let $\Sigma X = (\Sigma\{V_\alpha : \alpha \in X\}, \Sigma\{W_\alpha : \alpha \in X\})$.

Claim 1.7.

(1) For $\alpha < S$, $\alpha \in X_i$ iff $\alpha \cong \alpha^i$ and $V_\alpha \not\subseteq K$-span$(X_{<i})$.

(2) $\Sigma X_{<i} < S$.

(3) Assume $\beta_0, \ldots, \beta_n \in X_i$, $V_{\beta_0} \cap K$-span$(X_{<i} \cup \{\beta_1, \ldots, \beta_n\}) \neq 0$. Then $V_{\beta_0} \subseteq K$-span$(X_{<i} \cup \{\beta_1, \ldots, \beta_n\})$.

(4) Assume $\beta_1, \ldots, \beta_n \in X_i$. Then $\Sigma(X_{<i} \cup \{\beta_1, \ldots, \beta_n\}) < S$.

Proof. The proof is a modification of [DR, 4.2]. $\{1, e, f\}$ is a basis of K over L. We proceed by induction on i. For $i = 0$ the claim is easy. Also, (2) follows always from the induction hypothesis, and except for $i = 2$, $\dim(V_{\alpha^i}) = 1$, hence for $i \neq 2$, (1) and (3) are trivial. Let $i = 1$. We need to prove (4). We proceed by induction on n. Without loss of generality (wlog) K-span$(X_{<i})$ has dimension $k < \omega$. Let $\Sigma X_{<i} = (V_0, V_0)$, $\beta_j = (V_j, W_j)$. By (3) and the inductive hypothesis we can assume that $V_0, V_1, \ldots, V_n$ are independent. Let $W' = W \cap V'$ where $V' = V_0 + \cdots + V_n$. Suppose $W' \neq V_0 + W_1 + \cdots + W_n$. We know that $\dim_L(V_0 + W_1 + \cdots + W_n) = 3k + 2n$, hence $\dim_L(W') > 3k + 2n$, Thus $e^{-1}W' \cap f^{-1}W' \cap W'$ properly extends V_0. Let $u \in W' \cap e^{-1}W' \cap f^{-1}W' \setminus V_0$. Then $u, eu, fu \in W$, hence $(Ku, Ku) \in X_0$, contradicting $u \notin V_0$.

Now let $i = 2$.

(1) $\leftarrow$. Suppose $\Sigma X_0 = (V_0, V_0)$. Wlog K-span$(X_{<2})$ has finite dimension. Assume $\beta_1 = (V_1, W_1), \ldots, \beta_n = (V_n, W_n) \in X_1$ are independent (i.e., $V_t \cap (V_0 + \cdots + V_{t-1}) = 0$), $\alpha \in X_2$, $V_\alpha \cap (V_0 + \cdots + V_n) \neq 0$. Let $W' = W \cap V'$ where $V' = V_0 + \cdots + V_n + V_\alpha$, $\dim_K(V_0) = k$. So $\dim_L(W') \geq 3k + 2(n+1)$, and $\dim_L(V') = 3k + 3(n+1)$. $V_0 \subseteq W' \cap e^{-1}W'$ and $\dim_L(W' \cap e^{-1}W') \geq 3k + n + 1$. Let $u_t \in W_t \cap e^{-1}W'$, $t = 1, \ldots, n$, and $u_\alpha \in W' \cap e^{-1}W' \setminus (V_0 + L\text{-span}(u_1, \ldots, u_n))$.

If $u_\alpha \in V_0+\cdots+V_n$, we get a contradiction as in case 1. Hence $u_\alpha \notin V_0+\cdots+V_n$. Also, $u_\alpha, eu_\alpha \in W$, hence $(Ku_\alpha, Lu_\alpha + Leu_\alpha) \in X_1$, a contradiction.

We prove (3) and (4) simultaneously, by induction on n. As above, wlog K-span$(X_{<i})$ has finite dimension. For $n = 0$ we are done by (1). So we can assume that K-span$(X_{<i}), V_1, \ldots, V_n$ are independent, where $\beta_t = (V_t, W_t)$. Now if $V_0 \cap (K\text{-span}(X_{<i}) + V_1 + \cdots + V_n) \neq 0$ or $\Sigma(X_{<i} \cup \{\beta_1, \ldots, \beta_n\}) \not< S$ then as in the proof of (1) we get a $u \in (K\text{-span}(X_{<i}) + \cdots + V_n) \setminus K\text{-span}(X_{<i})$ such that $(Ku, Lu + Leu) \in X_1$, a contradiction.

Let $i = 3$. We need to prove (4). We can assume K-span$(X_{<i})$ has finite dimension. Let $\beta_n = (V_n, W_n), \Sigma X_{<i} = (V_0, W_0), V' = V_0 + \cdots + V_n$, $W' = W \cap V'$. Let $X = W_0 + W_1 + \ldots W_n$. We have $X + eX + fX = V'$. Suppose $X \neq W'$. Then we can choose $v \in W' \setminus X$, and $v = v_0 + ev_1 + fv_2$ for some $v_0, v_1, v_2 \in X$. Replacing v by $v - v_0$ we can assume $v_0 = 0$, and $v + ev_1 + fv_2$. If $v_1, v_2 \in W_0$, we get a contradiction with the inductive hypothesis (Claim 1.7(4)). If $v_1, v_2 \notin W_0$, then v, v_1, v_2 give rise to an $\alpha < S$ with $V_\alpha \cap V_0 = 0$ and $\alpha \in X_{<i}$, a contradiction. If, say, $v_1 \in W_0$ and $v_2 \notin W_0$ then $(Kv_1 + Kv_2, Lv_1 + Lv_2 + Lv) \in X_2$, a contradiction.

The case when $i = 4$ is trivial.

REMARK. The above claim is true as well when K, L are only division rings, and there is a central subfield F of both K and L such that $[K : F]$ and $[L : F]$ are finite.

Claim 1.7 justifies the following algorithm for decomposing $S = (V, W) \in \mathcal{K}(K, L)$ into a direct sum of (K, L)-structures of type $\alpha^0, \ldots, \alpha^4$. Suppose $Y_i = \{\beta_i^j : j < n_i\} \subseteq X_i$, $i < 5$, satisfy the condition: $V_{\beta_i^j} \not\subseteq K\text{-span}(X_{<i} \cup \{\beta_i^t : t < j\})$. Then $(V, W) = \oplus\{\beta_i^j : i < 5, j < n_i\}$. This algorithm enables us in the next section to get rid of the parameters $C \cup E$ added to the signature and determine $I(p, \aleph_0)$. More generally, if K, L are arbitrary, $R = \begin{pmatrix} K & R_0 \\ 0 & L \end{pmatrix}$ is of finite representation type, $\alpha_0, \ldots, \alpha_n$ are the indecomposable R-modules, and the counterpart of 1.7 holds then we can also get rid of the parameters $C \cup E$ and determine $I(p, \aleph_0)$. The reason for that is that the algorithm shown above is "cumulative." It is not clear to us if such an algorithm may be found for any ring R of finite type.

§2. Getting rid of the parameters. In this section we show how to prove that $I(p, \aleph_0)$ is countable in the case when K, L are fields, $n^* = 1$, and $[K : L] = 3$. So from now on as far as Theorem 2.5 we assume that $n^* = 1$, K is a field, and $[K : L] = 3$. By the discussion in the previous section we know that for some finite $C \cup E$, $I(p \restriction C \cup E, \aleph_0)$ is countable, and isomorphism types of sets of realizations of $p \restriction C \cup E$ in countable models of T correspond to isomorphism types of (K, L)-structures. The algorithm of decomposition of any (K, L)-structure into indecomposables from §1 enables us to find a decomposition of $p(M)$. More precisely we find essentially finitely many indiscernible sets $I_1, \ldots, I_k$ such that $p(M)$ is prime over $I_1 \cup \cdots \cup I_k$. The proof we give here is a variant of the reasoning from [Ne4, §4] (see Fact 4.8, Lemmas 4.9, 4.11, and Theorem 4.13 there).

We return now to the original meaning of Q, i.e., $Q = q(M)$ for some M, $\dim(Q) = \aleph_0$, q is a stationary, modular type over $\emptyset$, generic of a connected, type-definable over $\emptyset$ weakly minimal (w.m.) group $G = (G, +)$. K is the division ring of pseudo-endomorphisms of G, so that $Q \cup \{0\}$ is a vector space over K.

Let $\{E_n : n < \omega\}$ be an enumeration of $FE(\emptyset)$. Let $\mathrm{acl}_n(\emptyset) = \{a/E_k : a \in \mathfrak{C}, k \leq n\}$. Notice that $\mathrm{acl}_n(\emptyset)$ is finite. Assume A is a finite subset of Q, $R \subseteq P_A$ is finite, ACL-independent and such that $R \perp P_b$ for some (any) $b \in Q \setminus \mathrm{acl}(A)$ (by the local character of ACL it implies that $R \perp P_B$ for any $B \subseteq Q$ with $B \perp A$). Let B be a selector from $\{s(\mathfrak{C}) : s \in R\}$ and $r = \mathrm{tp}(B/A)$. By the transitivity of finite multiplicity ([PS] or [Sa, 1.5]), $\mathrm{Mlt}(BA/\emptyset)$ is finite. We say that r and (R, A) are n-determined if $\mathrm{tp}(BA/\mathrm{acl}_n(\emptyset))$ is stationary. Also, we say that (R, A) corresponds to $\mathrm{tp}(BA/\mathrm{acl}_n(\emptyset))$. This definition tacitly assumes an enumeration of A and R.

Fact 2.1. *(1) n-determined implies k-determined for $k > n$.*
(2) Every r is n-determined for some n.
(3) If $C \perp BA$ then every completion of r over $A \cup \mathrm{acl}_n(\emptyset)$ is weakly orthogonal to $\mathrm{tp}(C/A \cup \mathrm{acl}_n(\emptyset))$.

Proof. Easy.

Fix a finite $E \subseteq Q$ large enough, so that if R is a basis of P_E and C is a selector from $\{s(\mathfrak{C}) : s \in R\}$, then over $C \cup E$ the translation Φ of ACL-dependence on P_Q into L-dependence on Q works. Assume A is a finite subset of Q independent from E, $R \subseteq P_{AE} \setminus P_E$ is finite, ACL-independent, with $R \perp P_E$, and B is a selector from $\{s(\mathfrak{C}) : s \in\}$. We say that (R, A) and (B, A) are of type α^i if the (K, L)-structure (V, W) corresponding to $\mathrm{ACL}(R \cup P_E)$ (through Φ) is isomorphic to α^i (α^i, $i < 5$, are defined before Claim 1.7). This implies of course $\dim(A/E) \leq 2$. Let M be a countable model of T with $Q = q(M)$. For $A \subseteq Q$ let $P_A^M = \{r \in P_A : r \text{ is realized in } M\}$. Notice that P_A^M is ACL-closed in P_A. We define by induction sets X_i^M, $i < 5$, corresponding to sets X_i defined before Claim 1.7. Let $X_i^M = \{A \subseteq Q : \text{for some } R \subseteq P_{AE}^M, (R, A) \text{ is of type } \alpha^i \text{ and } A \perp X_{<i}^M(E)\}$, where $X_{<i}^M = \bigcup_{j<i} X_j^M$. Applying Claim 1.7 we get

Claim 2.2.
(1) For $A \subseteq Q$, $A \in X_i^M$ iff for some $R \subseteq P_{AE}^m$, (R, A) is of type α^i and $A \not\subseteq \mathrm{acl}(\bigcup X_{<i}^M \cup E)$.
(2) Choose for any j and $A \in X_j^M$ an $R_A \subseteq P_{AE}^M$ witnessing $A \in X_j^m$. Then $P_{EX_{<i}^M}^M = ACL(P_E^M \cup \bigcup\{R_A : A \in X_{<i}^M\}) \cap P_{EX_{<i}^M}$.
(3) Assume $A_0, \ldots, A_n \in X_i^M$, $A_0 \not\perp (X_{<i}^M \cup A_1 \cup \cdots \cup A_n)(E)$. Then $A_0 \subseteq \mathrm{acl}(X_{<i}^M \cup A_1 \cup \cdots \cup A_n \cup E)$.
(4) Assume $A_1, \ldots, A_n \in X_i^M$. Then $P_{EX_{<i}^M \cup A_1 \cup \cdots \cup A_n}^M \subseteq ACL(P_{EX_{<i}^M}^M \cup R_{A_1} \cup \cdots \cup R_{A_n})$, where R_A are chosen as in (2).

The next lemma corresponds to Lemmas 4.9 and 4.11 in [Ne4].

LEMMA 2.3. *Let* $R_n = ACL(P_E^M \cup \bigcup\{R$: *for some* A, (R, A) *witnesses* $A \in X_i$ *for some* $i < 5$, *and* (R, AE) *is* n*-determined*$\}$. *Then for some* $n < \omega$, $P_Q^M \subseteq R_n$.

Proof. Suppose not. Then we can choose n_i, $i < \omega$, so that n_i is the minimal k such that for $j = i - 1$, $R_k \neq R_j$. Of course, $P_Q^M \subseteq \bigcup_n R_n$. For $I < \omega$ choose $A_i \in X_j^M$ for some $j < 5$ and R_{A_i} witnessing $A_i \in X_j^M$ so that $(R_{A_i}, A_i E)$ is n_i-determined, and $R_{A_i} \not\subseteq R_{n_{i-1}}$.

For $X \subseteq \omega$ let $R_X = \mathrm{ACL}(P_E^M \cup \bigcup\{R_{A_i} : i \in X\})$. By the omitting types theorem we can find a model M' with $Q = q(M')$ and $P_Q^{M'} = R_X$ and (modulo Claims 1.7 and 2.2) as in the proof of [Ne4, 4.9], we can recover X from (M', E). This contradicts $I(T, \aleph_0) < 2^{\aleph_0}$.

Fix $n^0 < \omega$ with $P_Q^M \subseteq R_{n^0}$. Let $\{r_k : k < \omega\}$ be an enumeration of $\bigcup_{n<\omega} S_n(\mathrm{acl}_{n^0}(\emptyset))$. The proof of the next lemma is similar to that of Lemma 2.3.

LEMMA 2.4. *Let* $R_{n^0}^k = ACL(P_E^M \cup \bigcup\{R$: *for some* A, (R, A) *witnesses* $A \in X_i^M$ *(for some* $i < 5$*)*, (R, AE) *is* n^0*-determined and corresponds to some* r_i, $i < k\}$). *Then for some* $k < \omega$, $P_Q^M \subseteq_{n^0}^k$.

Fix $k^0 < \omega$ with $P_Q^M \subseteq R_{n^0}^{k^0}$.

THEOREM 2.5. *If* $n^* = 1$, K, L *are fields, and* $[K : L] = 3$ *then* $I(p, \aleph_0)$ *is countable.*

Proof. We can absorb E and $\mathrm{acl}_{n^0}(\emptyset)$ into the signature. We need to find a finite set of invariants determining $p(M)$ up to isomorphism. In order to characterize P_Q^M, we shall decompose it into indecomposables according to the algorithm from §1. We define by induction on $j < k^0$ sets $X_{ij}^M = \{A \in X_i^M$: for some R witnessing $A \in X_i^M$, (R, AE) is n^0-determined, corresponds to r_j, and $(*)$ $A \perp X_{<i}^M \cup X_{i,<j}^M(E)\}$, where $X_{i,<j}^M = \bigcup_{t<j} X_{it}^M$. By Claim 1.7, $(*)$ in the definition of X_{ij}^M is equivalent to $A \not\subseteq \mathrm{acl}(X_{<i}^M \cup X_{i,<j}^M \cup E)$. Now define by induction sets $A_{ij}^t \in X_{ij}^M$, $t < n_{ij} \leq \omega$, so that $A_{ij}^t \perp X_{<i}^M \cup X_{i<j}^M \cup A_{ij}^0 \cup \cdots \cup A_{ij}^{t-1}(E)$, and $X_{ij}^M \subseteq \mathrm{acl}(X_{<i}^M \cup X_{i,<j}^M \cup \bigcup\{A_{ij} : t < n_{ij}\} \cup E)$.

Let R_{ij}^t witness $A_{ij}^t \in X_{ij}^M$, and let B_{ij}^t be a selector from $\{s(\mathfrak{C}) : s \in R_{ij}^t\}$. Then by Fact 2.1 and Claims 1.7 and 2.2 we have:

(1) $P_Q^M = \mathrm{ACL}(P_E^M \cup \bigcup\{R_{ij}^t : i < 5, j < k^0, t < n_{ij}\})$,

(2) $\{B_{ij}^t A_{ij}^t E : t < n_{ij}\}$ is a Morley sequence in r_j,

(3) $\{B_{ij}^t A_{ij}^t E : i < 5, j < k^0, t < n_{ij}\}$ is independent.

It follows that the isomorphism type of $p(M)$ is determined by n^0, k^0, $n_{ij}(I < 5, j < k^0)$, the isomorphism type of P_E^M, and the dimension of $r(M)$ for any $r \in P_Q^M$. We see that that there are only countably many possibilities.

At the beginning we assumed that p_a is non-isolated. Now we will show how to omit this assumption, at least in some cases. So we assume now that

p is a stationary, non-isolated complete type over $\emptyset$ of U-rank 2 and p is not almost orthogonal to some $q \in S(\emptyset)$ of U-rank 1. For b satisfying p choose a realizing q with $a \in \mathrm{acl}(b)$, and let $p_a = \mathrm{tp}(b/a)$. Now we admit that p_a is possibly isolated. Again, we can dispose easily with some trivial cases, and after some manipulation we can assume that q is stationary, and p_a is non-trivial, properly weakly minimal. Again we may restrict ourselves to the case when $Q = q(M)$ has dimension $\aleph_0$, and now we focus our attention on P^* (defined in the introduction). If $p_a \perp p_b$ for $a, b \in Q$ with $a \mathop{\smile\!\!\!\!\!\!|}\, b$, then a standard argument (using ideas from [Bu2]) shows that either $I(T, \aleph_0) = 2^{\aleph_0}$ or $I(p, \aleph_0) \leq \aleph_0$. Indeed, for $a \in Q$ consider the set $S_a(M) = \{p_b(M) : b \in \mathrm{acl}(a) \cap Q\}$. This set is a countable union of w.m. sets and we can apply [Bu2] to show that there are (in T^{eq}) properly w.m. types $q_0, \ldots, q_{n-1}$ over a, of finite multiplicity, such that for any M with $Q = q(M)$, $S_a(M)$ is prime over $\{a\} \cup I_0 \cup \cdots \cup I_{n_1}$, where I_j is a Morley sequence in q_j. As for $a \mathop{\smile\!\!\!\!\!\!|}\, b$, $p_a \perp p_b$, the structures of $S_a(M)$ and $S_b(M)$ may be chosen independently, hence unless $n = 0$ and every $S_a(M)$ is prime just over a, we get $I(T, \aleph_0) = 2^{\aleph_0}$.

But in case when $S_a(M)$ is prime over a, $S_a(M)$ is unique up to isomorphism, hence we get $I(p, \aleph_0) \leq \aleph_0$.

So we can assume that $p_a \not\perp p_b$ for $a, b \in Q$ with $a \mathop{\smile\!\!\!\!\!\!|}\, b$. As for c realizing p_b, $b \in \mathrm{acl}(c)$ and $\mathrm{tp}(c/\emptyset)$ is stationary, we get easily that for all $a, b \in Q$, all stationarizations of p_a, p_b are non-orthogonal. Hence P^* is a family on non-orthogonal types.

FACT 2.6. *If $n(r) = 0$ for every $r \in P^*$ then for some finite $C \subseteq \mathrm{acl}(\emptyset)$ there are stationary $q_0, \ldots, q_n \in S(C)$ such that for every M with $Q = q(M)$, $P^*_Q(M) \subseteq \mathrm{acl}(I_0 \cup \cdots \cup I_n \cup C) \subseteq M$ for some Morley sequences $I_0, \ldots, I_n$ in $q_0, \ldots, q_n$ respectively.*

Proof. ACL-dimension of $P^*_\emptyset$ is finite (see [Ne4, 0.3] or [Bu2, 5.2(a)]).

Later we shall see how to conclude the computation of $I(p, \aleph_0)$ in case when $n(r) = 0$ for any $r \in P^*$. Now assume $n(r) > 0$ for some $r \in P^*$. Then by the proofs in [Bu4, §2], Q is locally modular. Of course we can assume that Q is non-trivial, hence again we can assume that q is modular and generic of a weakly minimal type-definable over $\emptyset$ group $G = (G, +)$. Let K be the division ring corresponding to q, and L be the division ring corresponding to any stationarization of p_a. We shall prove the following theorem.

THEOREM 2.7. *If K, L are finite and $DIM(P^*_a) > 1$ then $I(p, \aleph_0) \leq \aleph_0$.*

Proof. In [Ne4] we worked with P_Q, translating the ACL-dependence on P_Q into a linear dependence. But the same proofs work for P^*_Q as well, hence we get an $n^* = \max\{n(r) : r \in P^*\}$, and an embedding of L into the ring of matrices $M_{n^* \times n^*}(K)$ so that the ACL-dependence on P^* translates into L-dependence on $(Q \cup \{0\})^{n^*}$. Now by [DR], unless $n^* = 1$ and $[K : L] \leq 3$, $I(T, \aleph_0) = 2^{\aleph_0}$ (because $\mathcal{K}(K, L)$ is of infinite representation type then). If $n^* = 1$ and $[K : L] \leq 3$ then [Ne4, 4.13] and the proof of Theorem 2.5 show that there is a finite set $C \subseteq \mathrm{acl}(\emptyset)$, and finitely many stationary types $q_0, \ldots, q_n$ such that for every M with $Q = q(M)$ there are Morley sequences $I_0, \dots I_n \subseteq M$ in $q_0, \dots, q_n$ respectively such that $P^*(M) \cup Q \subseteq \mathrm{acl}(I_0 \cup \cdots \cup I_n)$. Notice that this is also the conclusion of Fact 2.6. As in [Bu2, §5] we prove the following claim which concludes the proof of the theorem and shows that when the assumptions of Claim 2.6 hold then $\dot{I}(p, \aleph_0) \leq \aleph_0$ as well.

Claim 2.8. *Assume $Q = q(M)$, $I_0, \dots, I_m \subseteq M$ are Morley sequences in $q_0, \dots, q_n$ respectively, and $P^*(M) \cup Q \subseteq acl(I_0 \cup \cdots \cup I_n)$. Then $Q \cup \bigcup_{a \in Q} p_a(M)$ is prime (i.e., atomic) over $I_0 \cup \cdots \cup I_n$.*

Proof. Let $N \subseteq M$ be a prime model over $I_0 \cup \cdots \cup I_n$ such that $\bigcup_{a \in Q} p_a(M)$ is maximal. We have to show $Q \cup \bigcup_{a \in Q} p_a(M) \subseteq N$. Suppose not. Take $b \in p_a(M) \setminus N$. As in the proof of [Bu2, 5.1], there is a finite $B \subseteq \bigcup_{a \in Q} p_a(N) \cup Q$ such that $\mathrm{tp}(b/B)$ is non-isolated. We can assume also that $a \in B$, and for every $c \in B \setminus Q$ there is $d \in B$ such that c realizes p_d. Let $D = B \cap Q$. Applying [Bu2, 4.1] to $T(D)$, we get an $r \in P^*_D$ such that $r \restriction B \not\perp^a \mathrm{stp}(b/B)$. So there is c realizing r with $c \not\perp b(B)$. Thus $c \in M \setminus N$. On the other hand, $c \in P^*(M) \subseteq \mathrm{acl}(I_0 \cup \cdots \cup I_n) \subseteq N$, a contradiction.

If T is superstable then in general there are no prime models over infinite sets. However, if in addition T is small then there are prime models over indiscernible sets. The first author conjectures that in case of a model M of a superstable T of finite U-rank, if $A \subseteq M$ is a skeleton of M, then, although M may not be prime over A, there are finitely many Morley sequences $I_0, \dots, I_n \subseteq M$ with $A \subseteq \mathrm{acl}(I_0 \cup \cdots \cup I_n)$ such that M is prime over $I_0 \cup \cdots \cup I_n$. This seems to work in case of classification of types of U-rank 2.

References

[Ba] J. T. Baldwin, ***Fundamentals of stability theory***, Springer 1987.

[BK] J. T. Baldwin, R. N. McKenzie, *Counting models in universal Horn classes*, ***Algebra universalis***, vol. 15 (1982), pp. 359–384.

[Bu1] S. Buechler, *The geometry of weakly minimal types*, ***The journal of symbolic logic***, vol. 50 (1985), pp. 1044–1054.

[Bu2] S. Buechler, *The classification of small weakly minimal sets I*, ***Classification theory, Chicago 1985***, ed. J. Baldwin, Springer 1987, pp. 32–71.

[Bu3] S. Buechler, *Classification of small weakly minimal sets III $\frac{1}{2}$*, manuscript.

[Bu4] S. Buechler, *Rank 2 types vis-a-vis Vaught's conjecture*, preprint 1989.

[DR] V. Dlab, C. M. Ringel, *On algebras of finite representation type*, ***Journal of algebra***, vol. 33 (1975), pp. 306–394.

[H] E. Hrushovski, *Locally modular regular types*, ***Classification theory, Chicago 1985***, ed. J. Baldwin, Springer 1987, pp. 132–164.

[Ne1] L. Newelski, *A proof of Saffe's conjecture*, ***Fundamenta Mathematicae***, to appear.

[Ne2] L. NEWELSKI, *Weakly minimal unidimensional formulas: a global approach*, ***Annals of pure and applied logic***, vol. 46 (1990), pp. 65–94.

[Ne3] L. NEWELSKI, *Classifying U-rank 2 types*, manuscript, February 1989.

[Ne4] L. NEWELSKI, *On U-rank 2 types*, ***Transactions of the American Mathematical Society***, submitted.

[P] A. PILLAY, ***Geometrical stability theory***, Oxford University Press, in preparation.

[PS] A. PILLAY, C. STEINHORN, *A note on non-multidimensional superstable theories*, ***The journal of symbolic logic***, vol. 50 (1985), pp. 1020–1024.

[PR] M. PREST, *Rings of finite representation type and modules of finite Morley rank*, ***Journal of algebra***, vol. 88 (1984), pp. 502–533.

[Sa] J. SAFFE, *On Vaught's conjecture for superstable theories*, preprint 1982.

[Sh] S. SHELAH, ***Classification theory***, North Holland 1978.

Steven Buechler
Department of Mathematics
University of Notre Dame
Notre Dame, IN 46556, USA

Ludomir Newelski
IM PAN
Kopernika 18
51-617 Wrocław
Poland

DEFINABILITY AND GLOBAL DEGREE THEORY[1]

S. BARRY COOPER

Gödel's work [Gö34] on undecidable theories and the subsequent formalisations of the notion of a recursive function ([Tu36], [Kl36] etc.) have led to an ever deepening understanding of the nature of the non-computable universe (which as Gödel himself showed, includes sets and functions of everyday significance). The nontrivial aspect of Church's Thesis (any function not contained within one of the equivalent definitions of recursive/Turing computable, cannot be considered to be effectively computable) still provides a basis not only for classical and generalised recursion theory, but also for contemporary theoretical computer science. Recent years, in parallel with the massive increase in interest in the computable universe and the development of much subtler concepts of 'practically computable,' have seen remarkable progress with some of the most basic and challenging questions concerning the non-computable universe, results both of philosophical significance and of potentially wider technical importance.

Relativising Church's Thesis, Kleene and Post [KP54] proposed the now standard framework of the *degrees of unsolvability* $\boldsymbol{\mathcal{D}}$ as the appropriate fine structure theory for ω^ω. A technical basis was found in the various equivalent notions of relative computability provided by Turing [Tu39], Kleene [Kl43], Post [Po43] and others. Within the study of $\boldsymbol{\mathcal{D}}$ it has become usual to distinguish (see [Sh81]) two approaches: that of *global degree theory*, based more or less on a number of general questions concerning the structure of the degrees first stated by Rogers in his book [Ro67]; and that of *local degree theory* with its emphasis on degree structure not far removed from the degree $\mathbf{0}$ of recursive functions (in particular the *recursively enumerable*—or *r.e.*—degrees and the degrees below $\mathbf{0}'$—the degree of the coded theorems of Peano arithmetic). Of course, there is an intimate relationship between the two approaches, and the aim here is to describe some recent results showing how even the most archetypal local degree theory can be used to resolve interesting and important global questions.

§1. Notation and terminology.

We use standard notation and terminology (see for example [So87]).

For instance, corresponding to the ith Turing machine, Φ_i denotes the ith partial recursive (p.r.) functional $2^\omega \to 2^\omega$. A set A is *Turing reducible to* a set B ($A \leq_T B$) if and only if $A = \Phi_i^B$ for some $i \in \omega$, and A, B are *Turing equivalent*

[1]Research partially supported March–June 1989 by a C.N.R. visiting professorship at the University of Siena. We are grateful to Andrea Sorbi and Bob di Paola for encouragement during the early stages of this work.

($A \equiv_T B$) if and only if $A \leq_T B$ and $B \leq_T A$. The *degree of unsolvability* or *Turing degree* of A is defined by

$$\deg(A) = \{X \in 2^\omega \mid A \equiv_T X\}.$$

We write $\leq$ for the partial ordering on $\mathcal{D}$, the set of all degrees, $\mathbf{0}$ for the least degree, consisting of all recursive sets of numbers, and $\boldsymbol{\mathcal{D}}$ for the structure $\langle \mathcal{D}, \leq \rangle$.

Kleene and Post [KP54] also defined the notion of *jump operator* on sets and degrees. Let $W_i^A = \operatorname{dom} \Phi_i^A$ denote the ith *recursively enumerable in* A (A*-r.e.*) set ($W_i = W_i^\emptyset$ being the ith r.e. set). Then the *jump* (*$n+1$th jump*) of a set A is defined by $A' = A^{(1)} = \{x \mid x \in W_x^A\}$ ($A^{(n+1)} = (A^{(n)})'$). This induces a *jump operator* on degrees defined by $\mathbf{a}' = \deg(A')$, $A \in \mathbf{a}$, with the special properties that $\mathbf{a} < \mathbf{a}'$, and $\mathbf{a}'$ is the least upper bound of the degrees of sets r.e. in $A \in \mathbf{a}$. Post's Theorem [Po48] that $X \in \Delta_{n+1}^A \Leftrightarrow X \leq_T A^{(n)}$ attaches special importance to the ascending sequence $\mathbf{a}, \mathbf{a}', \ldots, \mathbf{a}^{(n)}, \ldots$. We define the standard *ω-jump* of $\mathbf{a}$ by $\mathbf{a}^{(\omega)} = \deg(\oplus_{n \in \omega} A^{(n)})$, $A \in \mathbf{a}$. Kleene and Post were the first to investigate the structure $\boldsymbol{\mathcal{D}}' = \langle \mathcal{D}, \leq, ' \rangle$. They speculated ([KP54], p. 384) that the jump operator may not be capable of description within $\boldsymbol{\mathcal{D}}$ itself, a question returned to by many authors since then (for instance, in recent times, Simpson [Si77], Epstein [Ep79], Shore [Sh81] and Odifreddi [Od89]).

§2. Global degree theory.

How rich a structure is $\boldsymbol{\mathcal{D}}$ or $\boldsymbol{\mathcal{D}}'$? To what extent are degrees locally individuated? Are particular parts of the non-computable universe recognisable by their context in $\boldsymbol{\mathcal{D}}$ or $\boldsymbol{\mathcal{D}}'$? What mathematical concepts are describable within the degrees of unsolvability? Following Rogers [Ro67] global questions tend to be grouped under the following headings:

HOMOGENEITY. *For which* $\mathbf{a}, \mathbf{b} \in \mathcal{D}$ *is* $\boldsymbol{\mathcal{D}}(\geq \mathbf{a}) \equiv \boldsymbol{\mathcal{D}}(\geq \mathbf{b})$, *or* $\boldsymbol{\mathcal{D}}'(\geq \mathbf{a}) \equiv \boldsymbol{\mathcal{D}}'(\geq \mathbf{b})$?

STRONG HOMOGENEITY. *For which* $\mathbf{a}, \mathbf{b} \in \mathcal{D}$ *is* $\boldsymbol{\mathcal{D}}(\geq \mathbf{a}) \cong \boldsymbol{\mathcal{D}}(\geq \mathbf{b})$, *or* $\boldsymbol{\mathcal{D}}'(\geq \mathbf{a}) \cong \boldsymbol{\mathcal{D}}'(\geq \mathbf{b})$?

The Strong Homogeneity/Homogeneity Conjectures of Rogers [Ro67]/Yates [Ya70], respectively, refer to the special case when $\mathbf{b} = \mathbf{0}$.

AUTOMORPHISMS. *Are there any nontrivial automorphisms of* $\boldsymbol{\mathcal{D}}$ *or* $\boldsymbol{\mathcal{D}}'$?

That is, is the non-computable universe *rigid*?

DEFINABILITY. *What can we describe in terms of the structure of* $\boldsymbol{\mathcal{D}}$ *or* $\boldsymbol{\mathcal{D}}'$? *In particular:*

Is the jump $\mathbf{a}'$ *of* $\mathbf{a}$ *definable purely in terms of the structure of* $\boldsymbol{\mathcal{D}}(\geq \mathbf{a})$?

Rogers' [Ro67] chooses invariance under all automorphisms of $\boldsymbol{\mathcal{D}}$ (that is, the notion of being *order-theoretic*) as an alternative formalisation of the idea of a relation on degrees being fixed by the structure of $\boldsymbol{\mathcal{D}}$. For instance he asks (see also Question 5.12 of [Si77] and Q8. of [Ep79]):

Is the jump operator order-theoretic?

We briefly review what was previously known concerning the above questions. A more detailed discussion can be found in [Od89].

§3. Definability.

Initial segments and their relativisations provided the first source of expressive structure within the degrees of unsolvability. Lachlan [La68] used the embeddability of all countable distributive lattices as initial segments of $\mathcal{D}$ to show the undecidability and non-axiomatisability of Th $(\mathcal{D})$ (the first-order theory of $\mathcal{D}$). The best result in this direction was provided by Simpson [Si77], following Jockusch and Simpson's [JSi76] initial work on coding and definability results for $\mathcal{D}'$:

THEOREM 1 (Simpson [Si77]). *The degree of Th* $(\mathcal{D})$ = *the degree of Th* $(\mathcal{N})$ *(the theory of second-order arithmetic).*

Theorem 1 was originally proved using an ad hoc coding of Th $(\mathcal{N})$ into Th $(\mathcal{D})$, but was proved more directly by Nerode and Shore [NS80] using the countable distributive lattice initial segment embedding result.

In relation to the original question of Kleene and Post concerning the definability of the jump operator, we have:

THEOREM 2 (Shore [Sh82]). *Any relation on* $\mathcal{D}(\geq \mathbf{0}^{(3)})$ *which is definable in second-order arithmetic is definable in* $\mathcal{D}'$.

(The first result of this kind was proved by Simpson [Si77] with $\mathbf{0}^{(\omega)}$ in place of $\mathbf{0}^{(3)}$, the improvement to $\mathbf{0}^{(3)}$ emerging via $\mathbf{0}^{(7)}$ in Nerode and Shore [NS80], [NS80a], who also showed how to replace the jump by a parameter such as $\mathbf{0}''$ in many of these global results. There was already a natural definition of $\mathbf{0}^{(\omega)}$ in $\mathcal{D}'$, got by combining results of Enderton and Putnam [EP70] and Sacks [Sa71], as the least double-jump of an upper-bound for the arithmetical degrees.)

Without the jump much less could be said:

THEOREM 3 (Jockusch and Shore [JSh84]). *Any relation on the degrees above all the arithmetical ones is definable in* $\mathcal{D}$ *if and only if it is definable in second-order arithmetic. In particular,* $\mathbf{0}^{(\omega)}$, *and hence the* ω*-jump, is definable in* $\mathcal{D}$.

(The first part of Theorem 3 improves Harrington and Shore [HS81] by replacing 'hyperarithmetic' with 'arithmetical.')

There are also results concerning the existence of structural characterisations of the jump related classes of the high/low hierarchy, defined by

$$\text{High}_n = \{\mathbf{a} \leq \mathbf{0}' \mid \mathbf{a}^{(n)} = \mathbf{0}^{(n+1)}\},$$
$$\text{Low}_n = \{\mathbf{a} \leq \mathbf{0}' \mid \mathbf{a}^{(n)} = \mathbf{0}^{(n)}\}.$$

THEOREM 4 (Shore [Sh88]). *High*$_n$, *Low*$_n$ *are definable in* $\mathcal{D}(\leq \mathbf{0}')$ *for each* $n \geq 3$.

The above results are all proved using degree-theoretic codings, Theorems 2, 3 and 4 using developments of the Nerode/Shore coding methods. More recently, all such results have been derived using the simpler coding technique of Slaman and Woodin (see [SW86] and [OS*ta*]). The three quantifiers intrinsic to codings involving $\leq_T$ explain the best possible lower-bound $\mathbf{0}^{(3)}$ in these and other global results. Another striking definability result using these codings is:

THEOREM 5 (Slaman and Woodin [SW86]). *The set of r.e. degrees is definable in* $\mathcal{D}(\leq \mathbf{0}')$ *using a finite number of parameters.*

In another direction, there are a number of results concerning definability of particular jump ideals in $\mathcal{D}$ or $\mathcal{D}'$ (see [HS81], [NS80a] and [Sh81]). Jockusch and Shore [JSh84] obtain definability results using their theory of pseudo-jumps. For instance, with no use of codings they get:

THEOREM 6 (Jockusch and Shore [JSh84]). $\mathcal{A}$ *(= the set of arithmetical degrees) is definable in* $\mathcal{D}$.

§4. Homogeneity.

Here again we see a large gap between the situations with and without the jump. We single out the following from among the strongest previous results concerning strong homogeneity for $\mathcal{D}'$:

THEOREM 7 (Richter [Ri79]). *If* $\mathcal{D}'(\geq \mathbf{a}) \cong \mathcal{D}'(\geq \mathbf{b})$ *then* $\mathbf{a}^{(3)} = \mathbf{b}^{(3)}$.

This means, for instance, that for each $n \geq 1$, $\mathcal{D}' \not\cong \mathcal{D}'(\geq \mathbf{0}^{(n)})$. Theorem 7 essentially improves [JSol77] by a factor of one jump, and is the culmination of a sequence of results obtained by various authors, starting with Feiner's [Fe70] refutation of the Strong Homogeneity Conjecture for $\mathcal{D}'$ (see also [Ya72]). Yates [Ya70] formulated the Homogeneity Conjecture by replacing isomorphism with elementary equivalence in Rogers' original question, and Simpson [Si77] gave a negative answer. The strongest refutation is due to Shore:

THEOREM 8 (Shore [Sh81]). *If* $\mathcal{D}' \equiv \mathcal{D}'(\geq \mathbf{a})$ *then* $\mathbf{a}^{(3)} = \mathbf{0}^{(3)}$.

Without the jump information is much more difficult to obtain. Combining Theorem 4.7 from [NS80] concerning automorphisms of $\mathcal{D}$ with work of Jockusch and Soare [JSoa70] and Harrington and Kechris [HK75] on minimal covers, Shore [Sh79] disproved the Strong Homogeneity Conjecture for $\mathcal{D}$. Shore [Sh82] extended this result on isomorphisms of cones to one on elementary equivalence: *If the degree of Kleene's* $\mathcal{O} \leq \mathbf{a}^{(n)}$ *for some* n *then* $\mathcal{D} \not\equiv \mathcal{D}(\geq \mathbf{a})$, so disproving homogeneity without the jump. Improving Harrington and Shore [HS81] (by replacing 'hyperarithmetic' with 'arithmetic') we have:

THEOREM 9 (Jockusch and Shore [JSh84]). *1) If* $\mathcal{D}(\geq \mathbf{a}) \cong \mathcal{D}(\geq \mathbf{b})$ *then* $\mathbf{a}$ *and* $\mathbf{b}$ *are arithmetically equivalent.*
2) If $\mathcal{D}(\geq \mathbf{a}) \equiv \mathcal{D}$ *then* $\mathbf{a}$ *is arithmetical.*

So in particular, $\mathcal{D} \not\equiv \mathcal{D}(\geq \mathbf{0}^{(\omega)})$.

§5. Automorphisms.

Corresponding to Theorem 7 we have:

THEOREM 10 (Epstein [Ep79], Richter [Ri79]). *Let* f *be an automorphism of* $\mathcal{D}'$. *Then* $f(\mathbf{a}) = \mathbf{a}$ *for all* $\mathbf{a} \geq \mathbf{0}^{(3)}$.

Jockusch and Solovay [JSol77] were the first to prove that jump preserving automorphisms are the identity on a cone (with $\mathbf{0}^{(4)}$ instead of $\mathbf{0}^{(3)}$ as the base of the cone). Nerode and Shore [NS80] show how to replace the jump by the parameter $\mathbf{0}'$ in the above result.

Without the jump we have:

THEOREM 11 (Jockusch and Shore [JSh84]). *If ψ is an automorphism of $\mathcal{D}$ then $\psi(\mathbf{a}) = \mathbf{a}$ for all $\mathbf{a} \geq \mathbf{0}^{(\omega)}$.*

Theorem 4.7 of Nerode and Shore [NS80] was the first result of this kind for $\mathcal{D}$, improvements in the location of the base of the cone appearing in [HS81] and [Sh81].

Other restrictions on the possible automorphisms of $\mathcal{D}$ are provided by the notion of an ***automorphism base*** (see Jockusch and Posner [JP81]).

The situation described so far is one in which much stronger results, and simpler proofs, are possible for $\mathcal{D}'$ than for $\mathcal{D}$. The aim now is to reduce the theory for $\mathcal{D}$ to that for $\mathcal{D}'$.

§6. Pseudo-jump operators.

Jockusch and Shore [JSh83], [JSh84] observe that constructions in recursion theory relative to a set A produce a set $J(A)$ which is, from a formal point of view, very similar in definition to the jump A', or more generally αth jump $A^{(\alpha)}$ for suitable recursive ordinal α, of A. This leads them to abstract from this the notion of an ***α-REA operator***, and to mimic (in a nontrivial way) completeness and cupping theorems of Friedberg [Fr57], MacIntyre [Ma77] and Posner and Robinson [PR81] for the usual αth jump to produce cones of degrees with interesting structural properties. We need below the Jockusch and Shore pseudo-jump machinery for α finite (in fact, for $\alpha = 2$).

DEFINITION. We say that J^n is an ***n-REA operator*** if and only if there exist $j_0, j_1, \ldots, j_{n-1} \in \omega$ such that J^n is defined by

$$J^0(A) = A,$$
$$J^{k+1}(A) = J^k(A) \oplus W_{j_k}^{J^k(A)}, \ (k < n).$$

Natural examples of n-REA operators are given by

(1) Choose j_0 such that $W_{j_0}^A = \{x \mid x \in W_x^A\}$. Then the 1-REA operator J^1 defined by $J^1(A) = A \oplus W_{j_0}^A$ is Turing equivalent to A', the usual Turing jump of A.
(2) If $D = W_i - W_j$ is a *d-r.e. set* (a difference of two r.e. sets), we can define a 2-REA operator J^2 (see p. 1209 of [JSh84] for a detailed verification) by

$$J^2(A) = A \oplus (W_i^A - W_j^A).$$

(One can derive n-REA operators from n-r.e. sets in a similar way.)

We will need the following analogues of the Friedberg completeness and Posner–Robinson cupping theorems:

COMPLETENESS THEOREM FOR n-REA OPERATORS. *If J is an n-REA operator, for each C there is a set X^C such that $C \oplus \emptyset^n \equiv_T J(X^C)$.*

(The proof is a simplification of that of Theorem 2.3 of Jockusch and Shore [JSh84].)

CUPPING THEOREM FOR n-REA OPERATORS. *If J is an n-REA operator derived from an n-r.e. set, then if $D \geq_T \emptyset^{(n)} \oplus X$ and $X \not\leq_T \emptyset^{(n-1)}$, we can find an A such that*

$$X \oplus A \equiv_T D \equiv_T J(A).$$

(The proof is essentially contained in that of Theorem 3.2 of Jockusch and Shore [JSh84].)

In the next section we apply these theorems to a particular 2-REA operator.

§7. Some local degree theory and the definability of the jump.

We first outline how the construction of a d-r.e. degree with special properties yields the required 2-REA operator.

DEFINITION. Given $\mathbf{a}, \mathbf{b}, \mathbf{d}$, we say $\mathbf{d}$ is *unsplittable over* $\mathbf{a}$ *avoiding* $\mathbf{b}$ if and only if $\mathbf{a}, \mathbf{b} \leq \mathbf{d}$, $\mathbf{b} \not\leq \mathbf{a}$, and for all $\mathbf{d}_0, \mathbf{d}_1 < \mathbf{d}$, if $\mathbf{a} < \mathbf{d}_0, \mathbf{d}_1$ then *either* $\mathbf{b} \leq \mathbf{d}_0$ or $\mathbf{d}_1$, or $\mathbf{d} \neq \mathbf{d}_0 \cup \mathbf{d}_1$.

$\mathbf{d}$ is *relatively unsplittable* if and only if $\mathbf{d}$ is unsplittable over $\mathbf{a}$ avoiding $\mathbf{b}$, some $\mathbf{a}, \mathbf{b}$.

It is important to notice that, by the relativised Sacks Splitting Theorem (see [So87], p.124), there is no relatively unsplittable r.e. degree.

THE MAIN THEOREM. *There is a relatively unsplittable d-r.e. degree.*

That is, there is a d-r.e. set $D = W_i - W_j$ (say) and sets $A, B \leq_T D$ such that $\deg(D)$ is unsplittable over $\deg(A)$ avoiding $\deg(B)$.

Before outlining the construction we list some immediate applications of the main theorem.

We first notice that we can use the main theorem to get a 2-REA operator J such that for each B we have $J(B) = B \oplus (W_i^B - W_j^B)$ and $\deg(J(B))$ is unsplittable over some $\mathbf{a} \geq \deg(B)$. Then applying the Completeness Theorem for n-REA Operators to J we get:

THEOREM 12. *There is a cone of relatively unsplittable degrees with base* $\mathbf{0}''$. □

Using J with the Cupping Theorem for n-REA Operators we get:

THEOREM 13 (DEFINABILITY OF $\mathbf{0}'$). *$\mathbf{0}'$ is definable in $\boldsymbol{\mathcal{D}}$ as the largest degree satisfying*

$$(\dagger) \qquad \neg(\exists \mathbf{a}, \mathbf{b})[\mathbf{x} \cup \mathbf{a} \text{ is unsplittable over } \mathbf{a} \text{ avoiding } \mathbf{b}].$$

Proof. As previously remarked, each r.e. degree, including $\mathbf{0}'$, satisfies (†) by the relativised Sacks Splitting Theorem.

On the other hand, say $X \not\leq_T \emptyset'$. Then if $D \geq_T \emptyset'' \oplus X$ the Cupping Theorem gives us an A_1 for which $X \oplus A_1 \equiv_T D \equiv_T J(A_1)$, and $\deg(X \oplus A_1)$ is unsplittable over some degree $\mathbf{a} \geq \deg(A_1)$ avoiding some $\mathbf{b}$, where $\deg(X) \cup \deg(A_1) = \deg(X) \cup \mathbf{a}$. □

Relativising, this means that for each $\mathbf{a}$, $\mathbf{a}'$ is definable in $\mathcal{D}(\geq \mathbf{a})$, giving:

THEOREM 14 (DEFINABILITY OF THE JUMP). *The Turing jump is definable in $\mathcal{D}$.* □

This of course implies that the jump is order-theoretic, answering the question of Rogers referred to in section 2 above. Since $\mathbf{0}^{(n)}$ is definable for each n, we can get Theorem 6 (definability of the set of arithmetical degrees) as another corollary. And although Theorem 14 adds nothing to the known definability results below $\mathbf{0}'$, we can complement Theorem 4 above with:

THEOREM 15. *All the jump classes $High_n$ and Low_n, $n > 0$, are definable in $\mathcal{D}$.* □

We can use Theorem 14 to translate known results on definability, homogeneity and automorphisms for $\mathcal{D}'$ into dramatically improved results for $\mathcal{D}$. In place of Theorem 3 above we can restate Theorem 2:

THEOREM 2+. *Any relation on $\mathcal{D}(\geq \mathbf{0}^{(3)})$ which is definable in second-order arithmetic is definable in $\mathcal{D}$.*

Instead of Theorem 9 we get from Theorems 7 and 8:

THEOREM 7+. *If $\mathcal{D}(\geq \mathbf{a}) \cong \mathcal{D}(\geq \mathbf{b})$ then $\mathbf{a}^{(3)} = \mathbf{b}^{(3)}$.*

THEOREM 8+. *If $\mathcal{D} \equiv \mathcal{D}(\geq \mathbf{a})$ then $\mathbf{a}^{(3)} = \mathbf{0}^{(3)}$.*

And using Theorem 10 we can replace 'ω' by '3' in Theorem 11:

THEOREM10+. *Let f be an automorphism of $\mathcal{D}$. Then $f(\mathbf{a}) = \mathbf{a}$ for all $\mathbf{a} \geq \mathbf{0}^{(3)}$.*

§8. Proof of the main theorem.

The following sketch can be used as an introduction to the full proof in [Co*tal*].

Let $(\Theta_k, \Psi_k, \Phi_k, \widehat{\Phi}_k)$, $k \geq 0$, be a standard listing of all 4-tuples of p.r. functionals. We need to construct a d-r.e. set D and sets $A, B \leq_T D$ satisfying the requirements:

$$P_k: \quad B \neq \Theta_k^A,$$

$$Q_k: \quad D = \Psi_k(\Phi_k^D, \widehat{\Phi}_k^D) \Rightarrow B = \Gamma_k(\Phi_k^D, A) \vee B = \Lambda_k(\widehat{\Phi}_k^D, A),$$

$k \geq 0$, where Γ_k, Λ_k are p.r. functionals to be constructed. We also need an overall constraint that $A = \Omega^D$, Ω a p.r. functional to be defined during the construction. The Q-requirements will ensure that $B \leq_T D$.

The basic module is closely related to the Lachlan 'monster construction' [La75] of a r.e. degree which is relatively non-splitting within the r.e. degrees. We consider just two requirements P ($= P_{k'}$, say) and Q ($= Q_k$, say) in relation to each other, Q being of higher priority than P. We follow the convention of writing $\theta_k, \varphi_k, \psi_k, \gamma_k, \lambda_k$ etc. for the respective standard use functions of $\Theta_k, \Phi_k, \Psi_k, \Gamma_k, \Lambda_k$ etc.

The naive P-strategy: Look for an x with $\Theta^A(x)\downarrow$ for which we can *define* $B(x) \neq \Theta^A(x)$ and *restrain* $A \restriction \theta(x)$.

The naive Q-strategy: First try to implement the Γ-strategy: If P requires us to make a $B(x)$-change, try to produce a situation such that *either*
(a) $\gamma(x) > \theta(x)$ (so we can rectify the equation $B(x) = \Gamma(\Phi^D, A)(x)$ following the $B(x)$-change with an A-change bigger than $\theta(x)$), *or*
(b) $\gamma(x) > \psi(y)$, some y, and hope to get a $\Phi^D \restriction \gamma(x)$-change by using a $D(y)$-change to force a $\Phi^D \restriction \psi(y)$-change.

If it looks like we always get a $\widehat{\Phi}^D \restriction \psi(y)$ change in (b), start to implement the Λ-strategy.

We consider in detail some of the problems involved in reconciling the strategies for P and Q:

Some problems: Roughly speaking, our strategy for P and Q together is as follows. If $\ell(D, \Psi(\Phi^D, \widehat{\Phi}^D))$ (the standard length of agreement function at stage $s+1$) grows large, we follow the naive Q-strategy in initially implementing the Γ-strategy for making $B \leq_T \Phi^D \oplus A$. P will have associated with it followers x for which we hope to get $\Theta^A(x)\downarrow \neq B(x)$. This may conflict with the Γ-strategy in that changing $B(x)$ to disagree with $\Theta^A(x)$ at stage $s+1$ may not result in a $\Phi^D \restriction \gamma(x)$ change. This will require the $B(x)$ change to be signalled through a change in $A \restriction \gamma(x) \subset A \restriction \theta(x)$, resulting in a possible reassertion of the equation $B(x) = \Theta^A(x)$ at some later stage.

According to the naive Q-strategy, our first approach to a resolution of this conflict will be to try to make such a $\gamma(x) > \theta(x)$, so that the A-change is above the use of $\Theta^A(x)$. But in general we can only do this by also injuring the existing use of $\Theta^A(x)$, in the hope that our new larger $\gamma(x)$ will be greater than $\theta(x)$ when this becomes defined again. This process (basically Harrington's 'capricious destruction,' although the capriciousness is only apparent) may be repeated using A-changes on larger and larger numbers, along with them an ascending sequence of numbers needed for corresponding D-permissions.

There are various possible outcomes to this. We may succeed in obtaining a suitable relatively small use for Θ^A, make our choice of $B(x)$, and satisfy P. On the other hand, infinite repetition of this process will lead to $\Theta^A(x)\uparrow$ (P satisfied again), but (without further analysis) we will also end up with $\Gamma^{\Phi^D, A}$ not total so that the Γ-strategy fails. There is another possibility for avoiding this. Since we are only concerned about Q if Φ^D is a total function, we may permit the A-changes needed to move $\gamma(x)$ by the enumeration of an *agitator* y into D, but defer actually making those A-changes until at least $\Phi^D \restriction \gamma(x)$ has become redefined. This leaves open the possibility that we may get a completely new $\Phi^D \restriction \gamma(x)$ (that is, not containing as an initial segment any previously defined $\Phi^D \restriction \gamma(x)$) which can be used to permit $x \searrow B$ ('x entering B') via Γ without the need for the A-changes to be made. But then, assuming we have timed our enumeration of y into D to coincide with $\Theta^A(x)\downarrow = B(x) = 0$, we avoid disturbing $A \restriction \theta(x)$. Hence we get $\Theta^A(x) = 0 \neq B(x)$ following the above actions, so satisfying P and in the process leaving the Γ-strategy intact.

We can assist this outcome by using A to increase the likelihood of the new $\Phi^D \restriction \gamma(x)$ being usable to permit $x \searrow B$ via Γ. Following the monster construction we might try to ensure that whenever we define $\gamma(x)$ or $\lambda(x)$ previous to $y \searrow D$ we have $\psi(y)\downarrow \leq \gamma(x)$ or $\lambda(x)$ respectively, so that $y \searrow D$ will at least produce some sort of change in either $\Phi^D \restriction \gamma(x)$ or $\widehat{\Phi}^D \restriction \lambda(x)$. This is attempted via a process similar to Harrington's 'honestification,' whereby if $\gamma(x)$ or $\lambda(x) < \psi(y)$ when we require $y \searrow D$, we first produce an $A \restriction w$ change, with $w \leq \min\{\gamma(x), \lambda(x)\}$, redefining $\gamma(x), \lambda(x) \geq \psi(y)$ when $\psi(y)$ is next defined. In fact honestification is extended by making $w \leq \min\{\gamma(x), \lambda(x)\}$ for all such $\gamma(x), \lambda(x)$ defined since the last occurrence of honestification, thereby ensuring that previously defined $\Phi^D \restriction \gamma(x)$ or $\widehat{\Phi}^D \restriction \lambda(x)$ will not return in tandem with corresponding $A \restriction \gamma(x)$ or $A \restriction \lambda(x)$ to prevent us Φ^D- or $\widehat{\Phi}^D$-permitting B-changes via Γ or Λ respectively following $y \searrow D$.

There is a problem here (apart from that of not knowing whether we get a Φ^D- or $\widehat{\Phi}^D$-change following $y \searrow D$), in that honestification will very likely also involve an $A \restriction \theta(x)$ change, so that $y \searrow D$ will have to wait for $\theta(x)$ to become redefined, by which time the effects of honestification may have worn off, demanding renewed honestification. If this repetition developes into an infinite outcome, we get $\Gamma^{\Phi^D, A}, \Lambda^{\widehat{\Phi}^D, A}$ not total. But Q is then satisfied since we must have $\min\{\gamma(x), \lambda(x)\} < \psi(y)$ infinitely often so that $\Psi(\Phi^D, \widehat{\Phi}^D)(y)\uparrow$ also. And $\theta(x)\uparrow$ infinitely often, so P is satisfied through $\Theta^A(x)\uparrow$.

However, honestification as described will still not be sufficient to supply the ideal conditions for $y \searrow D$. This is because of a new complication resulting from the possibility of returns to strings $\Phi^D \restriction \gamma(x)$ (following $y \searrow D$) which appeared since the last occurrence of honestification for (P, Q). So before allowing $y \searrow D$ we further ask that $\Phi^D \restriction \psi(y), \widehat{\Phi}^D \restriction \psi(y)$ are unchanged at all stages since the previous occurrence of honestification, and if this condition is not met, we again honestify (even if $\psi(y) \leq \gamma(x), \lambda(x)$). If we never get to act on such a y but continue to honestify we still get $\Psi(\Phi^D, \widehat{\Phi}^D)(y)\uparrow$, and Q is again satisfied (along with P since $\Theta^A(x)\uparrow$).

Anticipating any consideration of how all this is to coexist with our actions on other requirements, we should mention one situation where we do need D to be d-r.e., and not just r.e. Say we have a P', of priority intermediate between that of Q and P, and that we get to enumerate y into D, achieving a suitable new $\Phi^D \restriction \gamma(x)$ which we restrain in order to be able to preserve $\Theta^A(x)\downarrow \neq B(x)$ following $x \searrow B$ while maintaining the Γ-strategy. It may happen at a later stage that we act on some y' through P', resulting in a loss of the new $\Phi^D \restriction \gamma(x)$ (replaced by a new $\widehat{\Phi}^D \restriction \lambda(x)$, presumably). It may not be possible now to rectify Γ by a suitable A-change, as this may conflict with the actions for P' (for instance). We then have no alternative but to extract x from B via an extraction from D. (There is still the possibility of the new $\Phi^D \restriction \gamma(x)$ which permitted $x \searrow B$ via Γ reasserting itself at a later stage, but in defining the corresponding $\gamma(x)$ we will have been able to have regard for higher priority P' to the extent that we can avoid having to re-enumerate x into B by making a suitable $A \restriction \gamma(x)$ change.)

It remains to follow through the consequences of infinitely many occurrences of agitators $y \searrow D$ for (P, Q) and producing no appropriate new strings $\Phi^D \upharpoonright \gamma(x)$. In this case we utilise the fact that following each $y \searrow D$ we get a new $\widehat{\Phi}^D \upharpoonright \lambda(x)$ to satisfy P, Q through the Λ-strategy. In fact, this outcome for (P, Q) gives us a successful Λ-strategy for each (P', Q) with P' ($= P_{k''}$ say) of lower priority than P, so we will not assume that the infinite set of y's we act on necessarily relates to P'.

We now assume that the Λ-strategy has its own set of followers $z \geq 0$ each with its own set of agitators $\widehat{y} \geq 0$, disjoint from any other set of followers or agitators. We act on each $\widehat{y}$ with the pre-knowledge that we get infinitely many $A \upharpoonright \gamma(x)$ changes through capricious destruction, and infinitely many usable $\widehat{\Phi}^D \upharpoonright \lambda(y)$ changes (or, more relevant, no usable $\Phi^D \upharpoonright \gamma(y)$ changes). This means we only bother to act in the interests of $B(z) \neq \Theta^A_{k''}(z)$ if $\theta_{k''}(z) < \gamma(x)$. Since $\gamma(x)$ goes to infinity, this will still provide sufficient space in which to satisfy $P_{k''}$.

In order to use an agitator $\widehat{y}$ we also need to obtain $\psi(\widehat{y}) \leq \psi(y)$ and $\psi(\widehat{y}) \leq \lambda(z)$, and to then act simultaneously to obtain $\widehat{y} \searrow D$ and $y \searrow D$ in the interests of obtaining a usable $\widehat{\Phi}^D \upharpoonright \lambda(z)$ change to permit $z \searrow B$ via Λ without the need to injure $\Theta^A_{k''}(z) \neq D(z)$ with an $A \upharpoonright \lambda(z)$ change. This requires its own honestification, which we can time to coincide with the honestification for (P, Q). Again, the honestification takes the stronger form described previously.

The strategies for the different requirements can be harmonised via a tree of outcomes, in a (by now) fairly standard $0'''$-priority context.

We now give a more formal description of the strategies for (P, Q).

The basic module for P confronted with one higher priority Q.

(All statements in the description below are assumed to relate to stage $s+1$ of the construction.)

Let

$$\begin{aligned} \ell(D, \Psi(\Phi^D, \widehat{\Phi}^D)) &= \mu z\,[D(z) \neq \Psi(\Phi^D, \widehat{\Phi}^D; z)] \text{ and} \\ \ell(B, \Theta^A) &= \mu z\,[B(z) \neq \Theta^A(z)] \qquad \text{(at stage } s+1). \end{aligned}$$

We associate with (P, Q) four disjoint infinite recursive sets $\xi, \eta, \widehat{\eta}$ and ζ.

We have two overall constraints on the construction relative to (P, Q):

(a) If $z \leq s$ then we must define $A(z) = \Omega^D(z)$ at all stages $s' + 1 > s$.

(b) If $\ell(D, \psi(\Phi^D, \widehat{\Phi}^D)) > z$ then we must define $\Gamma^{\Phi^D, A} \upharpoonright x$ or $\Lambda^{\widehat{\Phi}^D, A} \upharpoonright x$ whenever z is a D-agitator for x with $z \notin D$.

And if y, z are D-, $\widehat{D}$-agitators respectively for x with $y, z \notin D$ (at stage $s+1$) we ask that whenever we redefine $\Gamma^{\Phi^D, A} \upharpoonright x$ or $\Lambda^{\widehat{\Phi}^D, A} \upharpoonright x$ we choose $\gamma(x), \lambda(x)$ so that $\gamma(x) \geq \psi(y)$ or $\lambda(x) \geq \psi(z)$ respectively. Also, whenever we redefine $\Gamma^{\Phi^D, A}(w)$ or $\Lambda^{\widehat{\Phi}^D, A}(w)$, $w \geq 0$, we define $\Gamma^{\Phi^D, A}(w) = B(w)$ or $\Lambda^{\widehat{\Phi}^D, A}(w) = B(w)$, respectively.

Whenever we redefine values of $\Gamma^{\Phi^D,A}$ or of $\Lambda^{\widehat{\Phi}^D,A}$ in such a way that $\Gamma^{\Phi^D,A} \simeq B$ (that is, they agree on all values on which both are defined) or $\Lambda^{\widehat{\Phi}^D,A} \simeq B$, we say that we *rectify* Γ or Λ, respectively.

The basic module consists of the following phases together with the above overall constraints.

1) We *select* the least $x \in \xi - B$ to *follow* (P, Q).
2) We *select* the least $y \in \eta - D$, $y > x$, as a *D-agitator* for x.
3) We *select* the least $w \in \zeta - A$ as an *A-agitator* for x.
4) We *wait* for $\ell(D, \Psi(\Phi^D, \widehat{\Phi}^D))$ to grow bigger than y and *define* $\gamma(x) \geq w$.
5) And we *wait* for $\ell(B, \Theta^A)$ to grow bigger than x.
6) We then *check* if $\gamma(x)\downarrow > \theta(x)$.
   If so, we *enumerate* x into B, *restrain* $A \restriction \theta(x)$ and *rectify* Γ with an $A \restriction \gamma(x)$ change.
   Outcome: P is *satisfied* and ceases to interfere with Q.
7) (Honestification and capricious destruction combined.)
   Otherwise we *change* $A \restriction \gamma(x)$ using $w \searrow A$ (and a corresponding $D \restriction \omega(w)$ change), and *proceed* through phases 3), 4), 5) and then 8).
8) We now *check* if $\gamma(x) \geq \psi(y)$ and if $\Phi^D \restriction \psi(y)$, $\widehat{\Phi}^D \restriction \psi(y)$ are unchanged at all stages since they became redefined following the latest application of phase 7).
   (a) If so, we *enumerate* y into D, and *go to* 9).
   (b) If not, *return* to 3).
9) We *wait* for $\ell(D, \Psi^{\Phi^D,\widehat{\Phi}^D}) > y$, and then:
10) *Check* if $\Gamma^{\Phi^D,A}(x)\uparrow$.
    (a) If so we *define* $\Gamma^{\Phi^D,A}(x) = B(x) \neq \Theta^A(x)$, *restrain* $\Phi^D \restriction \gamma(x)$ and $A \restriction \theta(x)$.
    Outcome: P is *satisfied*, and Γ is *rectified*.
    (b) Otherwise we *return* to 2).

In the case of infinitely many returns to 2) on behalf of (P, Q), we need to describe the Λ-strategy. This is an auxiliary strategy that synchronises its activities with phases 2), 7) and 8) of the Γ-strategy. As mentioned before, it can relate to (P', Q) even if $P' \neq P$.

$\widehat{1}$) We *select* the least $x' \in \xi' - B$ to *follow* (P', Q) with $x' > x$.
$\widehat{2}$) (Simultaneous with 2).) We *select* the least $z \in \widehat{\eta} - D$, $z < y$, as a $\widehat{D}$*-agitator* for x' (if such a z exists).
$\widehat{8}$) (Simultaneous with 8).) We also *check* if $\lambda(x') \geq \psi(z)$.
   (a) If so and y has entered D, *enumerate* z into D, and *go to* $\widehat{9}$).
   (b) Otherwise we *return* to 3) as already described.
$\widehat{9}$) We *wait* for $\ell(D, \Psi^{\Phi^D,\widehat{\Phi}^D}) > z$, and then:
$\widehat{10}$) *Check* if $\gamma(x) > \theta'(x')$ and $\Lambda^{\widehat{\Phi}^D,A}(x')\uparrow$.
   (a) If so we *define* $\Lambda^{\widehat{\Phi}^D,A}(x') = B(x') \neq \Theta'^A(x')$, and *restrain* $\widehat{\Phi}^D \restriction \lambda(x')$ and $A \restriction \theta'(x')$.
   Outcome: P' is *satisfied* and Λ is *rectified*.

(b) Otherwise, we *return* to $\widehat{2}$).

When we consider more than two requirements the sequence of events (for instance the timing of the A-restraints) will need modifying, but the basic framework still holds.

Summary of outcomes of the Γ- *and* Λ-*strategies for* (P, Q), (P', Q).

The finite outcomes:

$\boxed{w_1}$: The strategy halts at 4). Then $D \neq \Psi(\Phi^D, \widehat{\Phi}^D)$ and Q is satisfied and ceases to interfere with P.

$\boxed{w_2}$: The strategy halts at 5). Then $B \neq \Theta^A$ and P is satisfied and ceases to interfere with Q.

$\boxed{s_1}$: $B(x) \neq \Theta^A(x)$, P is satisfied and ceases to interfere with Q, due to phase 6) applying.

$\boxed{w_1'}$: The strategy halts at 9). Outcome as for $\boxed{w_1}$.

$\boxed{s_2}$: Strategy halts at 10). $B(x) \neq \Theta^A(x)$, P is satisfied while maintaining $\Gamma^{\Phi^D, A} = D$ via a Φ^D-change.

$\boxed{\widehat{w}_1}$: The strategy halts at $\widehat{9}$). Outcome as for $\boxed{w_1}$ and $\boxed{w_1'}$.

$\boxed{\widehat{s}_2}$: Strategy halts at $\widehat{10}$). $B(x') \neq \Theta'^A(x')$, P' is satisfied while maintaining $\Lambda^{\widehat{\Phi}^D, A} = D$ via a $\widehat{\Phi}^D$-change.

The infinitary outcomes:

$\boxed{i_1}$: The stategy passes through phase 7) (but not 8)(a)) infinitely often.

Since we infinitely often pass through phases 3) and 4), $\gamma(x)$ goes to infinity. Since we never halt at 6), $\theta(x) \geq \gamma(x)$ infinitely often, so $\Theta^A(x) \uparrow$ and P is satisfied.

Since we go through 8)(b) infinitely often, either $\psi(y) > \gamma(x)$ infinitely often, or $\Phi^D \restriction \psi(y)$ or $\widehat{\Phi} \restriction \psi(y)$ changes infinitely often, so in either case $\Psi^{\Phi^D, \widehat{\Phi}^D}(y) \uparrow$ (possibly with $\psi(y)$ bounded but $\Phi^D(u)$ or $\widehat{\Phi}^D(u) \uparrow$, some $u \leq \psi(y)$), giving Q also satisfied.

$\boxed{i_2}$: The strategy passes through phase 10)(b) infinitely often. *Outcome:* We implement the Λ-strategy, P' is satisfied as in $\boxed{i_1}$.

$\boxed{\widehat{i}_2}$: The strategy passes through $\widehat{10}$)(b) infinitely often.

As for $\boxed{i_1}$ we get P' satisfied through $\Theta'^A(x') \uparrow$, while the Λ-strategy for (P', Q) is maintained. This is because, by the conditions of 8)(a) and $\widehat{8}$)(a) we must arrive at $\widehat{10}$)(a) with either $\Lambda^{\widehat{\Phi}^D, A}(x') \uparrow$ or $\Gamma^{\Phi^D, A}(x) \uparrow$. Since 10)(a) does not apply, we must have $\Lambda^{\widehat{\Phi}^D, A}(x') \uparrow$, so we get to rectify Λ before returning to 2)/ $\widehat{2}$).

The diagram shown relates the strategies to the outcomes.

The following diagram relates the strategies to the outcomes:

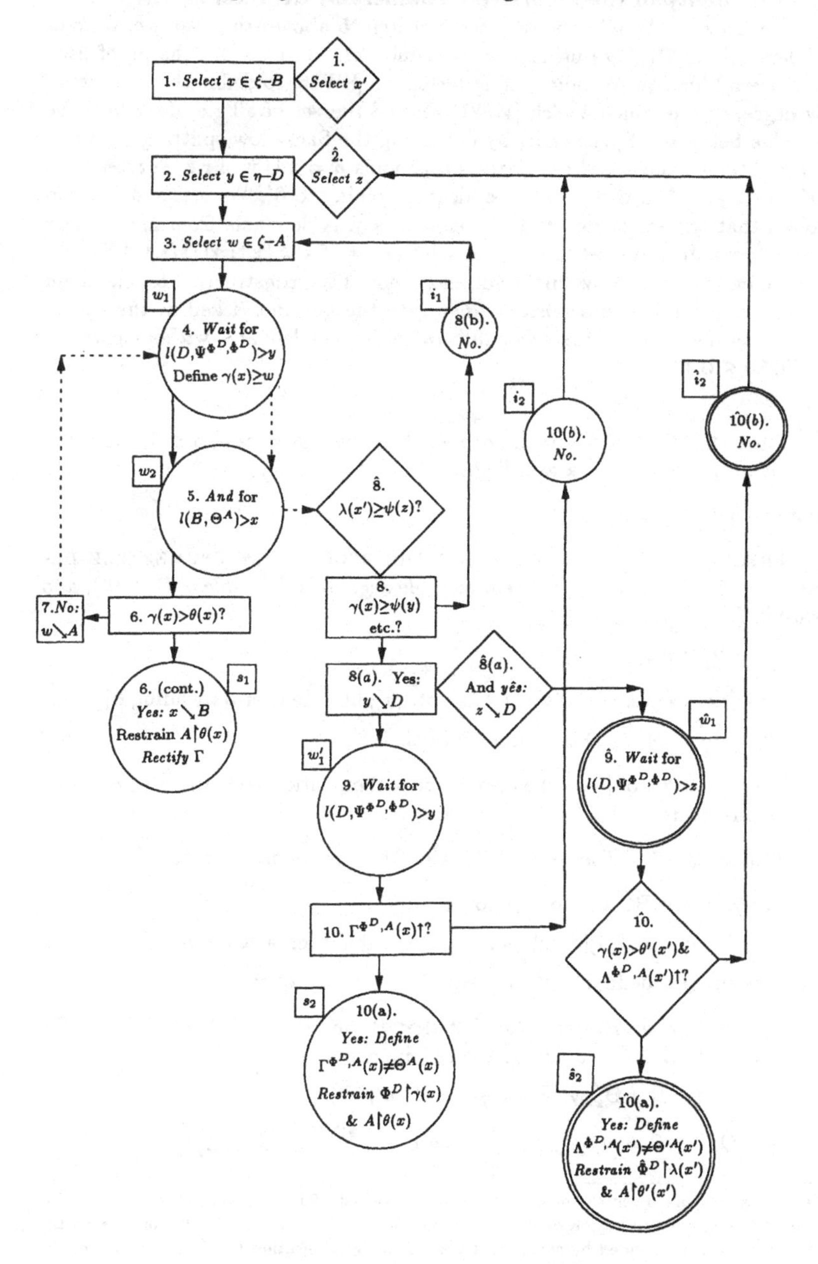

§9. Definability of the recursively enumerable degrees.

Slaman and Woodin showed (see Theorem 5 above) that the r.e. degrees are definable in $\mathcal{D}(\leq \mathbf{0}')$ using a finite number of parameters. The proof used the powerful Slaman–Woodin coding technique [SW86] to define the two sets of low degrees from which Welch [We81] showed the set of all r.e. degrees to be definable below $\mathbf{0}'$. Specifically, by extending the Sacks low splitting theorem, Welch obtained the set of r.e. degrees $\mathbf{a}$ as joins $\mathbf{a} = \mathbf{x} \cup \mathbf{y}$ of r.e. degrees $\mathbf{x}, \mathbf{y}$, $\mathbf{x} \leq \mathbf{c}$ and $\mathbf{y} \leq \mathbf{d}$, $\mathbf{c}, \mathbf{d}$ fixed low r.e. degrees joining to $\mathbf{0}'$. Slaman and Woodin showed that the set of r.e. degrees below $\mathbf{c}$, say, is definable from appropriate parameters $\mathbf{a}, \mathbf{b}$ as the set of minimal solutions $\mathbf{x}$ of $\mathbf{x} \not\leq (\mathbf{x} \cup \mathbf{a}) \cap (\mathbf{x} \cup \mathbf{b})$.

We now describe how further development of the construction for the main theorem above leads to a positive answer to the question asked at the end of [SW86]: *Is the set of recursively enumerable degrees definable without parameters in $\mathcal{D}(\leq \mathbf{0}')$?*

From:

THEOREM 16. *If $\mathbf{d} < \mathbf{0}'$ is not r.e., there exist degrees $\mathbf{a}, \mathbf{b} < \mathbf{0}'$ such that $\mathbf{a} \cup \mathbf{d}$ is unsplittable over $\mathbf{a}$ avoiding $\mathbf{b}$.*

we immediately get:

THEOREM 5+ (DEFINABILITY OF THE RECURSIVELY ENUMERABLE DEGREES). *The set of recursively enumerable degrees is definable in $\mathcal{D}(\leq \mathbf{0}')$, and hence in $\mathcal{D}$.*

Proof. If $\mathbf{d} \leq \mathbf{0}'$,

$$\mathbf{d} \text{ is r.e.} \Leftrightarrow (\forall \mathbf{a}, \mathbf{b} \leq \mathbf{0}')[\mathbf{a} \cup \mathbf{d} \text{ is not unsplittable over } \mathbf{a} \text{ avoiding } \mathbf{b}].$$

Since $\mathbf{0}'$ is definable in $\mathcal{D}$ the result follows. □

Rogers (p. 261 of [Ro67]) asked whether the relation *recursively enumerable in* is order-theoretic.

THEOREM 17. *The relation "$\mathbf{d}$ is $\mathbf{b}$-REA" is definable in $\mathcal{D}$.*

Proof. $\mathbf{d}$ is $\mathbf{b}$-REA $\Leftrightarrow \mathbf{d} \in [\mathbf{b}, \mathbf{b}']\ \&$

$$[\forall \mathbf{a}, \mathbf{c} \in [\mathbf{b}, \mathbf{b}'])[\mathbf{a} \cup \mathbf{d} \text{ is not unsplittable over } \mathbf{a} \text{ avoiding } \mathbf{c}].$$

Since the jump is definable in the degrees, the result follows.[2] □

Sketch proof of Theorem 16. Let $\mathbf{d} < \mathbf{0}'$ be not r.e., where $D \in \mathbf{d}$. We construct a set $A \in \Delta_2$ satisfying the requirements:

$$P_k: \quad B \neq \Theta_k^A \vee (\exists A^* \text{ co-r.e.})(A^* \equiv_T D),$$

$$Q_k: \quad D = \Psi_k(\Phi_k^{A,D}, \widehat{\Phi}_k^{A,D}) \Rightarrow B = \Gamma_k^{\Phi_k^{A,D},A} \vee B = \Lambda_k^{\widehat{\Phi}_k^{A,D},A},$$

[2] Following on from a remark of C. G. Jockusch (August 1991), we note that it is possible to extend the characterisation of "REA" to one of "r.e. in." But to do this one needs the extra information provided by replacing the standard relativisation to an upper-cone of the statement of Theorem 16 with an appropriate relativisation of the proof of the theorem.

$k \geq 0$, where $(\Theta_k, \Psi_k, \Phi_k, \widehat{\Phi}_k)$ is a standard list of all quadruples of p.r. functionals and Γ_k, Λ_k are partial recursive functionals to be constructed. As before, we get $B \leq_T A \oplus D$ from the satisfaction of the Q-requirements.

Roughly speaking, our strategy is as follows. We consider just two requirements $P = P_{k'}$ and $Q = Q_k$ in relation to each other, Q being of higher priority than P.

If (dropping reference to k and k' again) $\ell(D, \Psi(\Phi^{A,D}, \widehat{\Phi}^{A,D}))$ grows large, we first attempt to implement the Γ-strategy for making $B \leq_T \Phi^{A,D} \oplus A$. If carried through without help from $\Phi^{A,D}$ it may turn out that this leads to $B \leq_T A$. Our indication that this is happening will be some lower priority P producing a witness $\Theta^A = B$. We will use $\ell(B, \Theta^A)$ to monitor the extent to which this is happening. If $\ell(B, \Theta^A)$ grows large, we will try to prevent this initial segment of agreement from being disturbed by the enumeration of numbers into A. This happens when we have to enumerate x-traces into A on behalf of Γ (or Λ) following $x \searrow B$, some x's.

To avoid this, we select some x_0, and periodically try to use enumerations into A to move $\gamma(x_0)$ beyond the use of the current $\ell(B, \Theta^A)$. This process (of capricious destruction) may be successful in freeing larger and larger segments of agreement from the possibility of injury through some $y \searrow A$, but at the expense of the destruction of the Γ-strategy for Q through $\Gamma^{\Phi^{A,D},A}(x_0)\uparrow$. These initial segments may still be injured by extractions from A, and this is unavoidable, but the reason why the extractions are unavoidable is that they are linked to changes in an identifiable set of members of D through an appropriate $B = \Gamma_{k^*}^{A,D}$ or $\Lambda_{k^*}^{A,D}$. At the same time as $\ell(B, \Theta^A)$ grows large we code D into B up to $\ell(B, \Theta^A)$. As a result, if we find $B = \Theta^A$ then we will be able to modify A to a co-r.e. A^* which is $\equiv_T D$ due to $D \leq_T B = \Theta^A$ on the one hand, and due to the linking of extractions from A to D on the other. But then this outcome in which $\Gamma^{\Phi^{A,D},A}(x_0)\uparrow$ must be a pseudo-outcome. We are left with a result in which not only do we get the probability of $\Gamma^{\Phi^{A,D},A}(x_0)\uparrow$ but we also get $\Theta^A(z)\uparrow$ for some $z \geq x_0$, so satisfying P.

There is still a way of averting this outcome, the failure of which will provide a suitable replacement for the lost Γ-strategy. This consists in using suitable $\Phi^{A,D}$ changes to release us from the commitment to enumerate traces into A in order to A-permit B-changes via Γ. Since we are only concerned about Q if $\Phi^{A,D}$ is total, we may delay A-permissions via Γ for $B(w)$ changes (say) until at least $\Phi^{A,D} \upharpoonright \gamma(w)$ has become redefined. This leaves open the possibility that we may get a completely new $\Phi^{A,D} \upharpoonright \gamma(w)$ (that is, not containing as an initial segment any previously defined $\Phi^{A,D} \upharpoonright \gamma(w)$) which can be used to permit the $B(w)$ change via Γ without the need for A-changes to be made.

Moreover, we may get $\Phi^{A,D} \upharpoonright \gamma(y)$ to be new for other numbers y for which $B(y)$ changes are possible in the future. This will enable us to replace a trace for y which may need enumerating into A when we get a $B(y)$ change, with a trace (which can be chosen initially as large as we like) which need only be extracted from A to A-permit a $B(y)$ change via Γ. This indicates the possibility of ameliorating the effects of capricious destruction and satisfying P by exhibiting an A^* satisfying P while salvaging the Γ-strategy if we can obtain enough of the above $\Phi^{A,D} \upharpoonright \gamma(y)$ changes, $y \geq 0$.

We can assist this outcome by using A to increase the likelihood of our new $\Phi^{A,D} \upharpoonright \gamma(w)$ being usable to permit the $B(w)$ change via Γ, and, more important, enabling us to choose new traces for numbers y consistent with the A^*-strategy for P.

The main ingredient here is the process of 'honestification' whereby we use A changes to try to ensure that whenever we define $\gamma(w)$, some $w \geq x_0$, previous to a $D \upharpoonright w'$ change leading to a $B \upharpoonright w$ change (where we can assume $w \geq w'$) we have $\psi(w) \downarrow \leq \gamma(w), \lambda(w)$, so that the $D \upharpoonright w'$ change must be accompanied by some sort of change in either $\Phi^{A,D} \upharpoonright \gamma(w)$ or $\widehat{\Phi}^{A,D} \upharpoonright \lambda(w)$. In fact, we try to ensure via our A-changes for capricious destruction that for each y which is not yet accompanied by a 'positive trace' (that is, one $\in A$), we have $\gamma(y), \lambda(y) \geq \psi(w)$ for *all* $w < \ell(D, \Psi(\Phi^{A,D}, \widehat{\Phi}^{A,D}))$. Then if such a w occurs, its accompanying $\Phi^{A,D} \upharpoonright \gamma(w)$ or $\widehat{\Phi}^{A,D} \upharpoonright \lambda(w)$ change is usable by y for positive trace selection. The honestification has to be stronger than in the monster construction, as we are not just dealing with r.e. sets. If the honestification is attempted via an $A \upharpoonright u$ change, say, for the number y, then we ask that $u \leq \min\{\gamma(y), \lambda(y)\}$ for all such $\gamma(y), \lambda(y)$ defined since the last occurrence of honestification, thereby ensuring that previously defined $\Phi^{A,D} \upharpoonright \gamma(y)$ or $\widehat{\Phi}^{A,D} \upharpoonright \lambda(y)$ will not return in tandem with corresponding $A \upharpoonright \gamma(y)$ or $A \upharpoonright \lambda(y)$ following a $B \upharpoonright w$ change, $w < \ell(D, \Psi(\Phi^{A,D}, \widehat{\Phi}^{A,D}))$, to defeat the possibility of a positive trace selection for y.

There is a problem here (apart from that of not knowing whether we get a $\Phi^{A,D}$- or $\widehat{\Phi}^{A,D}$-change following the $A \upharpoonright u$ change, and of knowing whether we keep such a change), in that $B \upharpoonright w$ changes may occur while honestification is in progress, demanding renewed honestification. But this repetition can only develope into an infinite outcome, giving $\Gamma^{\Phi^{A,D},A}$, $\Lambda^{\widehat{\Phi}^{A,D},A}$ not total, if we have some w with $\min\{\gamma(y), \lambda(y)\} < \psi(w)$ infinitely often. In this case Q is satisfied since $\Psi(\Phi^{A,D}, \widehat{\Phi}^{A,D})(w) \uparrow$. And we still have $\theta(y) \uparrow$ infinitely often, so P is satisfied through $\Theta^A(y) \uparrow$.

However, honestification as described will still not be sufficient to supply the ideal conditions for positive trace definition for y. This is because of the possibility of returns to strings $\Phi^{A,D} \upharpoonright \gamma(y)$ (following a $B \upharpoonright w$ change) which appeared since the last occurrence of honestification for (P, Q). So before remitting honestification on behalf of y we further require that $\Phi^{A,D} \upharpoonright \psi(w)$, $\widehat{\Phi}^{A,D} \upharpoonright \psi(w)$ are unchanged at all stages since the previous occurrence of honestification, and if this condition is not met, we again honestify (even if $\psi(w) \leq \gamma(y), \lambda(y)$). If we never get a positive trace for y but continue to honestify we still get $\Psi(\Phi^{A,D}, \widehat{\Phi}^{A,D})(w) \uparrow$ for some w (by the non-recursiveness of D) and Q is again satisfied (along with P since $\Theta^A(x_0) \uparrow$).

There is now the possibility of a continuing successful honestification accompanied by infinitely many suitable $D \upharpoonright w'$ changes, but producing only finitely many new strings $\Phi^{A,D} \upharpoonright \gamma(y)$, $y \geq 0$. In this case we utilise the fact that following each suitable $D \upharpoonright w'$ change we get a new $\widehat{\Phi}^{A,D} \upharpoonright \lambda(y)$ to pursue the A^* strategy for P with while satisfying Q via the Λ-strategy. In fact, this outcome for (P, Q) gives a successful Λ-strategy for each $(P_{k''}, Q)$, $k'' > k'$. The Λ-strategy acts with the pre-knowledge that we get infinitely many $A \upharpoonright \gamma(x_0)$ changes, some x_0, through capricious destruction, and infinitely many usable $\widehat{\Phi}^{A,D} \upharpoonright \lambda(y)$ changes. This means we pursue the A^*-strategy for $P_{k''}$ related to Λ below $\gamma(x_0)$. Since $\gamma(x_0)$ goes to infinity, this will still provide sufficient space in which to satisfy $P_{k''}$. The Λ-strategy links up with the already initiated honestification process, and always gets its positive traces needed for the A^*-strategy for $P_{k''}$, without any need for additional A-changes in case the $\widehat{\Phi}^{A,D}$-changes do not live up to their promise (as they always do). There is still the possibility of the A^*-strategy failing through $\Theta^A_{k''}(w)\uparrow$, some w, but this is not the concern of the Λ-strategy, which is not disrupted, unlike the Γ-strategy.

As for the previous construction, one needs a tree of outcomes on which to reconcile the strategies for different pairs (P', Q'), and one of the standard frameworks for discussing $0'''$-injury priority arguments. It is worth mentioning here some of the special complications arising from the fact that D is not r.e. (or perhaps even d-r.e.) but Δ_2. An inevitable consequence is a certain amount of 'Δ_2-noise' in the construction, and the need for one or two nontrivial adjustments. The extra unpredictability of D- and hence B-changes does not in itself cause too many problems for A. With the help of redefinitions of γ and λ, A can cope with the corresponding demands of the Γ- and Λ-strategies. There are slightly more problems with the consequent lack of control over $\Phi^{A,D}$ and $\widehat{\Phi}^{A,D}$. We relied above, in certain situations, on $\Phi^{A,D}$- or $\widehat{\Phi}^{A,D}$-changes enabling us to avoid certain sorts of A-changes (by positive trace selection) in the interests of the A^*-strategy for the P-requirements. When these changes are in doubt, we will have to fall back on the undesired type of A-changes. However, temporary vacillations in $\Phi^{A,D}$ or $\widehat{\Phi}^{A,D}$ can be matched by a corresponding flexibility in the A-changes, and where ultimate outcomes for $\Phi^{A,D}$- or $\widehat{\Phi}^{A,D}$-changes are assured we will be able to maintain our aims in regard to the A^*-strategies. On the other hand, in reconciling the demands of different P-requirements, the A^*-strategies may demand negative A-changes where the immediate need may seem to be new positive A-changes. The key factor here is, of course, that the success of the A^*-strategy for P lies in producing $B \neq \Theta^A$, not in an infinitary outcome of $D \equiv_T$ a co-r.e. A^*.

See [Co*ta*2] for a more formal description of the basic module for (P, Q) and further discussion of problems in reconciling the strategies.

§10. Questions and further results.

More recently Slaman and Woodin [SW*ta*] (see [Sl91]) have extended their earlier results [SW86] to obtain new proofs (not involving the jump) of a number of global theorems concerning the degrees of unsolvability. In some cases improvements have been obtained. For instance, in a more general context including other common degree structures, they improve Theorem 10+:

THEOREM 10++ (Slaman and Woodin [SW*ta*]). *Let f be an automorphism of $\mathcal{D}$. Then $f(\mathbf{a}) = \mathbf{a}$ for all $\mathbf{a} \geq \mathbf{0}''$.*

Other results are proved making full use of Theorem 17 above. Their most dramatic result is:

THEOREM 18 (Slaman and Woodin [SW*ta*]). *The recursively enumerable degrees form an automorphism base for $\mathcal{D}$.*

This of course reduces the problem of showing that $\mathcal{D}$ is rigid to that of showing $\mathcal{R}$ (the structure of the r.e. degrees) to be rigid. In the other direction, Slaman [Sl91] asks whether every automorphism of the r.e. degrees can be extended to one of $\mathcal{D}(\leq \mathbf{0}')$ or of $\mathcal{D}$. Increasingly, not only does one find the more intractable and technically interesting questions of degree theory at the local level, but therein is seen to lie the key to the main outstanding problems of global degree theory.

We summarise some of the more important remaining open questions.

Homogeneity and automorphisms.

Jockusch [Jo81] has shown that there is a comeager set of degrees which are bases of elementarily equivalent cones of degrees. But:

QUESTION 1. *Do there exist degrees $\mathbf{a}, \mathbf{b}$, $\mathbf{a} \neq \mathbf{b}$, with $\mathcal{D}(\geq \mathbf{a}) \cong \mathcal{D}(\geq \mathbf{b})$?*

QUESTION 2. *How far can Theorems 2+, 7+, 8+ and 10++ be improved? For instance, is any automorphism of $\mathcal{D}$ the identity above $\mathbf{0}'$?*

QUESTION 3. *Do there exist nontrivial automorphisms of $\mathcal{D}(\leq \mathbf{0}')$ or of $\mathcal{R}$?*

Definability.

Shore [Sh88] has shown that all levels of the high/low hierarchy from the three level onwards are definable in $\mathcal{D}(\leq \mathbf{0}')$, and Shore and Slaman [SS*ta*] have succeeded in distinguishing between the high and low degrees within $\mathcal{D}(\leq \mathbf{0}')$.

QUESTION 4. *Are $High_1$, $High_2$, Low_1 or Low_2 definable in $\mathcal{D}(\leq \mathbf{0}')$?*

There are questions concerning the definability of particular degrees. For instance (questions first stated by Slaman and Harrington, respectively, in [Sa85]):

QUESTION 5. *Do there exist r.e. degrees other than $\mathbf{0}$ or $\mathbf{0}'$ definable in $\mathcal{D}(\leq \mathbf{0}')$ or $\mathcal{R}$? Are there any r.e. degrees not definable in $\mathcal{D}(\leq \mathbf{0}')$ or $\mathcal{R}$?*

Two particularly interesting questions concerning the definability of classes of degrees are:

QUESTION 6. *Define the class of n-r.e. degrees in $\mathcal{D}(\leq \mathbf{0}')$ for $n \geq 2$.*

QUESTION 7. *For which r.e. degrees $\mathbf{a} > \mathbf{0}$ can one define the r.e. degrees below $\mathbf{a}$ in $\mathcal{D}(\leq \mathbf{a})$?*

REFERENCES

[Cota1] S. B. Cooper, *On a conjecture of Kleene and Post*, to appear.

[Cota2] ———, *The recursively enumerable degrees are absolutely definable*, to appear.

[EP70] H. B. Enderton and H. Putnam, *A note on the hyperarithmetical hierarchy*, J. Symbolic Logic **35** (1970), 429–430.

[Ep79] R. L. Epstein, "Degrees of Unsolvability. Structure and Theory," Lecture Notes in Mathematics No. 759, Springer-Verlag, New York, 1979.

[Fe70] L. Feiner, *The strong homogeneity conjecture*, J. Symbolic Logic **35** (1970), 375–377.

[Fr57] R. M. Friedberg, *A criterion for completeness of degrees of unsolvability*, J. Symbolic Logic **22** (1957), 159–160.

[Gö34] K. Gödel, *On undecidable propositions of formal mathematical systems*, mimeographed notes, in "The Undecidable. Basic Papers on Undecidable Propositions, Unsolvable Problems, and Computable Functions," (M. Davis, ed.), Raven Press, New York, 1965, pp. 39–71.

[HK75] L. Harrington and A. Kechris, *A basis result for Σ_3^0 sets of reals with an application to minimal covers*, Proc. Amer. Math. Soc. **53** (1975), 445–448.

[HS81] L. Harrington and R. A. Shore, *Definable degrees and automorphisms of $\mathcal{D}$*, Bull. Amer. Math. Soc. (new series) **4** (1981), 97–100.

[Jo81] C. G. Jockusch, Jr., *Degrees of generic sets*, Lond. Math. Soc. Lect. Notes **45** (1981), 110–139.

[JP81] C. G. Jockusch, Jr. and D. Posner, *Automorphism bases for degrees of unsolvability*, Israel J. Math. **40** (1981), 150–164.

[JSh84] C. G. Jockusch, Jr. and R. A. Shore, *Pseudo jump operators II: Transfinite iterations, hierarchies, and minimal covers*, J. Symbolic Logic **49** (1984), 1205–1236.

[JSi76] C. G. Jockusch, Jr. and S. G. Simpson, *A degree theoretic definition of the ramified analytic hierarchy*, Ann. Math. Logic **10** (1976), 1–32.

[JSoa70] C. G. Jockusch, Jr. and R. I. Soare, *Minimal covers and arithmetical sets*, Proc. Amer. Math. Soc. **25** (1970), 856–859.

[JSol77] C. G. Jockusch, Jr. and R. M. Solovay, *Fixed points of jump preserving automorphisms of degrees*, Israel J. Math. **26** (1977), 91–94.

[Kl36] S. C. Kleene, *General recursive functions of natural numbers*, Math. Ann. **112** (1936), 727–742.

[Kl43] ———, *Recursive predicates and quantifiers*, Trans. Amer. Math. Soc. **53** (1943), 41–73.

[KP54] S. C. Kleene and E. L. Post, *The upper semi-lattice of degrees of recursive unsolvability*, Ann. Math. (2) **59** (1954), 379–407.

[La68] A. H. Lachlan, *Distributive initial segments of the degrees of unsolvability*, Z. Math. Logik Grundlag. Math. **14** (1968), 457–472.

[La75] ________, *A recursively enumerable degree which will not split over all lesser ones*, Ann. Math. Logic **9** (1975), 307–365.

[Ma77] J. M. MacIntyre, *Transfinite extensions of Friedberg's completeness criterion*, J. Symbolic Logic **42** (1977), 1–10.

[NS80] A. Nerode and R. A. Shore, *Second order logic and first order theories of reducibility orderings*, in "The Kleene Symposium," (J. Barwise et al., eds.), North-Holland, Amsterdam, 1980, pp. 181-200.

[NS80a] ________, *Reducibility orderings: theories, definability and automorphisms*, Ann. Math. Logic **18** (1980), 61–89.

[Od89] P. Odifreddi, "Classical Recursion Theory," North-Holland, Amsterdam, New York, Oxford, 1989.

[OSta] P. Odifreddi and R. A. Shore, *Global properties of local structures of degrees*, to appear.

[PR81] D. Posner and R. W. Robinson, *Degrees joining to* $\mathbf{0}'$, J. Symbolic Logic **46** (1981), 714–722.

[Po44] E. L. Post, *Recursively enumerable sets of positive integers and their decision problems*, Bull. Amer. Math. Soc. **50** (1944), 284–316.

[Po48] ________, *Degrees of recursive unsolvability*, Bull. Amer. Math. Soc. **54** (1948), 641–642.

[Ri79] L. J. Richter, *On automorphisms of the degrees that preserve jumps*, Israel J. Math. **32** (1979), 27–31.

[Ro67] H. Rogers, Jr., "Theory of recursive functions and effective computability," McGraw-Hill, New York, 1967.

[Sa71] G. E. Sacks, *Forcing with perfect closed sets*, in "Axiomatic Set Theory I," (D. Scott, ed.), Proc. Symp. Pure Math. *13*, Los Angeles, 1967, Amer. Math. Soc., Providence, R.I., 1971, pp. 331-355.

[Sa85] ________, *Some open questions in recursion theory*, in "Recursion Theory Week," (H. D. Ebbinghaus et al., eds.), Lecture Notes in Mathematics No. 1141, Springer-Verlag, Berlin, Heidelberg, New York, 1985, pp. 333-342.

[Sh79] R. A. Shore, *The homogeneity conjecture*, Proc. Natl. Acad. Sci. U.S.A. **76** (1979), 4218–4219.

[Sh81] ________, *The degrees of unsolvability: global results*, in "Logic Year 1979-80: University of Connecticut," (eds. M. Lerman et al.), Lecture Notes in Mathematics No. 859, Springer-Verlag, Berlin, Heidelberg, New York, 1981, pp. 283–301.

[Sh82] ________, *On homogeneity and definability in the first order theory of the Turing degrees*, J. Symbolic Logic **47** (1982), 8–16.

[Sh88] ________, *Defining jump classes in the degrees below* $\mathbf{0}'$, Proc. Amer. Math. Soc. **104** (1988), 287–292.

[SSta] R. A. Shore and T. A. Slaman, *Working below a high recursively enumerable degree*, to appear.

[Si77] S. G. Simpson, *First order theory of the degrees of unsolvability*, Ann. of Math. (2) **105** (1977), 121–139.

[Sl91] T. A. Slaman, *Degree structures*, in "Proc. Int. Congress of Math., Kyoto, 1990," Springer-Verlag, Tokyo, 1991, pp. 303–316.

[SW86] T. A. Slaman and W. H. Woodin, *Definability in the Turing degrees*, Illinois J. Math. **30** (1986), 320–334.

[SWta] ———, *Definability in degree structures*, to appear.

[So87] R. I. Soare, "Recursively enumerable sets and degrees," Springer-Verlag, Berlin, Heidelberg, London, New York, 1987.

[Tu36] A. M. Turing, *On computable numbers, with an application to the Entscheidungsproblem*, Proc. London Math. Soc. **42** (1936), 230–265.

[Tu39] ———, *Systems of logic based on ordinals*, Proc. London Math. Soc. **45** (1939), 161–228.

[We81] L. V. Welch, *A Hierarchy of Families of Recursively Enumerable Degrees and a Theorem on Bounding Minimal Pairs*, Ph.D. Dissertation, University of Illinois at Urbana-Champaign, 1981.

[Ya70] C. E. M. Yates, *Initial segments of the degrees of unsolvability, Part I: A survey*, in "Mathematical Logic and Foundations of Set Theory," (Y. Bar-Hillel, ed.), North-Holland, Amsterdam, 1970, pp. 63–83.

[Ya72] ———, *Initial segments and implications for the structure of degrees*, in "Conference in Mathematical Logic, London, 1970," (W. Hodges, ed.), Lecture Notes in Mathematics No. 255, Springer-Verlag, Berlin, Heidelberg, New York, 1972, pp. 305–335.

University of Leeds
Leeds LS2 9JT
England

ABOUT THE IRREFLEXIVITY HYPOTHESIS FOR FREE LEFT DISTRIBUTIVE MAGMAS

PATRICK DEHORNOY *

Abstract. An important combinatorial statement about free left distributive structures, the irreflexivity hypothesis, has been proved by R. Laver using a large cardinal axiom. We discuss here another approach that could, if completed, lead to a new proof independent of any set theoretical assumption.

A left distributive magma—or LD-magma—will be any set endowed with a binary law satisfying the left distributivity identity

$$x(yz) = (xy)(xz).$$

The interest for (free) LD-magmas was emphasized by the study of the iterations of an elementary embedding of a rank into itself in set theory and the conjecture that the structure obtained in this way is actually a free (monogenic) LD-magma. This conjecture has been proved in 1989 by Richard Laver ([La]); an alternative proof in given in [De4]. Both proofs make an intensive use of the relation of being a left factor in LD-magmas.

DEFINITION. Let $\mathfrak{g}$ be a LD-magma, and x, y belong to $\mathfrak{g}$; write $x <_L^{\mathfrak{g}} y$ if, and only if, there exists a (positive) integer p and a finite sequence $z_1, z_2, \ldots, z_p$ in $\mathfrak{g}$ such that y is equal to $(\cdots((xz_1)z_2)\cdots)z_p$.

The statement we shall discuss here is the

IRREFLEXIVITY HYPOTHESIS (IH). *Let $\mathfrak{f}$ be the free monogenic LD-magma; then $<_L^{\mathfrak{f}}$ is an irreflexive relation.*

This property proved to be crucial in the study of free LD-magmas. In particular, the following was proved independently in [De2], and in [La1] (for the monogenic case):

PROPOSITION. *If IH is true, then for any set Σ, the word problem for the free LD-magma generated by Σ is decidable; also every free LD-magma admits left cancellation.*

---

* This work was supported in part by a CNRS grant PRC mathématiques et informatique.

The point in the proofs is that $<_L^{\mathfrak{f}}$ is a linear ordering on $\mathfrak{f}$. However the irreflexivity property is clearly preserved under projection, so that if $\mathfrak{g}$ is any LD-magma such that $<_L^{\mathfrak{g}}$ is irreflexive, then $<_L^{\mathfrak{f}}$ must be irreflexive too. It follows moreover that (if $\mathfrak{g}$ is monogenic) it must be free as well, and this is the way R. Laver proves that the iterations of an elementary embedding make a free LD-magma.

PROPOSITION (Laver). *Assume that j is an elementary embedding of a rank V_λ into itself; let $\mathfrak{a}^j$ be the structure generated by j using the operation $ik := \bigcup_{\alpha<\lambda} i(k \upharpoonright V_\alpha)$; then $<_L^{\mathfrak{a}^j}$ is irreflexive—and therefore $\mathfrak{a}^j$ is free, and* IH *is true.*

Let EE be the large cardinal axiom "there exists an elementary embedding of a rank into itself". The irreflexivity hypothesis is thus proved under EE—and therefore such natural questions about free LD-magmas as decidability of the word problem or left cancellation are proved under EE. There is a surprising contrast between the (very) large set theoretical hypothesis EE and the algebraic and, more or less, finitistic properties of LD-magmas that are, up to now, established under EE too. Though no metamathematics is known to force it, it seems likely that the axiom EE could be dropped from the proofs—and, at first, from the proof of the key statement IH.

Owing to the preservation of irreflexivity under projection, the most natural way for proving IH would be to exhibit some particular LD-magma $\mathfrak{g}$ such that $<_L^{\mathfrak{g}}$ is irreflexive—or, at least, satisfies $x \not<_L^{\mathfrak{g}} x$ for sufficiently many x's. Unfortunately, little is known about monogenic LD-magmas : most of the (numerous) examples of LD-magmas are in fact idempotent, so they give rise only to trivial monogenic structures. In [De3], a nontrivial (infinite) monogenic LD-magma $\mathfrak{d}$ is constructed that satisfies "1-irreflexivity": $x = xz_1$ is impossible in $\mathfrak{d}$. But no extension to 2-irreflexivity ("$x = (xz_1)z_2$ is impossible") or more is known.

The aim of this paper is to develop the scheme of a proof for IH that is connected with the study of a certain structural monoid associated with the distributive structures. This proof is *not* complete, so that the conjecture is still open. Nevertheless we hope that the reduced form we obtain for IH in this way can be considered a progress toward a complete solution.

1. General framework.

The general setting will be the one of [De1]. We start with any nonempty set Σ (we shall assume that Σ has at least two elements), and let $\mathcal{T}(\Sigma)$ be the free magma generated by Σ, i.e., the set of all terms constructed from Σ using some fixed binary operator, say $*$. It will be convenient to use here the ***right Polish notation***, so that the product of two terms S, T is denoted by $ST*$. Now the free LD-magma generated by Σ is the quotient of $\mathcal{T}(\Sigma)$ under the least congruence $\equiv$ that forces the left distributivity condition, i.e., that satisfies, for every S, T, U in $\mathcal{T}(\Sigma)$,

$$STU{*}{*} \equiv ST{*}SU{*}{*}.$$

The main tool introduced in [De1] to analyse this congruence $\equiv$ is a monoid ϑ generated by some elementary term transformations. To describe it easily in this nonassociative framework, it is convenient to think to terms as to *binary trees* (with leaves indexed by Σ): thus the subterms of a given term can be specified using any system of addresses in a binary tree, for instance finite sequences of 0's and 1's describing the path in the corresponding tree to reach the current node from the root of the tree (0 meaning "going to the left," 1 meaning "going to the right").

DEFINITION. Let $\mathbf{S}$ be the set of all finite sequences of 0's and 1's; for w in $\mathbf{S}$, and S in $\mathcal{T}(\Sigma)$, $S_{/w}$ denotes the subterm of S whose root has address w (if it exists); the set of all addresses w such that $S_{/w}$ exists is called the *support* of S and written $\mathbf{Supp}(S)$; the set of all addresses w such that $S_{/w}$ exists is denoted by $\mathbf{Supp}^+(S)$.

EXAMPLE. Let S be $abcd{*}{*}{*}$; then $S_{/0}$ is a, $S_{/1}$ is $bcd{*}{*}$; $\mathbf{Supp}(S)$ is $\{0, 10, 110, 111\}$, while $\mathbf{Supp}^+(S)$ is $\{0, 10, 110, 111, 11, 1, \Lambda\}$, where Λ denotes the empty sequence (the address of the root in any tree). Notice that $\mathbf{Supp}^+ S$ is (in any case) the closure of $\mathbf{Supp} S$ under word prefixing.

DEFINITION. i) For w in $\mathbf{S}$, define a partial mapping w^+ of $\mathcal{T}(\Sigma)$ into itself by:

S is in $\mathbf{Dom} w^+$ if, and only if, $wu0, w10, w11$ are in $\mathbf{Supp}(S)$, and, in this case, the value of w^+ on S is obtained from S by replacing $S_{/w}$ (that is $S_{/w0} S_{/w10} S_{/w11}{*}{*}$) by $S_{/w0} S_{/w10} {*} S_{/w0} S_{/w11}{*}{*}$.

ii) The monoid generated by all w^+'s for w in $\mathbf{S}$ with reverse composition is denoted by $\boldsymbol{\vartheta}^+$. The inverse mapping of w^+ is denoted by w^-; the monoid generated by all w^+'s and w^-'s is denoted by $\boldsymbol{\vartheta}$.

The action of w^+ is clear: it consists in expanding according to left distributivity "at w." We use reverse composition to make reading more natural, and to avoid ambiguity write $\mathbf{val}(S, \varphi)$ to denote the image of S under φ (the value of φ at S).

EXAMPLE. A picture should make clear that $\mathbf{val}(abcd{*}{*}{*}, \Lambda^+)$ is $ab{*}acd{*}{*}{*}$, while $\mathbf{val}(abcd{*}{*}{*}, 1^+)$ is $abc{*}bd{*}{*}{*}$.

With this definition, $\boldsymbol{\vartheta}$ acts on $\mathcal{T}(\Sigma)$, and $\equiv$ is exactly the equivalence relation attached to this action: $S{\equiv}T$ holds if, and only if, T is the image of S under some element of $\boldsymbol{\vartheta}$. Notice that, owing to the above definition, $\boldsymbol{\vartheta}$ could depend on Σ; in fact it is proved in [De1] that it does not, at least if Σ has strictly more than one element, what we shall assume henceforth.

The monoid $\boldsymbol{\vartheta}$ is closely connected with the infinite braid group B_∞. The existence of canonical operation of braid groups on distributive structures has already been noticed and used (see [Br]). The "structural" monoid $\boldsymbol{\vartheta}$ introduced above is in fact an extension of B_∞. Let us describe B_∞ as the group generated by an infinite family of generators $(\sigma_i)_{i=0,1,\ldots}$ under the relations

$$\sigma_i \sigma_{i+1} \sigma_i = \sigma_{i+1} \sigma_i \sigma_{i+1}$$

$$\sigma_i \sigma_j = \sigma_j \sigma_i \quad \text{if} \quad |\, i - j \,| \geq 2;$$

then the mapping π defined by

$$\pi(w^+) := \begin{cases} \sigma_i & \text{if } w \text{ is } 1^i \text{ for some } i \\ 1 & \text{otherwise} \end{cases} \qquad \pi(w^-) := \begin{cases} \sigma_i^{-1} & \text{if } w \text{ is } 1^i \text{ for some } i \\ 1 & \text{otherwise} \end{cases}$$

is an epimorphism of ϑ onto B_∞. The process described below can be seen as extending some work about normal forms for the elements of B_∞: for instance, proposition 1 below extends the existence of Garside form (see for instance [Bi]). However ϑ appears as much more complicated than B_∞ to handle, and the results in B_∞ that can be derived using π from the ones established here in ϑ have in general (much) more simple direct proofs.

With the present notations, the left subterms of a term S are the various $S_{/0^p}$ for $p = 1, 2, \ldots$, and the irreflexivity conjecture can be stated as:

IRREFLEXIVITY HYPOTHESIS (IH'). *There cannot exist S in $\mathcal{T}(\Sigma)$, φ in ϑ and a positive integer p such that φ maps S to $S_{/0^p}$.*

A first reduction of the question can be obtained by focusing on the positive terms in ϑ, i.e., the terms in ϑ^+. Let us introduce the following notations.

DEFINITION. For S, T in $\mathcal{T}(\Sigma)$, write $S \longrightarrow T$ (respectively, $S \longrightarrow^n T$) if, and only if, T is the image of S under some element of ϑ^+ (respectively, under the product of at most n successive w^+'s).

The main result about $\longrightarrow$ is its confluent character proved in [De1]; it immediately implies

PROPOSITION 1. *Let S, T be arbitrary terms in $\mathcal{T}(\Sigma)$; the following are equivalent:*

i) $S \equiv T$ holds;

ii) there exist U such that both $S \longrightarrow U$ and $T \longrightarrow U$ hold.

We can therefore rewrite IH as

IRREFLEXIVITY HYPOTHESIS (IH''). *There cannot exist S, T in $\mathcal{T}(\Sigma)$ and a positive integer p such that both $S \longrightarrow T$ and $S_{/0^p} \longrightarrow T$ hold.*

2. Progressive sequences.

The main obstruction for a direct proof of conjecture IH'' is the *lack of uniqueness* in the writing of the elements in ϑ^+ as products of a sequence of w^+ transformations. For instance, the sequences $\Lambda^+1^+\Lambda^+$, $1^+\Lambda^+0^+1^+$ and $1^+\Lambda^+1^+0^+$ represent the same element of ϑ^+. The idea is to distinguish particular sequences that enjoy some uniqueness property, so that we reach a contradiction in assuming both $S \longrightarrow T$ and $S_{/0^p} \longrightarrow T$ because the canonical sequences witnessing for these relations should be both equal (by uniqueness) and different (since they represent 1-1 mappings having different arguments and the same image).

It will be convenient in the sequel to use as "atomic terms" no longer the w^+ transformations, but some simple products. Also we shall use an ordering on $\mathbf{S}$. So we put

NOTATION. i) For w in $\mathbf{S}$ and r a positive integer, set $w^{(r)} := w^{+}(w0)^{+}\cdots(w0^{r-1})^{+}$;
ii) $\prec$ is the natural linear ordering on $\mathbf{S}$ for which $0\prec 1\prec \Lambda$ holds, i.e., $u\prec v$ holds either if v is a prefix of u (that is $u = vw$ for some w) or u is on the left of v (that is $w0$ is a prefix of u and $w1$ is a prefix of v for some w);
iii) $\overline{\mathbf{S}}$ will be $\mathbf{S}$ enlarged with a new element 0^{∞} that will be considered minimal for $\prec$.

The aim of this paper is to introduce the following refinement of the relation $\longrightarrow$.

DEFINITION. i) The relation $\Longrightarrow^{1}_{\bullet}$ on $\mathcal{T}(\Sigma)\times\overline{\mathbf{S}}$ is defined by:

$(S,u)\Longrightarrow^{1}_{\bullet}(T,v)$ holds if, and only if, for some w in $\mathbf{S}$ and some integer r one has $T = \mathbf{val}(S, w^{(r)})$, $u\preceq w10^{r}$ and $w0^{r}\preceq v$.

The reflexive-transitive closure of $\Longrightarrow^{1}_{\bullet}$ is denoted by $\Longrightarrow_{\bullet}$, and the projection of $\Longrightarrow_{\bullet}$ on $\mathcal{T}(\Sigma)$ is denoted by $\Longrightarrow$.
ii) A sequence $\langle w_1^{(r_1)}\dots w_n^{(r_n)}\rangle$ in ϑ^{+} is said to be *progressive* if, and only if, $w_i0^{p_i}\preceq w_{i+1}10^{p_{i+1}}$ holds for $i = 0,1,\dots,n-1$. The elements of ϑ^{+} that can be written (in at least one way) as the product of a progressive sequence are called *progressive*; the set of all progressive elements in ϑ^{+} is written $\vartheta^{+}_{\mathbf{prog}}$.

The connection between progressive transformations and $\Longrightarrow$ is easy: the second coordinate in $\Longrightarrow_{\bullet}$ is used to witness for the progressivity of the sequence of elementary transformations applied to the first coordinate, so that one has

LEMMA 1. *Let S,T belong to $\mathcal{T}(\Sigma)$; then $S\Longrightarrow T$ holds if, and only if, T is the image of S under some progressive element of ϑ^{+}.*

The set $\vartheta^{+}_{\mathbf{prog}}$ is a strict subset of ϑ^{+}, i.e., $\Longrightarrow$ is a strict refinement of $\longrightarrow$. Indeed, while $\Longrightarrow_{\bullet}$ is designed to be transitive, $\Longrightarrow$ is *not* transitive: for instance, Λ^{+} and 00^{+} are in $\vartheta^{+}_{\mathbf{prog}}$ (so is every atomic element $w^{(r)}$), but the product $\Lambda^{+}00^{+}$ is not (the sequence $\langle\Lambda^{+},00^{+}\rangle$ does not satisfy the combinatorial criterion since $0\preceq 0010$ is false, and one can easily show that it is the only possible decomposition of $\Lambda^{+}00^{+}$ as a product of positive terms). Notice that the corresponding projections on B_{∞} behave nicely and give rise to a simple "unique normal form" result: every positive term in B_{∞} is progressive, and has exactly one progressive writing, since $\langle\sigma_i,\sigma_j\rangle$ satisfies the progressivity assumption exactly when $j\leq i+1$ holds.

We prove now that progressive sequences enjoy the required uniqueness properties.

DEFINITION. i) Recall that the terms in $\mathcal{T}(\Sigma)$ are considered as finite sequences from $\Sigma\cup\{*\}$; we shall treat such a sequence say S as a mapping of the integer interval $1..|S|$ to $\Sigma\cup\{*\}$. Then we denote by $\mathbf{add}(.,S)$ the increasing bijection of $(1..|S|,<)$ onto $(\mathbf{Supp}^{+}S,\prec)$ (the *address* in S).
ii) Let S,T be two distinct terms in $\mathcal{T}(\Sigma)$; the *divergence* of S and T, written $\mathbf{div}(S,T)$, is the least integer p such that $S(p+1)$ and $T(p+1)$ are not equal (i.e., either both are defined and have different values, or one is defined while the other is not).

EXAMPLE. Let S be $abc{*}{*}$ and T be $ab{*}ac{*}{*}$; $|T|$ is 7, and one has e.g., $T(1) = a$, $T(3) = *$. Since the $\prec$-increasing enumeration of $\mathbf{Supp}(T)$ is $00, 01, 0, 10, 11, 1, \Lambda$, $\mathbf{add}(1,T)$ is 00 while $\mathbf{add}(3,T)$ is 0. Finally $\mathbf{div}(S,T)$ is 2, since $S(3)$ is a and $T(3)$ is $*$.

One will easily verify that $u = \mathbf{add}(p, S)$ means that p is the rank in S of the last occurrence of a character coming from the subterm $S_{/u}$.

LEMMA 2. *Assume $S \Longrightarrow T$; then the first term of any progressive sequence such that T is the image of S under the product of this sequence is $w^{(r)}$ where $w10^r$ is $\mathbf{add}(\mathbf{div}(S,T),S)$.*

Proof. A direct computation shows for $T = \mathbf{val}(S, w^{(r)})$ the following equalities

$$\mathbf{add}(\mathbf{div}(S,T),S) = w10^r$$
$$\mathbf{add}(\mathbf{div}(S,T)+1,T) = w0^r.$$

It follows that if we start with a progressive sequence

$$(S_0,u_0) \Longrightarrow^1_\bullet (S_1,u_1) \Longrightarrow^1_\bullet \cdots \Longrightarrow^1_\bullet (S_n,u_n),$$

then the integers $\mathbf{div}(S_0,S_1), \mathbf{div}(S_1,S_2), \ldots, \mathbf{div}(S_{n-1},S_n)$ make a strictly increasing sequence: indeed if $w_i^{(r_i)}$ maps S_{i-1} to S_i, then the inequality $\mathbf{div}(S_{i-1},S_i) + 1 \leq \mathbf{div}(S_i,S_{i+1})$ is equivalent to $\mathbf{add}(\mathbf{div}(S_{i-1},S_i)+1,S_i) \preceq \mathbf{add}(\mathbf{div}(S_i,S_{i+1}),S_i)$, hence to $w_i0^{r_i} \preceq w_{i+1}10^{r_{i+1}}$, which is the progressivity assumption. It follows that $\mathbf{div}(S_0,S_n)$ is equal to $\mathbf{div}(S_0,S_1)$, and, therefore, that $\mathbf{add}(\mathbf{div}(S_0,S_n),S_0)$ is $w_110^{r_1}$, and this determines uniquely both w_1 and r_1. ∎

The preceding proof makes the notion of progressive sequence clear: a sequence is progressive when the divergences produced by successive application of its terms appear in strictly increasing order (from the left to the right). We deduce

LEMMA 3. *i) Every member of $\vartheta^+_{\mathrm{prog}}$ has exactly one progressive decomposition.*

ii) There cannot exist S,T in $T(\Sigma)$ and a positive integer p such that both $S \Longrightarrow T$ and $S_{/0^p} \Longrightarrow T$ simultaneously hold.

Proof. i) is an obvious iteration of lemma 2. In fact, we get in this way an *algorithm* that produces when starting with two terms S,T the unique progressive sequence $\langle w_1^{(r_1)}, \ldots, w_n^{(r_n)} \rangle$ such that T is the image of S under $w_1^{(r_1)} \cdots w_n^{(r_n)}$ if such a sequence exists. Indeed start with S, get $w_1^{(r_1)}$ from $\mathbf{div}(S,T)$, replace S by $\mathbf{val}(S, w_1^{(r_1)})$ and loop until equality with T is obtained.

ii) Assume $S_{/0^p} \Longrightarrow T$: the algorithm above running on $S_{/0^p}$ and T provides φ in $\vartheta^+_{\mathrm{prog}}$ such that T is $\mathbf{val}(S_{/0^p}, \varphi)$. Now remember that $S_{/0^p}$ is just a prefix of S (when viewed as words on $\Sigma \cup \{*\}$): then for every $p < |S_{/0^p}|$, $\mathbf{add}(p,S)$ is $0^p\mathbf{add}(p, S_{/0^p})$, so that the same algorithm running on S and T will produce "$0^p\varphi$" (same as φ but add 0^p before each term) after scanning $S_{/0^p}$, and the current value of the first term at this step will be TX if S was $S_{/0^p}X$. It is then clearly impossible that the algorithm continues on TX and T and succeeds, since $\mathbf{add}(\mathbf{div}(TX,T),TX)$ is Λ: so $S \Longrightarrow T$ is impossible. ∎

Comparing lemma 3ii) with conjecture IH″ suggests a way for proving IH, namely to replace $\longrightarrow$ by its (strict) refinement $\Longrightarrow$ in the confluency results.

PROGRESSIVITY HYPOTHESIS (PH*). *Let S, T be arbitrary terms in $\mathcal{T}(\Sigma)$; the following are equivalent:*

i) $S \equiv T$ holds;

ii) there exist U such that both $S \Longrightarrow U$ and $T \Longrightarrow U$ hold.

It is clear that PH* implies IH. Due to the lack of transitivity of $\Longrightarrow$, it will be convenient to introduce a more technical statement that seems easier to prove than PH* and nevertheless implies IH. To do that, we shall first recall the construction of the derivation operation on $\mathcal{T}(\Sigma)$, which is the key tool for proving the confluency of $\longrightarrow$. The problem there lies in the fact that $\longrightarrow$ is not a noetherian relation, i.e., there are infinitely long nontrivial sequences $S_0 \longrightarrow S_1 \longrightarrow S_2 \longrightarrow \cdots$, and that therefore the easy local confluency does not imply the global one. The solution given in [Del] introduces a kind of "local noetherianity" by constructing for every term S an infinite sequence $S, \partial S, \partial^2 S, \ldots$, so that some lower bound phenomenon appears, from which proposition 1.1 easily follows.

DEFINITION. i) The binary operation **dist** on $\mathcal{T}(\Sigma)$ is inductively given by

$$\mathbf{dist}(S,T) := \begin{cases} ST* & \text{if } T \text{ is in } \Sigma, \\ \mathbf{dist}(S, T_{/0})\mathbf{dist}(S, T_{/1})* & \text{otherwise.} \end{cases}$$

ii) The unary operation ∂ on $\mathcal{T}(\Sigma)$ is inductively given by

$$\partial S := \begin{cases} S & \text{if } S \text{ is in } \Sigma, \\ \mathbf{dist}(\partial(S_{/0}), \partial(S_{/1})) & \text{otherwise.} \end{cases}$$

The operation **dist** is a "complete" distribution: $\mathbf{dist}(S,T)$ is obtained from T by replacing every variable a in T by $Sa*$; ∂ corresponds to recursively applying **dist** to every subterm of its argument. The key lemma for proving the confluency of $\longrightarrow$ is the following

LEMMA 4. *For any S, T in $\mathcal{T}(\Sigma)$ and any integer n, $S \longrightarrow^n T$ implies $T \longrightarrow \partial^n S$.*

The convenient refinement of this result we wish to set as a reachable conjecture is the following.

PROGRESSIVITY HYPOTHESIS (PH). *For any S, T in $\mathcal{T}(\Sigma)$ and any integer n, $S \longrightarrow^n T$ implies $T \Longrightarrow \partial^n S$.*

LEMMA 5. PH *implies* IH.

Proof. In order to prove IH″, assume that $S \longrightarrow T$ and $S_{/0^p} \longrightarrow T$ hold for some S, T and positive p. An easy induction on ϑ^+ shows that, if $S \longrightarrow T$ holds, then for every $p > 0$ (such that $S_{/0^p}$ exist) there exists $q \geq p$ such that $S_{/0^p} \longrightarrow T_{/0^q}$ holds. Let n be large enough so that $S_{/0^p} \longrightarrow^n T$ and $S_{/0^p} \longrightarrow^n T_{/0^q}$ hold. If PH is true, this implies $T \Longrightarrow \partial^n(S_{/0^p})$ and $T_{/0^q} \Longrightarrow \partial^n(S_{/0^p})$, a contradiction to lemma 3 since $q \geq p \geq 1$. ∎

NOTATION. Let PH_n, PH'_m be the following statements:

PH_n: For any S, T in $\mathcal{T}(\Sigma)$, $S \longrightarrow^n T$ implies $T \Longrightarrow \partial^n S$.

PH'_m: For any S, T, U in $\mathcal{T}(\Sigma)$, $S \longrightarrow^1 T$ and $S \Longrightarrow^m U$ imply $T \Longrightarrow \partial U$ (where $\Longrightarrow_\bullet^m$ is the m-th power of $\Longrightarrow_\bullet^1$, and $\Longrightarrow^m$ the projection of $\Longrightarrow_\bullet^m$).

An immediate induction shows that, if PH'_m is true for every m, then PH_n is also true for every n, i.e., PH is true. The following section gives a proof of PH_1, that is also PH'_0. The best result presently proved in this direction is PH'_1. The main improvement brought by replacing IH by PH seems to be that IH is a negative statement while PH is positive and, due to the uniqueness of progressive decomposition, the "double arrow" whose existence is claimed is in fact completely determined, so that proving PH (if it is true!) appears as a kind of (very complicated) verification only.

3. A proof of PH_1.

In order to prove that $T \Longrightarrow \partial S$ holds for every T satisfying $S \longrightarrow^1 T$ (and, at first, that $S \Longrightarrow \partial S$ holds), it will be necessary to define a family of terms that contains S, ∂S, all T's satisfying $S \longrightarrow^1 T$, but also all intermediate terms appearing in the progressive transformations $T \Longrightarrow \partial S$.

DEFINITION. Assume that $\mathcal{X}$ is a subset of $\mathcal{T}(\Sigma)$;

i) write $(U, u) \overset{\mathcal{X}}{\Longrightarrow}_\bullet (V, v)$ whenever $(U, u) \Longrightarrow_\bullet (V, v)$ holds and all intermediate terms appearing in the progressive transformation (including U and V themselves) are in $\mathcal{X}$; use $U \overset{\mathcal{X}}{\Longrightarrow} V$ in the same way;

ii) say that $\mathcal{X}$ is *S-directed* if, and only if, $U \overset{\mathcal{X}}{\Longrightarrow} S$ holds for every U in $\mathcal{X}$.

For every S in $\mathcal{T}(\Sigma)$, $\{U \in \mathcal{T}(\Sigma); U \Longrightarrow S\}$ is the maximal S-directed set. Clearly a subset of $\mathcal{T}(\Sigma)$ can be S-directed for at most one term S. In order to describe the progressive transformations toward terms written as "**dist**," we introduce a machinery that controls partial distribution.

NOTATION. If f is any mapping with domain included in Σ and u is in Σ, write $f_{/u}$ for the new mapping defined by $f_{/u}(w) = x$ if, and only if, $f(uw) = x$. Also, we write $\mathbf{Dom}^+ f$ for the prefix closure of $\mathbf{Dom} f$: w is in $\mathbf{Dom}^+ f$ if, and only if, there is some v in $\mathbf{Dom} f$ such that w is a prefix of v.

DEFINITION. Assume that $\mathcal{X}$, $\mathcal{Y}$ are subsets of $\mathcal{T}(\Sigma)$ and T is any term in $\mathcal{T}(\Sigma)$;

i) an *$\mathcal{X}$-graft* is a mapping whose domain is a support (i.e., is $\mathbf{Supp}(S)$ for some term S in $\mathcal{T}(\Sigma)$), and whose range is included in $\mathcal{X}$; an $\mathcal{X}$-graft is said to be *T-suitable* if its domain is included in $\mathbf{Supp}^+ T$. If Γ is a T-suitable $\mathcal{X}$-graft, a new term $\langle \Gamma, T \rangle$ is defined inductively on $\mathbf{Dom}\Gamma$ by

$$\langle \Gamma, T \rangle := \begin{cases} \Gamma(\Lambda) T * & \text{if } \mathbf{Dom}\Gamma \text{ is } \{\Lambda\}, \\ \langle \Gamma_{/0}, T_{/0} \rangle \langle \Gamma_{/1}, T_{/1} \rangle * & \text{otherwise.} \end{cases}$$

ii) The set $\{\langle \Gamma, T \rangle; \Gamma \text{ is an } \mathcal{X}\text{-graft for } T\}$ is denoted by $\mathcal{D}ist(\mathcal{X}, T)$, and $\bigcup_{T \in \mathcal{Y}} \mathcal{D}ist(\mathcal{X}, T)$ by $\mathcal{D}ist(\mathcal{X}, \mathcal{Y})$.

Point i) in the definition makes sense since if $\mathbf{Dom}\Gamma$ is not $\{\Lambda\}$, $\mathbf{Supp}T$ is not $\{\Lambda\}$, so that $T_{/0}$ and $T_{/1}$ exist and moreover $\mathbf{Dom}(\Gamma_{/e})$ is included in $\mathbf{Supp}^+T_{/e}$ for $e = 0, 1$.

EXAMPLE. Let $\mathcal{X}$ be $\{a, ab*\}$ and T be $cde**$; set $\Gamma = \{(0, ab*), (1, a)\}$; then Γ is an T-suitable $\mathcal{X}$-graft, and $\mathbf{dist}(\Gamma, T)$ is $ab*c*ade***$: $\langle\Gamma, T\rangle$ is obtained from T by "grafting" some members from $\mathcal{X}$ in T at the places prescribed by $\mathbf{Dom}\Gamma$.

The key result of this section will be the following

PROPOSITION 1. *Assume that $\mathcal{X}$ is S-directed and $\mathcal{Y}$ is T-directed; then $\mathcal{D}\mathbf{ist}(\mathcal{X},\mathcal{Y})$ is $\mathbf{dist}(S,T)$-directed.*

LEMMA 2. *Assume that Γ is a T-suitable graft;*
i) for u in $\mathbf{Dom}^+\Gamma$, one has

$$\langle\Gamma, T\rangle_{/u} = \langle\Gamma_{/u}, T_{/u}\rangle;$$

ii) for u in $\mathbf{Dom}\Gamma$ and w short enough, one has

$$\langle\Gamma, T\rangle_{/u} = \langle\Gamma(u), T_{/u}\rangle\ ;\ \langle\Gamma, T\rangle_{/u0w} = \Gamma(u)_{/w}\ ;\ \langle\Gamma, T\rangle_{/u1w} = T_{/uw}.$$

The proof is an easy induction on $\mathbf{Dom}\Gamma$. Notice that, for any terms S, T in $\mathcal{T}(\Sigma)$, $\mathbf{dist}(S,T)$ is $\langle\Gamma, T\rangle$ where Γ is the constant $\{S\}$-graft with domain $\mathbf{Supp}T$ and value S.

NOTATION. For u in Σ, we write $\{u\}^\sim$ for the least support that contains u: $\{u\}^\sim$ can be defined inductively by $\{\Lambda\}^\sim := \{\Lambda\}$, $\{0u\}^\sim := 0\{u\}^\sim \cup \{1\}$ and $\{1u\}^\sim := \{0\} \cup 1\{u\}^\sim$.

LEMMA 3. *Assume that u is in $\mathbf{Supp}T$; the following are equivalent:*
i) the term U is $\langle\Gamma, T\rangle$ for some T-suitable $\mathcal{X}$-graft Γ such that u is in $\mathbf{Dom}^+\Gamma$;
ii) for every w in $\{u\}^\sim$, the term $U_{/w}$ is in $\mathcal{D}\mathbf{ist}(\mathcal{X}, T_{/w})$.

Proof. Induction on u. If u is Λ, $\{u\}^\sim$ is $\{\Lambda\}$ and both i) and ii) say that U is in $\mathcal{D}\mathbf{ist}(\mathcal{X}, T)$. Otherwise, assume $u = eu'$ with $e = 0$ or $e = 1$. Use $\bar{e}$ to mean 1 (respectively, 0) if e is 0 (respectively, 1). Assume i). As u is in $\mathbf{Dom}^+\Gamma$, hence in $\mathbf{Supp}T$, T is not in Σ, and $\langle\Gamma, T\rangle$ is $\langle\Gamma_{/0}, T_{/0}\rangle\langle\Gamma_{/1}, T_{/1}\rangle*$. Now u' is in $\mathbf{Supp}^+(T_{/e})$, so $U_{/e_{/w}}$ must be in $\mathcal{D}\mathbf{ist}(\mathcal{X}, T_{/e_{/w}})$ for every w in $\{u'\}^\sim$. Moreover, $U_{/\bar{e}}$ is in $\mathcal{D}\mathbf{ist}(\mathcal{X}, T_{/\bar{e}})$, so ii) is proved. Now assume ii). Then $U_{/e_{/w}}$ is in $\mathcal{D}\mathbf{ist}(\mathcal{X}, T_{/e_{/w}})$ for every w in $\{u'\}^\sim$, so (induction hypothesis) $U_{/e}$ is $\langle\Gamma_e, T_{/e}\rangle$ for some $\mathcal{X}$-graft Γ_e with $u' \in \mathbf{Dom}^+(\Gamma_e)$. Moreover $U_{/\bar{e}}$ is in $\mathcal{D}\mathbf{ist}(\mathcal{X}, T_{/\bar{e}})$, i.e., is $\langle\Gamma_{\bar{e}}, T_{/\bar{e}}\rangle$ for some $\mathcal{X}$-graft $\Gamma_{\bar{e}}$. Then U is $\langle\Gamma, T\rangle$ where Γ is, with obvious notations, $0\Gamma_0 \cup 1\Gamma_1$, and u is in $\mathbf{Dom}^+\Gamma$. ∎

DEFINITION. i) For T in $\mathcal{T}(\Sigma)$ and u in $\mathbf{S}$, write $\mathbf{Supp}_u T$ for $\{w \in \mathbf{Supp}T; w \preceq u\}$.
ii) Assume that $\mathcal{X}$ is S-directed; we say that Γ is a (T,u)-*complete* $\mathcal{X}$-graft whenever Γ is a T-suitable $\mathcal{X}$-graft, $\mathbf{Dom}\Gamma$ includes $\mathbf{Supp}_u T$, and for w in the latter set, $\Gamma(w)$ is equal to S.

LEMMA 4. *Assume that $\mathcal{X}$ is S-directed and u is in* **Supp**T*; the following are equivalent:*

i) U is $\langle\Gamma, T\rangle$ for some (T, u)-complete $\mathcal{X}$-graft Γ;

ii) for every w in $\{u\}^\sim$, the term $U_{/w}$ is in $\mathcal{D}ist(\mathcal{X}, T_{/w})$, and moreover it is equal to $\mathbf{dist}(S, T_{/w})$ whenever $w \preceq u$ holds.

Proof. Induction on u. If u is Λ, then Γ is (T, u)-complete if, and only if, $\mathbf{Dom}\Gamma$ is equal to $\mathbf{Supp}T$ and $\Gamma(w)$ is equal to S for every w in $\mathbf{Dom}\Gamma$, so if, and only if, $\langle\Gamma, T\rangle$ is $\mathbf{dist}(S, T)$. Otherwise let u be eu' with $e = 0$ or $e = 1$. Assume i). By lemma 3, we know that $U_{/w}$ is in $\mathcal{D}ist(\mathcal{X}, T_{/w})$ for w in $\{u\}^\sim$. Assume $e = 0$. Then Γ is (T, u)-complete if, and only if, $\Gamma_{/0}$ is $(T_{/0}, u')$-complete, since $\mathbf{Supp}_u T$ is exactly $0\mathbf{Supp}_{u'}(T_{/0})$. This implies $U_{/0_{/w}} = \mathbf{dist}(S, T_{/0_{/w}})$ for w in $\mathbf{Supp}_{u'}(T_{/0})$, and proves ii). Conversely if ii) holds, then "ii) holds" as well for $U_{/0}$ with respect to u', so (induction hypothesis) $U_{/0}$ is $\langle\Gamma_0, T_{/0}\rangle$ for some $(T_{/0}, u')$-complete $\mathcal{X}$-graft Γ_0. Moreover $U_{/1}$ is assumed to be $\langle\Gamma_1, T_{/1}\rangle$ for some $T_{/1}$-suitable $\mathcal{X}$-graft Γ_1. So finally U is $\langle\Gamma, T\rangle$ where Γ is $0\Gamma_0 \cup 1\Gamma_1$, and Γ is (T, u)-complete. Now assume $e = 1$: $\mathbf{Dom}_u T$ is $0\mathbf{Supp}(T_{/0}) \cup 1\mathbf{Supp}_{u'}(T_{/1})$, so Γ is (T, u)-complete if, and only if, $\Gamma_{/0}$ is $(T_{/0}, \Lambda)$-complete and $\Gamma_{/1}$ is $(T_{/1}, u')$-complete, if, and only if, $\langle\Gamma_0, T_{/0}\rangle$ is $\mathbf{dist}(S, T_{/0})$ and $\langle\Gamma_1, T_{/1}\rangle_{/w} = \mathbf{dist}(S, T_{/1w})$ holds for $w \preceq u'$ in $\{u'\}^\sim$. This is exactly ii). The converse direction is proved as above. ∎

LEMMA 5. *Assume that $\mathcal{X}$ is S-directed, that T' is $\mathbf{val}(T, w^{(r)})$ and Γ is a $(T, w10^r)$-complete $\mathcal{X}$-graft; then there exists a $(T', w0^r)$-complete $\mathcal{X}$-graft, say Γ', such that $\langle\Gamma', T'\rangle$ is $\mathbf{val}(\langle\Gamma, T\rangle, w^{(r)})$, and, therefore $(\langle\Gamma, T\rangle, w10^r) \Longrightarrow_\bullet (\langle\Gamma', T'\rangle, w0^r)$ holds.*

Proof. Let $u_1, \ldots, u_p$ (respectively, $v_1, \ldots, v_q$) be the elements of $\{w\}^\sim$ such that $u_i \prec w$ (respectively, $w \prec v_j$) holds. Let U' be $\mathbf{val}(\langle\Gamma, T\rangle, w^{(r)})$ (which exists since $w10^r$ is in $\mathbf{Supp}U$). By lemma 4, we get

$$
\begin{aligned}
U'_{/x} &= U_{/x} = \mathbf{dist}(S, T_{/x}) = \mathbf{dist}(S, T'_{/x}) \qquad \text{for } x = u_1, \ldots, u_p,\\
U'_{/w0^r} &= U_{/w0} U_{/w10^r} *\\
&= \mathbf{dist}(S, T_{/w0}) \mathbf{dist}(S, T_{/w10^r}) *\\
&= \mathbf{dist}(S, T_{/w0} T_{/w10^r} *)\\
&= \mathbf{dist}(S, T'_{/w0^r}),\\
U'_{/x} &= U_{/x} \in \mathcal{D}ist(\mathcal{X}, T_{/x}) = \mathcal{D}ist(\mathcal{X}, T'_{/x}) \qquad \text{for } x = v_1, \ldots, v_q,\\
U'_{/w0^k1} &= U_{/w0} U_{/w10^k1} *\\
&\in \mathcal{D}ist(\mathcal{X}, T_{/w0}) \mathcal{D}ist(\mathcal{X}, T_{/w10^k1}) *\\
&\subseteq \mathcal{D}ist(\mathcal{X}, T_{/w0} T_{/w10^k1} *)\\
&= \mathcal{D}ist(\mathcal{X}, T'_{/w0^k1}) \qquad \text{for } k = 0, \ldots, r-1.
\end{aligned}
$$

By lemma 4 again, this shows that U' is $\langle\Gamma', T'\rangle$ for some $(T', w0^r)$-complete $\mathcal{X}$-graft Γ'. ∎

LEMMA 6. *Assume that $\mathcal{X}$ is S-directed, $\overline{S}$ is in $\mathcal{X}$ and 0^r is in* **Supp**T*; then there exists a $(T, 0^r)$-complete $\mathcal{X}$-graft Γ such that one has*

$$(\overline{S}T*, 0^\infty) \overset{\mathcal{D}ist(\mathcal{X}, T)}{\Longrightarrow_\bullet} (\langle\Gamma, T\rangle, 0^r).$$

Proof. First $\overline{S}\overset{\mathcal{X}}{\Longrightarrow}S$ holds, and therefore $(\overline{S}T*, 0^\infty)\overset{\mathcal{D}ist(\mathcal{X},T)}{\Longrightarrow_\bullet}(ST*, 0)$ holds as well. Now 10^r is in $\mathbf{Supp}(ST*)$, so $ST*$ is in $\mathbf{Dom}\Lambda^{(r)}$. Let U be $\mathbf{val}(ST*, \Lambda^{(r)})$: we have $(ST*, 10^r)\overset{\mathcal{D}ist(\mathcal{X},T)}{\Longrightarrow_\bullet}(U, 0^r)$ and

$U_{/0^r} = S(T_{/0^r})* = \mathbf{dist}(S, T_{/0^r})$ (since $T_{/0^r}$ is in Σ)

$U_{/0^k1} = S(T_{/0^k1})* \in \mathcal{D}ist(\mathcal{X}, T_{/0^k1})$ for $k = 0, \ldots, r-1$.

So U is $\langle\Gamma, T\rangle$ for some $(T, 0^r)$-complete $\mathcal{X}$-graft Γ . We are done, since $0 \prec 10^r$ holds and $\overset{\mathcal{D}ist(\mathcal{X},T)}{\Longrightarrow_\bullet}$ is a transitive relation. ■

LEMMA 7. *Assume that $\mathcal{X}$ is S-directed, U is in $\mathcal{D}ist(\mathcal{X},T)$ and 0^r is in* $\mathbf{Supp}T$; *then there exists a $(T, 0^r)$-complete $\mathcal{X}$-graft Γ such that one has*

$$(U, 0^\infty)\overset{\mathcal{D}ist(\mathcal{X},T)}{\Longrightarrow_\bullet}(\langle\Gamma, T\rangle, 0^r).$$

Proof. Either U is $\overline{S}T*$ for some $\overline{S}$ in $\mathcal{X}$, and lemma 6 applies, or U is U_0U_1* with U_e in $\mathcal{D}ist(\mathcal{X}, T_{/e})$. If r is 0, only the first case may occur, so that the induction starts. Assume the second case. Since 0^{r-1} is in $\mathbf{Supp}(T_{/0})$, there exists by induction hypothesis some $(T_{/0}, 0^{p-1})$-complete $\mathcal{X}$-graft Γ_0 such that the following holds

$$(U_0, 0^\infty)\overset{\mathcal{D}ist(\mathcal{X},T_{/0})}{\Longrightarrow_\bullet}(\langle\Gamma_0, T_{/0}\rangle, 0^{r-1}),$$

and so does

$$(U, 0^\infty)\overset{\mathcal{D}ist(\mathcal{X},T)}{\Longrightarrow_\bullet}(\langle\Gamma_0, T_{/0}\rangle U_1*, 0^r).$$

Since U_1 is in $\mathcal{D}ist(\mathcal{X}, T_{/1})$, by lemma 4 again, $\langle\Gamma_0, T_{/0}\rangle U_1*$ is $\langle\Gamma, T\rangle$ for some $(T, 0^r)$-complete $\mathcal{X}$-graft Γ ■

LEMMA 8. *Assume that $\mathcal{X}$ is S-directed, u, v are points in* $\mathbf{Supp}^+(T)$ *satisfying $u \preceq v$ and Γ is a (T, u)-complete $\mathcal{X}$-graft; then there exists a (T, v)-complete $\mathcal{X}$-graft Δ such that one has*

$$(\langle\Gamma, T\rangle, u)\overset{\mathcal{D}ist(\mathcal{X},T)}{\Longrightarrow_\bullet}(\langle\Delta, T\rangle, v).$$

Proof. If v is u, there is nothing to prove (take $\Delta := \Gamma$). Otherwise, by transitivity we can assume that v is the immediate successor of u in $\mathbf{Supp}^+T$ (with respect to $\prec$). If u is Λ, there is nothing to prove. If u is $w1$ for some w, then w is the successor of u, and (from lemma 4) the (T, u)-completeness of a graft implies its (T, w)-completeness. So assume $u = w0$. Then, for some positive integer r, v is $w10^r$, and, in this case, v is in $\mathbf{Supp}T$. We argue inductively on w. Assume $w = \Lambda$. Since Γ is $(T, 0)$-complete, $\langle\Gamma, T\rangle_{/0}$ is equal to $\mathbf{dist}(S, T_{/0})$ and $\langle\Gamma, T\rangle_{/1}$ is in $\mathcal{D}ist(\mathcal{X}, T_{/1})$. Applying lemma 7 to $T_{/1}$, we get a $(T_{/1}, 0^r)$-complete $\mathcal{X}$-graft Δ_1 such that

$$(\langle\Gamma, T\rangle_{/1}, 0^\infty)\overset{\mathcal{D}ist(\mathcal{X},T_{/1})}{\Longrightarrow_\bullet}(\langle\Delta_1, T_{/1}\rangle, 0^r)$$

holds, and so does

$$(\langle\Gamma, T\rangle, 0)\overset{\mathcal{D}ist(\mathcal{X},T)}{\Longrightarrow_\bullet}(\langle\Gamma, T\rangle_{/0}\langle\Delta_1, T_{/1}\rangle*, 10^r).$$

Since $\langle\Gamma, T\rangle_{/0}$ is equal to $\mathbf{dist}(S, T_{/0})$, lemma 4 guarantees that $\langle\Gamma, T\rangle_{/0}\langle\Delta_1, T_{/1}\rangle*$ is $\langle\Delta, T\rangle$ for some $(T, 10^r)$-complete $\mathcal{X}$-graft.

Assume now $w = ew'$ with $e = 0$ or $e = 1$; $\langle\Gamma, T\rangle$ is $\langle\Gamma_{/0}, T_{/0}\rangle\langle\Gamma_{/1}, T_{/1}\rangle*$, and Γ_e is $(T_{/e}, w')$-complete. So (induction hypothesis)

$$((\Gamma_{/e}, T_{/e}), w'0) \overset{\mathcal{D}ist(\mathcal{X}, T_{/e})}{\Longrightarrow_{\bullet}} ((\Delta_e, T_{/e}), w'10^r)$$

holds for some $(T_{/e}, w'10^r)$-complete $\mathcal{X}$-graft Δ_e. It follows that

$$(\langle\Gamma, T\rangle, w0) \overset{\mathcal{D}ist(\mathcal{X}, T)}{\Longrightarrow_{\bullet}} (\langle\Delta, T\rangle, w10^r)$$

holds, where Δ is the $(T, w10^r)$-complete $\mathcal{X}$-graft defined by $\Delta = 0\Delta_0 \cup 1\Gamma_{/1}$ if e is 0, and by $\Delta = 0\Gamma_{/0} \cup 1\Delta_1$ if e is 1. ∎

We are now ready to prove proposition 1. Let U be an arbitrary member of $\mathcal{D}ist(\mathcal{X}, \mathcal{Y})$: U is $\langle\Gamma_0, T_0\rangle$ for some T_0 in $\mathcal{Y}$ and some T_0-suitable $\mathcal{X}$-graft Γ_0. Let $T_0, T_1, \ldots, T_n = T$ be the intermediate terms witnessing for $T_0 \Longrightarrow T$. Introduce for $\ell = 1, \ldots, n$ the elements w_ℓ, r_ℓ such that $w_\ell^{(r_\ell)}$ maps $T_{\ell-1}$ to T_ℓ. Using lemma 7 to start, and then lemma 8, get a $(T_0, w_1 10^{r_1})$-complete $\mathcal{X}$-graft Δ_0 such that one has

$$(\langle\Gamma_0, T_0\rangle, 0^\infty) \overset{\mathcal{D}ist(\mathcal{X}, \mathcal{Y})}{\Longrightarrow_{\bullet}} (\langle\Delta_0, T_0\rangle, w_1 10^{r_1})$$

Using lemma 5, get a $(T_1, w_1 0^{r_1})$-complete $\mathcal{X}$-graft Γ_1 such that one has

$$(\langle\Delta_0, T_0\rangle, w_1 10^{r_1}) \overset{\mathcal{D}ist(\mathcal{X}, \mathcal{Y})}{\Longrightarrow_{\bullet}} (\langle\Gamma_1, T_1\rangle, w_1 0^{r_1})$$

Using lemma 8, get a $(T_0, w_1 10^{r_1})$-complete $\mathcal{X}$-graft Δ_1 such that one has

$$(\langle\Gamma_1, T_1\rangle, w_1 0^{r_1}) \overset{\mathcal{D}ist(\mathcal{X}, \mathcal{Y})}{\Longrightarrow_{\bullet}} (\langle\Delta_1, T_1\rangle, w_2 10^{r_2})$$

Using alternatively lemma 5 and lemma 8, one continues and finally gets some $(T_n, w_n 0^{r_n})$-complete $\mathcal{X}$-graft Γ_n. A last call to lemma 8 provides a (T_n, Λ)-complete $\mathcal{X}$-graft Δ_n such that one has

$$(\langle\Gamma_n, T_n\rangle, w_n 0^{r_n}) \overset{\mathcal{D}ist(\mathcal{X}, \mathcal{Y})}{\Longrightarrow_{\bullet}} (\langle\Delta_n, T_n\rangle, \Lambda),$$

and, by transitivity, one deduces

$$(\langle\Gamma_0, T_0\rangle, 0^\infty) \overset{\mathcal{D}ist(\mathcal{X}, \mathcal{Y})}{\Longrightarrow_{\bullet}} (\langle\Delta_n, T_n\rangle, \Lambda).$$

But T_n is T, and the Λ-completeness of Δ_n means that $\langle\Delta_n, T_n\rangle$ is $\mathbf{dist}(S, T)$. So the proof is complete. ∎

It is now very easy to conclude this section.

DEFINITION. For S in $\mathcal{T}(\Sigma)$, define $\mathcal{E}xtS$ inductively by

$$\mathcal{E}xtS := \begin{cases} \{S\} & \text{if } S \text{ is in } \Sigma, \\ \mathcal{D}ist(\mathcal{E}xt(S_{/0}), \mathcal{E}xt(S_{/1})) & \text{otherwise.} \end{cases}$$

PROPOSITION 9. *For every S in $\mathcal{T}(\Sigma)$, $\mathcal{E}xtS$ is ∂S-directed.*

Proof. Induction on S. If S is in Σ, S is the only member of $\mathcal{E}xtS$, and is equal to ∂S. Otherwise, assuming that $\mathcal{E}xt(S_{/e})$ is $\partial(S_{/e})$-directed, we apply proposition 1 to conclude that $\mathcal{E}xtS$ is $\mathbf{dist}(\partial(S_{/0}), \partial(S_{/1}))$-directed, that is ∂S-directed. ∎

COROLLARY. *The statement* PH_1 *is true.*

Proof. Assume $S\longrightarrow^1 T$ (or even $S\Longrightarrow^1 T$): we claim that T is in $\mathcal{E}xtS$, and this, by the latter proposition, implies that $T\Longrightarrow\partial S$ holds. So assume that $w^{(r)}$ maps S to T. We argue inductively on S, and then, for a given S, inductively on w. If S is in Σ, the result is vacuously true. Assume it proved for $S_{/e}$ ($e = 0, 1$). If w is Λ, then T is in $\mathcal{D}ist(\{S_{/0}\}, \{S_{/1}\})$, hence in $\mathcal{E}xtS$. And if w is ew', $T_{/e}$ is in $\mathcal{E}xt(S_{/e})$ (induction hypothesis), while $T_{/\bar{e}}$ is $S_{/\bar{e}}$, therefore belongs to $\mathcal{E}xt(S_{/\bar{e}})$: so T is in $\mathcal{E}xt(S_{/0})\mathcal{E}xt(S_{/1})*$, hence in $\mathcal{D}ist(\mathcal{E}xt(S_{/0}), \mathcal{E}xt(S_{/1}))$, that is $\mathcal{E}xtS$. ■

4. The syntactical approach.

All statements considered so far deal in fact with the study of particular expressions for the members of ϑ. For instance, the confluency of $\longrightarrow$ (proposition 1.1) can be stated as the equality $\vartheta = \vartheta^+\vartheta^-$ since it claims that every member of ϑ has an expression made by a block of positive generators followed by a block of negative generators. In the same way, the statement PH* can be stated as $\vartheta = \vartheta^+_{\mathrm{prog}}\vartheta^-_{\mathrm{prog}}$ (with the obvious meaning of $\vartheta^-_{\mathrm{prog}}$), and analog forms exist for PH, PH_n and PH'_m. In each case, the point is to prove that certain terms in ϑ are progressive, i.e., can be written as the products of progressive sequences.

Such results have been established above in a *semantical* way, in so far as we used the operation of ϑ on $\mathcal{T}(\Sigma)$, and proved the progressivity of a given term φ by showing that $S\Longrightarrow\mathbf{val}(S, \varphi)$ holds for some (any) term S in $\mathcal{T}(\Sigma)$. Another approach consists in directly guessing a progressive sequence and proving that φ can be written as the product of this sequence by means of the commutation relations that are known to hold in ϑ^+. This type of argument can be called *syntactical* since it only uses the relations in ϑ, but not the operation of ϑ on $\mathcal{T}(\Sigma)$.

DEFINITION. Let $(\mathbf{S}^*, \bullet)$ be the free monoid generated by $\mathbf{S}$, and denote by $\approx$ the congruence on $\mathbf{S}^*$ generated by the following pairs:

- all pairs $(u0v\bullet u1w\,,\, u1w\bullet u0v)$ ("$\perp$-pairs");
- all pairs $(u1\bullet u\bullet u0\bullet u1\,,\, u\bullet u1\bullet u)$ ("1-pairs");
- all pairs $(u0v\bullet u\,,\, u\bullet u00v\bullet u10v)$ ("0-pairs");
- all pairs $(u10v\bullet u\,,\, u\bullet u01v)$ ("10-pairs");
- all pairs $(u11v\bullet u\,,\, u\bullet u11v)$ ("11-pairs").

It is easily verified that the pairs above correspond to equal elements in ϑ^+, so that if ρ is the canonical projection of $\mathbf{S}^*$ onto ϑ^+ that maps w to w^+, ρ factorizes through $\approx$. Therefore any relation involving $\approx$ yields an equality when projected to ϑ^+. We quote below a few results in this direction; the proofs are rather painful, so they will be omitted.

DEFINITION. i) For w in $\mathbf{S}$, and $e = 0, 1$, we let $|w|_e$ be the number of e's occurring in w; we let $|w|^{\mathrm{fin}}_e$ be the number of *final* e's in w, i.e., the maximal integer r such that w can be written as $w'e^r$; finally set $|w|^{\mathrm{nfin}}_e = |w|_e - |w|^{\mathrm{fin}}_e$.

ii) Let w be in $\mathbf{S}$ and n be an integer; put

$$w^{[n]} := \begin{cases} w\cdot w0\cdot\cdots\cdot w0^{n-1} & \text{if } n>0,\\ \varepsilon & \text{otherwise,}\end{cases} \quad , \quad w^{\times} := w'^{[r]},$$

$$w^{!} := \begin{cases} (0^p1^{q-1})^{[r]}\cdot(0^p1^{q-2})^{[r]}\cdot\cdots\cdot(0^p)^{[r]} & \text{if } q\geq 1,\\ \varepsilon & \text{otherwise,}\end{cases}$$

where $p := |w|_0^{\mathrm{nfin}}, q := |w|_1, r := |w|_0^{\mathrm{fin}}, w = w'0^r$ and ε is the empty sequence in $\mathbf{S}^*$.

iii) For A included in $\mathbf{S}$, we set

$$A^{\times} := \prod_{w\in A}^{\prec} w^{\times} \quad \text{and} \quad A^{!} := \prod_{w\in A}^{\prec} w^{!}.$$

EXAMPLE. Let w be 101100; then the parameters p, q, r are respectively 1,3,2, so that $w^{\times}$ is 1011•10110 and $w^{!}$ is 011•0110•01•010•0•00.

Two important technical results are the following

LEMMA 1. *Assume that α is in $\mathbf{S}^*$ and that $\rho\alpha$ maps S to T; then one has*

$$(\mathbf{Supp}S)^{\times}\cdot\alpha \approx 1\alpha\cdot(\mathbf{Supp}T)^{\times},$$

where for any sequence $\alpha = w_1\cdot\cdots\cdot w_n$, $u\alpha$ means $uw_1\cdot\cdots\cdot uw_n$.

LEMMA 2. *Assume that A is a support and w is in $\mathbf{S}$; then one has*

$$(wA)^{!} \approx w^{!}\cdot 0^m1^{q-1}A^{\times}\cdot 0^m1^{q-2}A^{\times}\cdot\cdots\cdot 0^mA^{\times}\cdot 0^mA^{!},$$

where m is $|w|_0$ and q is $|w|_1$.

A typical (and self-contained) step toward the proof of such results is the following

CLAIM. *For $k\geq 1$ and $m\geq 0$, $\Lambda^{[k]}\cdot 1^{[m+1]}\cdot\Lambda \approx 1\cdot\Lambda^{[k+1]}\cdot 1^{[k]}\cdot 01^{[m]}$ holds.*

Proof. Induction on $m\geq 0$. Assume $m = 0$ and use induction on $k\geq 1$. For $k = 1$, we have $\Lambda\cdot 1\cdot\Lambda \approx 1\cdot\Lambda^{[2]}\cdot 1$ since the two members make a 1-pair. Now assume $k>1$, we have:

$$\begin{aligned}
\Lambda^{[k]}\cdot 1\cdot\Lambda &= \Lambda^{[k-1]}\cdot 0^{k-1}\cdot 1\cdot\Lambda \\
&\approx \Lambda^{[k-1]}\cdot 1\cdot 0^{k-1}\cdot\Lambda && (\perp\text{-pair})\\
&\approx \Lambda^{[k-1]}\cdot 1\cdot\Lambda\cdot 0^{k}\cdot 10^{k-1} && (0\text{-pair})\\
&\approx 1\cdot\Lambda^{[k]}\cdot 1^{[k-1]}\cdot 0^{k}\cdot 10^{k-1} && (\text{induction hypothesis})\\
&\approx 1\cdot\Lambda^{[k]}\cdot 0^{k}\cdot 1^{[k-1]}\cdot 10^{k-1} && (\perp\text{-pair})\\
&= 1\cdot\Lambda^{[k+1]}\cdot 1^{[k]}.
\end{aligned}$$

Now suppose $m\geq 1$ and the formula is proved for $m-1$ (and all k); we have:

$$\begin{aligned}
\Lambda^{[k]}\cdot 1^{[m+1]}\cdot\Lambda &= \Lambda^{[k]}\cdot 1^{[m]}\cdot 10^{m}\cdot\Lambda \\
&\approx \Lambda^{[k]}\cdot 1^{[m]}\cdot\Lambda\cdot 010^{m-1} && (10\text{-pair})\\
&\approx 1\cdot\Lambda^{[k+1]}\cdot 1^{[k]}\cdot 01^{[m-1]}\cdot 010^{m-1} && (\text{induction hypothesis})\\
&= 1\cdot\Lambda^{[k+1]}\cdot 1^{[k]}\cdot 01^{[m]}. \ \blacksquare
\end{aligned}$$

From lemmas 1 and 2, one deduces

PROPOSITION 3. *For any S in $\mathcal{T}(\Sigma)$, $\rho(\mathbf{Supp}S)^{!}$ maps S to ∂S.*

Proof. First, an induction on T shows that $\rho(\mathbf{Supp}T)^{\times}$ maps (for every term S) $ST*$ to $\mathbf{dist}(S,T)$. If T is in Σ, $ST*$ is equal to $\mathbf{dist}(S,T)$, and $\rho\Lambda^{\times}$ is the identity mapping. Otherwise, T is $T_{/0}T_{/1}*$, and the definition of $w^{\times}$ yields

$$(\mathbf{Supp}T)^{\times} = (0\mathbf{Supp}T_{/0})^{\times}\centerdot(1\mathbf{Supp}T_{/1})^{\times} = \Lambda\centerdot 0(\mathbf{Supp}T_{/0})^{\times}\centerdot 1(\mathbf{Supp}T_{/1})^{\times}.$$

So, if $\rho(\mathbf{Supp}T_{/e})^{\times}$ maps $ST_{/e}*$ to $\mathbf{dist}(S,T_{/e})$, $\rho(\mathbf{Supp}T)^{\times}$ maps $ST*$ to

$$\mathbf{val}(ST_{/0}*, \rho(\mathbf{Supp}T_{/0})^{\times})\mathbf{val}(ST_{/1}*, \rho(\mathbf{Supp}T_{/1})^{\times})*,$$

that is $\mathbf{dist}(S,T_{/0})\mathbf{dist}(S,T_{/1})*$, i.e., $\mathbf{dist}(S,T)$.
Now we argue inductively on S. The result is obvious for S in Σ. Otherwise, we have $(\mathbf{Supp}S)^{!} = (0\mathbf{Supp}S_{/0})^{!}\centerdot(1\mathbf{Supp}S_{/1})^{!}$, and lemma 2 gives $(0\mathbf{Supp}S_{/0})^{!} \approx 0(\mathbf{Supp}S_{/0})^{!}$ and $(1\mathbf{Supp}S_{/1})^{!} \approx (\mathbf{Supp}S_{/1})^{\times}\centerdot(\mathbf{Supp}S_{/1})^{!}$. Assume (induction hypothesis) that $\rho(\mathbf{Supp}S_{/e})^{!}$ maps $S_{/e}$ to $\partial S_{/e}$. Then by lemma 1 we have

$$(\mathbf{Supp}S_{/1})^{\times}\centerdot(\mathbf{Supp}S_{/1})^{!} \approx 1(\mathbf{Supp}S_{/1})^{!}\centerdot(\mathbf{Supp}\partial(S_{/1}))^{\times},$$

so that we deduce

$$(\mathbf{Supp}S)^{!} \approx 0(\mathbf{Supp}S_{/0})^{!}\centerdot 1(\mathbf{Supp}S_{/1})^{!}\centerdot(\mathbf{Supp}\partial(S_{/1}))^{\times}.$$

Now starting from S, i.e., $S_{/0}S_{/1}*$, $\rho(0(\mathbf{Supp}S_{/0})^{!})$ maps S to $(\partial S_{/0})S_{/1}*$, then $\rho(1(\mathbf{Supp}S_{/1})^{!})$ maps this term to $(\partial S_{/0})(\partial S_{/1})*$, and, finally, $\rho((\mathbf{Supp}\partial(S_{/1}))^{\times})$ maps this later term to $\mathbf{dist}(\partial S_{/0}, \partial S_{/1})$, that is ∂S. ∎

Since it is easily verified that, for any support A, $A^{\times}$ and $A^{!}$ are progressive sequences, we conclude from the proposition above that $S \Longrightarrow \partial S$ holds, and, moreover, we get the explicit progressive sequence witnessing for this property: this expression evaluates the exact contribution of each point w in $\mathbf{Supp}S$ to this sequence, namely the terms denoted by $\rho w^{!}$. Further computations could be made, for instance toward a syntactical proof of PH_1.

As a last question, let us mention the "completeness conjecture" that claims that ϑ^{+} is in fact isomorphic to $\mathbf{S}^{*}/\approx$, i.e., that the set of pairs used in the definition of $\approx$ is exactly a presentation of ϑ^{+}. If this conjecture is true, any semantical proof can be converted into a syntactical proof. However, it seems likely that any proof of the completeness conjecture will require a lot of results about $\approx$ first, and therefore the computations above are not useless in any case.

REFERENCES

[Bi] BIRMAN, J., *Braids, links, and mapping class groups*, Annals of Math. Studies **82**, Princeton Univ. Press (1975).

[Br] BRIESKORN, E., *Automorphic sets and braids and singularities*, Braids, Contemporary Math. **78**, Amer. Math. Soc. (1988) 45–117.

[De1] DEHORNOY, P., *Free distributive groupoids*, Journal of Pure and Applied Algebra, **61** (1989) 123–146.

[De2] DEHORNOY, P., *Sur la structure des gerbes libres*, Comptes-rendus de l'Acad. des Sciences de Paris, **309-I** (1989) 143–148.

[De3] DEHORNOY, P., *Algebraic properties of the shift mapping*, Proc. Amer. Math. Soc., **106-3** (1989) 617–623.

[De4] DEHORNOY, P., *An alternative proof of Laver's results on the algebra generated by elementary embeddings*, preprint 1989.

[De5] DEHORNOY, P., *Preuve de la conjecture d'irréflexivité pour les structures distributives libres*, Comptes-rendus de l'Acad. des Sciences des Paris, **314-I** (1992) 333–336.

[Do] DOUGHERTY, R., *On critical points of elementary embeddings*, preprint 1988.

[Ke] KEPKA, P., *Notes on left distributive groupoids*, Acta universitatis Carolinae, Mathematica et physica, **22-2** (1981) 23–37.

[Jo] JOYCE, D., *A classifying invariant of knots: the knot quandle*, Journal of Pure and Applied Algebra, **23** (1982) 37–65.

[La1] LAVER, R., *On the left distributive law and the freeness of an algebra of elementary embeddings*, Advances in Mathematics, **91-2** (1992), 209–231.

[La2] LAVER, R., *A division algorithm for the free left distributive algebra*, this volume.

Groupe de recherche algorithmique et logique
Université de Caen
14032 Caen-cédex, France
dehornoy@geocub.greco-prog.fr

ON ω_1-COMPLETE FILTERS

HANS-DIETER DONDER

Let us start with a definition. For an uncountable cardinal κ set

$$\mu(\kappa) = \min\{|H| \mid H \text{ is a set of } \omega_1\text{-complete uniform filters on } \kappa \text{ and } \forall A \subseteq \kappa\, \exists F \in H (A \in F \text{ or } \kappa - A \in F)\}$$

Clearly, $1 \leq \mu(\kappa) \leq 2^\kappa$.

A classical result of Ulam says that κ must be very large, if $\mu(\kappa) = 1$. On the other hand, by definition we have that $\mu(\kappa) = 1$ if κ is bigger than some strongly compact cardinal. Only recently (see [3]) Gitik has shown that $\mu(\kappa) \leq \omega$ implies that $\mu(\kappa) = 1$.

Can $\mu(\kappa)$ be small for small cardinals κ? Using a huge cardinal, Magidor showed in [4] that $\mu(\omega_3) \leq \omega_3$ is consistent. Shelah constructed a model of $\mu(\omega_1) = \omega_1$ starting with many supercompact cardinals (see [6]). With an almost huge cardinal Woodin produced a model where $\mu(\omega_1) = \omega_1$ is witnessed by normal filters. It seems to be an open problem whether $\mu(\omega_2) = \omega_1$ is consistent.

In this note we treat the following question. Is there always some κ such that $\mu(\kappa) \leq \kappa$? Prikry showed in [5] that $\mu(\omega_1) > \omega_1$ is consistent. Jensen showed later that the appropriate combinatorial principle holds in L which implies that $\mu(\omega_1) > \omega_1$ is true in L. We shall show:

THEOREM 1. *Assume $V = L$. Then $\mu(\kappa) > \kappa$ for all regular $\kappa > \omega$.*

To prove this we reduce the problem to a purely combinatorial question. So let us introduce the following principle. Let $\kappa > \omega$ be regular. Then Q_κ denotes the following property:

> There is some $G \subseteq \{f \mid f : \kappa \to 2\}$ such that $|G| > \kappa$ and for all $G^* \subseteq G$ such that $|G^*| > \kappa$ there is a countable $\overline{G} \subseteq G^*$ such that $\{\alpha < \kappa \mid \forall f, g \in \overline{G}, f(\alpha) = g(\alpha)\}$ is nonstationary.

This principle is closely related to some properties discussed in [7]. So the interested reader might also consult that paper. Now we have:

LEMMA 1. *Let $\kappa > \omega$ be regular and assume that Q_κ holds. Then $\mu(\kappa) > \kappa$.*

Proof. Assume not. Let $\mu(\kappa) \leq \kappa$ be given by H. By a result of Taylor (see [8]) we may assume that all $F \in H$ contain the club filter on κ. Let Q_κ be given by G. For each $f \in G$ choose $F_f \in H$ and $i_f < 2$ such that $\{\alpha < \kappa \mid f(\alpha) = i_f\} \in F_f$. Then there are $G^* \subseteq G$, $i < 2$, $F \in H$ such that

$|G^*| > \kappa$ and $F_f = F$, $i_f = i$ for all $f \in G^*$. Choose some countable $\overline{G} \subseteq G^*$ as in Q_κ. Then $\bigcap\{\alpha < \kappa \mid \forall f \in \overline{G} f(\alpha) = i\} \in F$ by ω_1-completeness and is nonstationary. This is a contradiction. □

So in order to prove Theorem 1 we only need to show:

PROPOSITION 1. *Assume $V = L$. Then Q_κ holds for all regular $\kappa > \omega$.*

Proof. We shall use the natural $(\kappa, 1)$-morass and the natural $\square_\infty$-sequence in L. The reader should look at [1] for the basic definitions. We use the standard notations. So for example $S = \{\nu \mid \nu > \omega,\ \nu \text{ p.r. closed},\ \nu \text{ singular}\}$, $\langle C_\nu \mid \nu \in S\rangle$ is the $\square_\infty$-sequence, $\prec$ is the morass tree, $\pi_{\bar{\nu}\nu}$ are the morass maps. Set $E = \{\nu \in S \cap \kappa^+ \mid C_\nu = \emptyset\}$. So we have

(1) (a) E is stationary in κ^+

(b) for all singular τ, $E \cap \tau$ is not stationary in τ

(c) if $\bar{\nu} \prec \nu, \bar{\nu} \in E$ and $\pi_{\bar{\nu}\nu}$ is cofinal, then $\nu \in E$.

Set $E_0 = \{\nu < \kappa \mid \nu \in S^+ \cap E,\ \nu \text{ is minimal in } \prec\}$. We also need:

(2) There is a sequence $\langle X_\eta \mid \eta \in E_0\rangle$ such that

(a) $\text{otp}(X_\eta) = \omega$, $X_\eta \subseteq \eta$ is cofinal in η

(b) for all unbounded $X \subseteq \kappa^+$ there are $\nu \in S_\kappa$ and $\eta \in E_0$ such that $\eta \prec \nu$ and $\pi_{\eta\nu}$"$X_\eta \subseteq X$.

The proof of this is very similar to the argument used in §3 of [1]. So we only give a sketch. We define $\langle X_\eta \mid \eta \in E_0\rangle$ by recursion. Given $\eta \in E_0$ let Z_η be the $<_L$-least unbounded subset of η such that there are no $\nu \in S_{\alpha_\eta}$ and $\tau \in E_0$ such that $\tau \prec \nu$ and $\pi_{\tau\nu}$"$X_\tau \subseteq Z_\eta$. Then choose $X_\eta \subseteq Z_\eta$ such that $\text{otp}(X_\eta) = \omega$ and $\sup X_\eta = \eta$. Note that every element of E_0 has cofinality ω. This will do it.

Now using (1) we easily get:

(3) For $\alpha < \kappa$ and $\mu < \alpha^+$ there is a function $h^\mu_\alpha : \mu \to 2$ such that for all $\nu \in S_\alpha \cap \mu$, $\eta \prec \nu$, $\eta \in E_0$ we have that $h^\mu_\alpha \restriction \pi_{\eta\nu}$"$X_\eta$ is not eventually constant.

Now let $\nu \in S_\kappa$. Set $A_\nu = \{\alpha_\tau \mid \tau \prec \nu\}$. We define a function $f_\nu : A_\nu \to \kappa$ such that $f_\nu(\alpha) < \alpha^+$ as follows. Let $\tau \prec \nu$, $\alpha = \alpha_\tau$ and $\pi = \pi_{\tau\nu}$. Here we regard π as a map from L_τ to L_ν. Set $U = \{X \subseteq \alpha \mid X \in L_\tau,\ \alpha \in \pi(X)\}$. Define a sequence $\langle \tau_i \mid i \leq \gamma\rangle$ as follows. Set $\tau_0 = \alpha + 1$. If $\tau_i > \tau$, then set $\gamma = i$ and stop. If $\tau_i \leq \tau$, then let τ_{i+1} be the least ordinal Θ such that $U \cap L_{\tau_i} \in L_\Theta$. If λ is a limit ordinal, set $\tau_i = \sup\{\tau_i \mid i < \lambda\}$. Because we are in L it is easy to see that $\gamma \leq \omega + 1$. Set $f_\nu(\alpha) = \tau_\gamma$.

We are now ready to define the set of functions G which will give us Q_κ. It suffices that every element of G is defined on a club subset of κ. So let $\nu \in S_\kappa$. We define $g_\nu : A_\nu \to 2$ by $g_\nu(\alpha) = h^\mu_\alpha(\tau)$ where $\mu = f_\nu(\alpha)$ and τ is the unique $\tau \prec \nu$ such that $\alpha = \alpha_\tau$. Then set $G = \{g_\nu \mid \nu \in S_\kappa\}$. Finally, we show that G satisfies Q_κ. So let $X \subseteq S_\kappa$ be unbounded. By (2)(b) choose $\nu_0, \nu_1 \in S_\kappa, \eta_0, \eta_1 \in E_0$ such that $\nu_0 < \nu_1$ and $\eta_i \prec \nu_i, \pi_{\eta_i\nu_i}$"$X_\eta \subseteq X$ for $i < 2$. Set $Y_i = \pi_{\eta_i\nu_i}$"X_{η_i} and $Y = Y_0 \cup Y_1$. It suffices to show that there is a club $C \subseteq \kappa$ such that for all $\alpha \in C$ there are $\tau_0, \tau_1 \in Y$ such that $F_{\tau_0}(\alpha) = 0$ and $f_{\tau_1}(\alpha) = 1$. For this let $\alpha \in A_{\nu_0} \cap A_{\tau_1}$ be sufficiently large. Let $\tau_i \prec \nu_i$ such that $\alpha_{\tau_i} = \alpha$. Set $\pi_i = \pi_{\tau_i \nu_i}$. Then $\pi_0 \subseteq \pi_1$. Looking at the definition of the functions f_ν we see that the sequence $\langle f_\nu(\alpha) \mid \nu \in Y_0 \rangle$ or the sequence $\langle f_\nu(\alpha) \mid \nu \in Y_1 \rangle$ is eventually constant. So (3) gives us what we need. □

We conjecture that $\mu(\kappa) \leq \kappa$ implies that there is an inner model with a measurable cardinal. Let us mention that in Theorem 1 we can replace the assumption $V = L$ by $V = K$, where K denotes the Dodd–Jensen core model. We now indicate a proof of a very special case of our conjecture.

THEOREM 2. *Assume $\mu(\omega_1) = \omega_1$. Then there is an inner model with a measurable cardinal.*

For this we use a result of Taylor (see [8]). He showed that $\mu(\omega_1) > \omega_1$ is true if every ω_1-complete filter on ω_1 containing the club filter possesses an almost disjoint family of sets of positive F-measure of size ω_2. Now let $\langle f_\nu \mid \nu < \omega_2 \rangle$ be the sequence of canonical functions for ω_1. By Taylor's result Theorem 2 follows from the following proposition.

PROPOSITION 2. *Let F be an ω_1-complete filter on ω_1 which contains every club subset of ω_1. Assume that for every $f : \omega_1 \to \omega_1$ there is some $\nu < \omega_2$ such that $\{\alpha < \omega_1 \mid f(\alpha) < f_\nu(\alpha)\} \in F$. Then there is an inner model with a measurable cardinal.*

Proof. This just uses the method applied in the proof of Theorem 2 in [2]. So we build the same system of embeddings as there. It is well known that we may assume that for all $\nu \in E$ and $\alpha \in C_\nu$ that $f_\nu(\alpha) = \nu_\alpha$. So by our assumption on F for all $f : \omega_1 \to \omega_1$ there is some $\nu \in E$ such that $\{\alpha \in C_\nu \mid f(\alpha) < \nu_\alpha\} \in F$. So we can easily construct $X \subseteq E$ such that $\text{otp}(X) = \omega^2$ and $S_{\nu\tau} = \{\alpha \mid [\nu_\alpha, \tau_\alpha] \cap I_\alpha \neq \emptyset\} \in F$ for all $\nu, \tau \in X$, $\nu < \tau$. Then $S = \bigcap\{S_{\nu\tau} \mid \nu, \tau \in X, \nu < \tau\} \in F$. So S is stationary. Now we argue exactly as in [2]. □

REFERENCES

[1] H.-D. Donder, *Another look at gap-1 morasses*, in **Recursion theory** (A. Nerode and R. Shore, editors), Proc. symp. in pure math. 42, American Mathematical Society, 1985, pp. 223–236.

[2] H.-D. Donder, *Families of almost disjoint functions*, **Contemporary mathematics**, vol. 31 (1984), pp. 71–78.

[3] M. Gitik and S. Shelah, *Forcing with ideals and simple forcing notions*, **Israel journal of mathematics**, vol. 68 (1989), pp. 129–160.

[4] M. Magidor, *On the existence of non-regular ultrafilters and the cardinality of ultrapowers*, **Transactions of the American Mathematical Society**, vol. 223 (1979), pp. 97–111.

[5] K. Prikry, *On a problem of Erdős, Hajnal and Rado*, **Discrete mathematics**, vol. 2 (1972), pp. 51–59.

[6] S. Shelah, *Iterated forcing and normal ideals on ω_1*, ***Israel journal of mathematics***, vol. 60 (1987), pp. 345–380.

[7] J. Steprāns and S. Watson, *Extending ideals*, ***Israel journal of mathematics***, vol. 54 (1986), pp. 201–226.

[8] A. Taylor, *On saturated sets of ideals and Ulam's Problem*, ***Fundamenta mathematicae***, vol. 109 (1980), pp. 37–53.

Mathematisches Institut
Universität München
D-80333 München
Germany

LABELLED DEDUCTIVE SYSTEMS: A POSITION PAPER

D. M. GABBAY

§1. Labelled deductive systems in context.

The purpose of this paper is to introduce a general notion of a logical system, namely that of a *Labelled Deductive System* (*LDS*), and show that many logical systems, new and old, monotonic and non-monotonic, all fall within this new framework. This research will eventually be published as a book, and this paper is based on Chapter 1 of [19].

We begin with the traditional view of what is a logical system.

Traditionally, to present a logic **L**, we need to present first the set of well-formed formulas of that logic. This is the *language* of the logic. We specify the sets of atomic formulas, connectives, quantifiers and the set of well-formed formulas. Secondly, we mathematically define the notion of consequence, that is, for sets of formulas Δ and formulas Q, we define the consequence relation $\Delta \vdash_{\mathbf{L}} Q$, which is read "$Q$ follows from Δ in the logic **L**".

The consequence relation is required to satisfy the following intuitive properties: (Δ, Δ' abbreviates $\Delta \cup \Delta'$).

Reflexivity

$$\Delta \vdash Q \text{ if } Q \in \Delta$$

Monotonicity

$$\frac{\Delta \vdash Q}{\Delta, \Delta' \vdash Q}$$

1.1. Transitivity (cut)

$$\frac{\Delta \vdash A; \Delta, A \vdash Q}{\Delta \vdash Q}$$

If you think of Δ as a database and Q as a query, then reflexivity means that the answer "yes" is given for any Q which is already listed in the database Δ. Monotonicity reflects the accumulation of data, and transitivity is nothing but lemma generation, namely, if $\Delta \vdash A$, then A can be used as a lemma to derive B from Δ.

These three properties have appeared to constitute minimal and most natural for a logical system, given that the main applications of logic were in mathematics and philosophy.

The above notions were essentially put forward by Tarski [8] in 1936 and is referred to as Tarski consequence. Scott [7], inspired by Gabbay [16], generalised the notion to allow Q to be a set of formulas Γ. The basic relation is then of the form $\Delta \vdash \Gamma$, satisfying:[1]

Reflexivity

$$\Delta \vdash \Gamma \text{ if } \Delta \cap \Gamma \neq \varnothing$$

Monotonicity

$$\frac{\Delta \vdash \Gamma}{\Delta, \Delta' \vdash \Gamma}$$

1.2. Transitivity (cut)

$$\frac{\Delta, A \vdash \Gamma; \Delta' \vdash A, \Gamma'}{\Delta, \Delta' \vdash \Gamma, \Gamma'}$$

Scott further showed that for any Tarski consequence relation $\vdash$ there exist two Scott consequence relations (a maximal one and a minimal one) that agree with it, namely, that $\Delta \vdash A$ (Tarski) iff $\Delta \vdash \{A\}$ (Scott) (see Gabbay [2]).

The above notions are monotonic. However, the increasing use of logic in theoretical computer science and artificial intelligence has given rise to logical systems which are not monotonic, i.e., to systems in which the axiom of monotonicity is not satisfied. There are many such systems, satisfying a variety of conditions and presented in a variety of ways. Furthermore, some are characterized in a proof-theoretical and some in a model-theoretical manner. All these different presentations give rise to some notion of consequence $\Delta \vdash Q$, but they only seem to all agree on reflexivity. The essential difference between these logics (commonly called *non-monotonic logics*) and the more traditional logics (now referred to as *monotonic logics*) is the fact that $\Delta \vdash A$ holds in the monotonic case because of some $\Delta_A \subseteq \Delta$, while in the non-monotonic case the entire set Δ is somehow used to derive A. Thus if Δ is increased to Δ', there is no change in the monotonic case, while there may be a change in the non-monotonic case.

The above describes the situation current in the early 1980's. We have had a multitude of systems generally accepted as "logics" without a unifying underlying theory and many had semantics without proof theory or vice-versa, though almost all of them were based on some sound intuitions of one form or another. Clearly there was the need for a general unifying framework. An early attempt at classifying non-monotonic systems was Gabbay [3]. It was put forward that basic axioms for a Tarski type consequence relation should be *reflexivity, cut* (version 1.2 above) and *restricted monotonicity*, namely:

Restricted monotonicity (cumulativity)

$$\frac{\Delta \vdash A; \Delta \vdash B}{\Delta, A \vdash B}$$

[1]The similarity with Gentzen sequents is obvious. A sequent $\Delta \vdash \Gamma$ is a relation between Δ and Γ. Such a relation can either be defined axiomatically (as a consequence relation) or be generated via closure conditions like $A \vdash A$ (initial) and other generating rules. The generating rules correspond to Gentzen rules. In many logics we have $\Delta \vdash \Gamma$ iff $\emptyset \vdash \bigwedge \Delta \to \bigvee \Gamma$, which gives an intuitive meaning to $\vdash$.

A variety of systems seem to satisfy this axiom. Further results were obtained (Lehmann [11, 12]), (Wojcicki [9, 10]), (Makinson [5, 6]) and the area was called "axiomatic theory of the consequence relation" by Wojcicki. A recent general theory is presented in Gabbay [20, 21].

Although some sort of classification was obtained and semantical results were proved, the approach does not seem to be strong enough. Many systems do not satisfy restricted monotonicity. Other systems such as relevance logic, do not even satisfy reflexivity. Others have a richness of their own which is lost in a simple presentation as an axiomatic consequence relation. Obviously a different approach is needed, one which would be more sensitive to the variety of features of the systems in the field. Fortunately, developments in a neighbouring area, that of automated deduction, seem to be of help. New automated deduction methods were developed for non-classical logics, and resolution was generalised and modified to be applicable to these logics. In general, because of the value of these logics in theoretical computer science and artificial intelligence, a greater awareness of the computational aspects of logical systems was developing and more attention was being devoted to proof-theoretical presentations. It became apparent to us that a key feature in the proof-theoretic study of these logics is that a slight natural variation in an automated or proof-theoretic system of one logic (say $\mathbf{L}_1$), can yield another logic (say $\mathbf{L}_2$).

Although $\mathbf{L}_1$ and $\mathbf{L}_2$ may be conceptually far apart (in their philosophical motivation, and mathematical definitions) when it comes to automated techniques and proof-theoretical presentation, they turn out to be brother and sister. This kind of relationship is not isolated and seems to be widespread. Furthermore, non-monotonic systems seem to be obtainable from monotonic ones through variations on some of their monotonic proof-theoretical formulation, thus giving us a handle on classifying non-monotonic systems.

This phenomena has prompted Gabbay [4, 15] to put forward the view that a logical system $\mathbf{L}$ is not just the traditional consequence relation $\vdash$ (monotonic or non-monotonic) but a pair $(\vdash, \mathbf{S}_\vdash)$, where $\vdash$ is a mathematically defined consequence relation (i.e., the set of pairs (Δ, Q) such that $\Delta \vdash Q$) satisfying whatever minimal conditions on a consequence relation one happens to agree on, and $\mathbf{S}_\vdash$ is an algorithmic system for generating all those pairs. Thus according to this definition classical logic $\vdash$ perceived as a set of tautologies together with a Gentzen system $\mathbf{S}_\vdash$ is not the same as classical logic together with the two-valued truth table decision procedure $\mathbf{T}_\vdash$ for it. In our conceptual framework,$(\vdash, \mathbf{S}_\vdash)$ is *not the same logic* as $(\vdash, \mathbf{T}_\vdash)$.

To illustrate and motivate our way of thinking, observe that it is very easy to move from $\mathbf{T}_\vdash$ for classical logic to a truth table system $\mathbf{T}_\vdash^n$ for Lukasiewicz n-valued logic. It is not so easy to move to an algorithmic system for intuitionistic logic. In comparison, for a Gentzen system presentation, exactly the opposite is true. Intuitionistic and classical logics are neighbours, while Lukasiewicz logics seem completely different. In fact, some of the results of this book show proof-theoretic similarities between Lukasiewicz's infinite valued logic and Girard's Linear Logic, which in turn is proof-theoretically similar to intuitionistic logic.

There are many more such examples among temporal logics, modal logics, defeasible logics and others. Obviously, there is a need for a more unifying framework. The question is then whether we can adopt a concept of a logic where the passage from one system to another is natural, and along predefined acceptable modes of variation? Can we put forward a framework where the computational aspects of a logic also play a role? Is it possible to find a common home for a variety of seemingly different techniques introduced for different purposes in seemingly different intellectual logical traditions?

To find an answer, let us ask ourselves what makes one logic different from another? How is a new logic presented and described and compared to another? The answer is obvious. These considerations are usually dealt with on the meta-level. Most logics are based on modus ponens and the quantifier rules are formally the same anyway and the differences between them are meta-level considerations on the proof theory or semantics. If we can find a mode of presentation of logical systems where meta-level features can reside side by side with object-level features then we can hope for a general framework. We must be careful here. In the logical community the notions of object-level vs. meta-level are not so clear. Most people think of *naming* and *proof predicates* in this connection. This is not what we mean by meta-level here. We need a more refined understanding of the concept. There is a similar need in computer science. In [19] we devote a chapter to these considerations. See also [25].

We found that the best framework to put forward is that of a *Labelled Deductive System, LDS*. Our notion of what constitutes a logic will be that of a pair $(\vdash, \mathbf{S}_\vdash)$ where $\vdash$ is a set-theoretic (possibly non-monotonic) consequence relation on a language $\mathbf{L}$ and $\mathbf{S}_\vdash$ is an *LDS*, and where $\vdash$ is essentially required to satisfy no more than *Identity* (i.e., $\{A\} \vdash A$) and *Surgical Cut* (see below and [20, 21]). This is a refinement of our concept of a logical system mentioned above and first presented in Gabbay [4]. We now not only say that a logical system is a pair $(\vdash, \mathbf{S}_\vdash)$, but we are adding that $\mathbf{S}_\vdash$ itself has a special presentation, that of an *LDS*.

An *LDS* system is a triple $(\mathbf{L}, \Gamma, \mathbf{M})$, where $\mathbf{L}$ is a logical language (connectives and wffs) and Γ is an algebra (with some operations) of labels and $\mathbf{M}$ is a discipline of labelling formulas of the logic (from the algebra of labels Γ), together with deduction rules and with agreed ways of propagating the labels via the application of the deduction rules. The way the rules are used is more or less uniform to all systems. In the general case we allow Γ, the algebra of labels, to be an *LDS* system itself! Furthermore, if our view of a logical system is that the declarative unit is a pair, a formula and a label, then we can also label the pair itself and get multiple labelling.

The perceptive reader may feel resistance to this idea at this stage. First be assured that you are not asked to give up your favourite logic or proof theory nor is there any hint of a claim that your activity is now obsolete. In mathematics a good concept can rarely be seen or studied from one point of view only and it is a sign of strength to have several views connecting different concepts. So the traditional logical views are as valid as ever and add strength to the new point of view. In fact, a closer examination of [19] would reveal that manifestations of our *LDS* approach already exist in the literature in various forms (see [1] and [18] and the references there), however, they were locally regarded as convenient tools and there was not the realisation that there is a general framework to be studied and developed. None of us is working in a vacuum and we build on each others' work. Further, the existence of a general framework in which any particular case can be represented does not necessarily mean that the best way to treat that particular case is within the general framework. Thus if some modal logics can be formulated in *LDS*, this does not mean that in practice we should replace existing ways of treating the logics by their *LDS* formulation. The latter may not be the most efficient for those particular logics. It is sufficient to show how the *LDS* principles specialise and manifest themselves in the given known practical formulation of the logic.

The reader may further have doubts about the use of labels from the computational point of view. What do we mean by a unifying framework? Surely a Turing machine can simulate any logic, is that a unifying framework? The use of labels is powerful, as we know from computer science, are we using labels to play the role of a Turing machine? The answer to the question is twofold. First that we are not operating at the meta-level, but at the object-level. Second, there are severe restrictions on the way we use *LDS*. Here is a preview:

1. The only rules of inference allowed are the traditional ones, modus ponens and some form of deduction theorem for implication, for example.

2. Allowable modes of label propagation are fixed for all logics. They can be adjusted in agreed ways to obtain variations but in general the format is the same. For example, it has the following form for implications:
   $(A \to B)$ gets label t iff $\forall x \in \Gamma_1$ [If A is labelled x then B can be proved with labels $t + x$], where Γ_1 is a set of labels characterising the implication in that particular logic. For example Γ_1 may be all atomic labels or related labels to t, or variations. The freedom that different logics have is in the choice of Γ_1 and the properties of "+". For example we can restrict the use of modus ponens by a wise propagation of labels.

3. The quantifier rules are the same for all logics.

4. Meta-level features are implemented via the labelling mechanism, which is object language.

The reader who prefers to remain within the traditional point of view of:

assumptions (data) proving a conclusion

can view the labelled formulas as another form of data.

There are many occasions when it is most intuitive to present an item of data in the form $t : A$, where t is a label and A is a formula. The common underlying reason for the use of the label t is that t represents information which is needed to modify A or to supplement (the information in) A which is not of the same type or nature as (the information represented by) A itself. A is a logical formula representing information declaratively, and the additional information of t can certainly be added declaratively to A to form A', however, we may find it convenient to put forward the additional information through the label t as part of a pair $t : A$.

Take for example a source of information which is not reliable. A natural way of representing an item of information from that source is $t : A$, where A is a declarative presentation of the information itself and t is a number representing its reliability. Such expert systems exist (e.g. Mycin) with rules which manipulate both t and A as one unit, propagating the reliability values t_i through applications of modus ponens. We may also use a label naming the source of information and this would give us a qualitative idea of its reliability.

Another area where it is natural to use labels is in reasoning from data and rules. If we want to keep track, for reasons of maintaining consistency and/or integrity constraints, where and how a formula was deduced, we use a label t. In this case, the label t in $t : A$ can be the part of the data which was used to get A. Formally in this case t is a formula, the conjunction of the data used. We thus get pairs of the form $\Delta_i : A_i$, where A_i are formulas and Δ_i are the parts of the database from which A_i was derived.

A third example where it is natural to use labels is time stamping of data. Where data is constantly revised and updated, it is important to time stamp the data items. Thus the data items would look like $t_i : A_i$, where t_i are time stamps. A_i itself may be a temporal formula. Thus there are two times involved, the logical time s_i in $A_i(s_i)$ and the time stamping t_i of A_i. For reasons of clarity, we may wish to regard t_i as a label rather than incorporate it into the logic (by writing for example $A^*(t_i, s_i)$).

To summarise then, we replace the traditional notion of consequence between formulas of the form $A_1, \ldots, A_n \vdash B$ by the notion of consequence between labelled formulas

$$t_1 : A_1, t_2 : A_2, \ldots, t_n : A_n \vdash s : B$$

Depending on the logical system involved, the intuitive meaning of the labels varies. In querying databases, we may be interested in labelling the assumptions so that when we get an answer to a query, we can record, via the label of the answer, from which part of the database the answer was obtained. Another area where labelling is used is temporal logic. We can time stamp assumptions as to when they are true and query, given those assumptions, whether a certain conclusion will be true at a certain time. Thus the consequence notion for labelled deduction is essentially the same as that of any logic: given assumptions does a conclusion follow?

Whereas in the traditional logical system the consequence is defined by using proof rules on the formulas, in the *LDS* methodology the consequence is defined by using rules on both formulas and their labels. Formally we have formal rules for manipulating labels and this allows for more scope in decomposing the various features of the consequence relation. The meta features can be reflected in the algebra or logic of the labels and the object features can be reflected in the rules of the formulas.

The notion of a database or of a "set of assumptions" also has to be changed. A database is a configuration of labelled formulas. The configuration depends on the labelling discipline. For example, it can be a linearly ordered set $\{a_1 : A_1, \ldots, a_n : A_n\}, a_1 < a_2 < \cdots < a_n$. The proof discipline for the logic will specify how the assumptions are to be used. We need to develop the notions of the Cut Rule and the Deduction Theorem in such an environment. This we do in a later section.

The next two sections will give many examples of *LDS* disciplines featuring many known monotonic and non-monotonic logics. It is of value to summarise our view listing the key points involved:

- The unit of declarative data is a labelled formula of the form $t : A$, where A is a wff of a language $\mathbf{L}$ and t is a label. The labels come from an algebra (set) of labels.
- A database is a set of labelled formulas.
- An *LDS* discipline is a system (algorithmic) for manipulating both formulas and their labels. Using this discipline the statement $\Delta \vdash \Gamma$ is well defined for the two databases Δ and Γ. Especially $\Delta \vdash t : A$ is well defined.
- $\vdash$ must satisfy the minimal conditions, namely

  1.3. Identity

  $$\{t : A\} \vdash t : A$$

  1.4. Surgical cut

  $$\frac{\Delta \vdash t : A, \Gamma[t : A] \vdash s : B}{\Gamma[\Delta] \vdash s : B}$$

  where $\Gamma[t : A]$ means that $t : A$ is contained/occurs somewhere in the structure Γ and $\Gamma[\Delta]$ means that Δ replaces A in the structure.

- A logical system is a pair $(\vdash, \mathbf{S}_\vdash)$, where $\vdash$ is a consequence relation and $\mathbf{S}_\vdash$ is an *LDS* for it.

§2. Examples from monotonic logics.

To motivate our approach we study several known examples in this section.

Example 2.1 below shows a standard deduction from Relevance Logic. The purpose of the example is to illustrate our point of view. There are many such examples in Anderson and Belnap [1]. Example 2.3 below considers a derivation in modal logic. There we use labels to denote essentially possible worlds. The objective of the example is to show the formal similarities to the relevance logic case in Example 2.1. Section 2.4 can reap the benefits of the formal similarities of the first two examples and introduce, in the most natural way, a system of relevant modal logic. The objective of Example 2.4 is to show that the labels in Example 2.1 and Example 2.3 can be read as determining the meta-language features of the logic and can therefore be combined "declaratively" to form the new system of Example 2.4. Example 2.5 considers strict implication. This example shows that for strict **S4** implication one can read the labels either as relevance labels or as possible world labels. Examples 2.6, 2.7 show how labels can interact with quantifiers in modal logic.

EXAMPLE 2.1. *Relevance and linear logic.*

Consider a propositional language with implication "$\rightarrow$" only. The forward elimination rule is modus ponens. From the theorem proving view, modus ponens is an object language consideration. Thus a proof of $\vdash (B \rightarrow A) \rightarrow ((A \rightarrow B) \rightarrow (A \rightarrow B))$ can proceed as follows:

Assume $a_1 : B \rightarrow A$ and show $(A \rightarrow B) \rightarrow (A \rightarrow B)$. Further assume $a_2 : A \rightarrow B$ and show $A \rightarrow B$. Further assume $a_3 : A$ and show B. We thus end up with the following problem:

Assumptions

1. $a_1 : B \rightarrow A$

2. $a_2 : A \rightarrow B$

3. $a_3 : A$

Derivation

4. $a_2a_3 : B$ by modus ponens from lines (2) and (3).

5. $a_1a_2a_3 : A$ from (4) and (1).

6. $a_2a_1a_2a_3 : B$ from (5) and (2).

7. $a_2a_1a_2 : A \rightarrow B$ from (3) and (6).

8. $a_2a_1 : (A \rightarrow B) \rightarrow (A \rightarrow B)$ from (2) and (7).

9. $a_2 : (B \rightarrow A) \rightarrow ((A \rightarrow B) \rightarrow (A \rightarrow B))$ from (1) and (8).

The meta aspect of this proof is the annotation of the assumptions and the keeping track of what was used in the deduction. A meta-level condition would determine the logic involved.

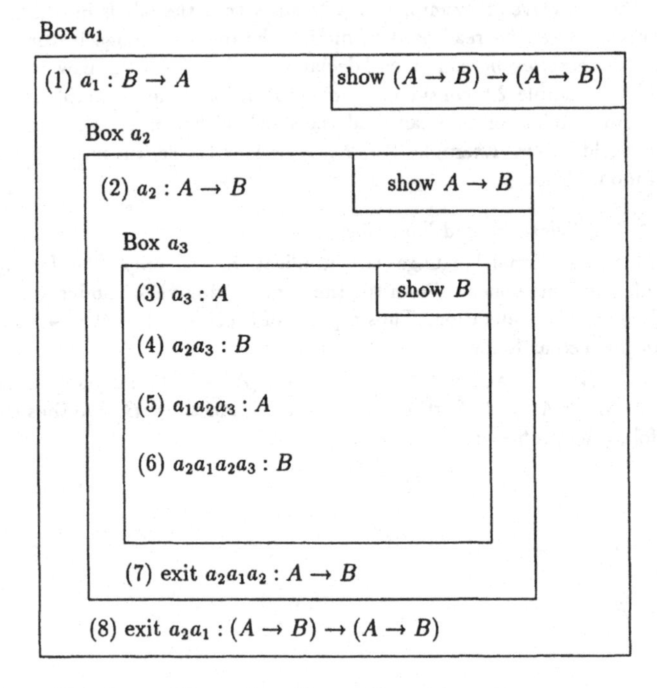

Box a_1
(1) $a_1 : B \to A$
show $(A \to B) \to (A \to B)$
Box a_2
(2) $a_2 : A \to B$
show $A \to B$
Box a_3
(3) $a_3 : A$
show B
(4) $a_2a_3 : B$
(5) $a_1a_2a_3 : A$
(6) $a_2a_1a_2a_3 : B$
(7) exit $a_2a_1a_2 : A \to B$
(8) exit $a_2a_1 : (A \to B) \to (A \to B)$

(9) exit $a_2 : (B \to A) \to ((A \to B) \to (A \to B)$

Figure 1

A formal definition of the labelling discipline for this class of logics is given in [19]. For this example it is sufficient to note the following three conventions:

1. Each assumption is labelled by a new atomic label.

   An ordering on the labels can be imposed, namely $a_1 < a_2 < a_3$. This is to reflect the fact that the assumptions arose from our attempt to prove $(B \to A) \to ((A \to B) \to (A \to B))$ and not for example from $(A \to B) \to ((B \to A) \to (A \to B))$ in which case the ordering would be $a_2 < a_1 < a_3$. The ordering can affect the proofs in certain logics.

2. If in the proof, A is labelled by the multiset α and $A \to B$ is labelled by β then B can be derived with a label $\alpha \cup \beta$ where "$\cup$" denotes multiset union.

3. If B was derived using A as evidenced by the fact that the label α of A is a submultiset of the label β of B $(\alpha \subseteq \beta)$ then we can derive $A \to B$ with the label $\beta - \alpha$ ("$-$" is multiset subtraction).

The derivation can be represented in a more graphical way.

To show $(B \to A) \to ((A \to B) \to (A \to B))$: See figure 1.

Figure 1 is the *meta-box* way of representing the deduction. Note that in line 8, multiset subtraction was used and only one copy of the label a_2 was taken out. The other copy of a_2 remains and cannot be cancelled. Thus this formula is not a theorem of linear logic, because the outer box does not exit with label $\varnothing$. In relevance logic, the discipline uses sets and not multisets. Thus the label of line 8 in this case would be a_1 and that of line 9 would be $\varnothing$. The above deduction can be made even more explicit as follows:

$(B \to A) \to ((A \to B) \to (A \to B))$ follows with a label from Box a_1.

Box a_1

a_1 :	$B \to A$ assumption
$a_2 a_1$:	$(A \to B) \to (A \to B)$ from Box a_2

Box a_2

a_2 :	$A \to B$ assumption
$a_2 a_1 a_2$:	$A \to B$ from Box a_3

Box a_3

$a_3 : A$	assumption
$a_2 : A \to B$	reiteration from box a_2
$a_2 a_3 : B$	by modus ponens
$a_1 : B \to A$	reiteration from box a_1
$a_1 a_2 a_3 : A$	modus ponens from the two preceding lines
$a_2 : A \to B$	repetition of an earlier line
$a_2 a_1 a_2 a_3 : B$	modus ponens from the two preceding lines

The following meta-rule was used:

We have a systems of partially ordered meta-boxes $a_1 < a_2 < a_3$. Any assumption in a box a can be reiterated in any box b provided $a < b$.

REMARK 2.2.

a. The above presentation of the boxes makes them look more like possible worlds. The labels are the worlds and formulas can be exported from one world to another according to some rules. The next example 2.3 describes modal logic in just this way.

b. Note that different meta-conditions on labels and meta-boxes correspond to different logics.

   The following table gives intuitively some correspondence between meta-conditions and logics.

Meta-condition:	*Logic*
ignore the labels	intuitionistic logic
accept only the derivations which use all the assumptions	relevance logic
accept derivations which use all assumptions exactly once	linear logic

The meta-conditions can be translated into object conditions in terms of axioms and rules. If we consider a Hilbert system with modus ponens and substitution then the additional axioms involved are given below:

Linear Logic
$A \to A$
$(A \to (B \to C)) \to (B \to (A \to C))$
$(C \to A) \to ((B \to C) \to (B \to A))$
$(C \to A) \to ((A \to B) \to (C \to B))$

Relevance Logic
Add the schema below to linear logic
$(A \to (B \to C)) \to ((A \to B) \to (A \to C))$

Intuitionistic Logic
Add the schema below to relevance logic:
$A \to (B \to A)$

The reader can note that the following axiom (Peirce Rule) yields classical logic. Further note that for example, we can define "Linear Classical Logic" by adding the Peirce Rule to linear logic. A new logic is obtained.

Classical Logic
Add the schema below to intuitionistic logic:
$((A \to B) \to A) \to A$.

EXAMPLE 2.3. This example shows the meta-level/object-level division in the case of modal logic. Modal logic has to do with possible worlds. We thus think of our basic database (or assumptions) as a finite set of information about possible worlds. This consists of two parts. The configuration part, the finite configuration of possible worlds for the database, and the assumptions part which tells us what formulas hold in each world. The following is an example of a database:

	Assumptions	*Configuration*
(1)	$t : \Box\Box B$	$t < s$
(2)	$s : \Diamond(B \to C)$	

The conclusion to show (or query) is:

$$t : \Diamond\Diamond C.$$

The derivation is as follows:

3. From (2) create a new point r with $s < r$ and get $r : B \to C$.

We thus have

Assumptions	*Configuration*
(1), (2), (3)	$t < s < r$

4. From (1), since $t < s$ we get $s : \Box B$.
5. From (4) since $s < r$ we get $r : B$.
6. From (5) and (3) we get $r : C$.
7. From (6) since $s < r$ we get $s : \Diamond C$.
8. From (7) using $t < s$ we get $t : \Diamond\Diamond C$.

Discussion:
The object rules involved are:
$\Box E$ *Rule:*

$$\frac{t < s; t : \Box A}{s : A}$$

$\Diamond I$ *Rule:*

$$\frac{t < s, s : B}{t : \Diamond B}$$

$\Diamond E$ *Rule:*

$$\frac{t : \Diamond A}{\text{create a new point } s \text{ with } t < s \text{ and deduce } s : A}$$

Note that the above rules are not complete. We do not have rules for deriving, for example, $\Box A$. Also, the rules are all for intuitionistic modal logic.

The meta-level consideration may be properties of $<$,
e.g. transitivity: $t < s \wedge s < r \rightarrow t < r$ or
e.g. linearity: $t < s \vee t = s \vee s < t$ etc.

EXAMPLE 2.4. The reader can already see the benefit of separating the meta-level (the handling of possible worlds, i.e., labels) and the object-level (i.e., formulas) features. We can combine both the meta-level features of Examples 2.1 and 2.3 to create for example a modal relevance logic in a natural way. Each assumption has a relevance label as well as world label. Thus the proof of the previous example becomes the following:

	Assumptions	*Configuration*
(1)	$(a_1, t) : \Box\Box B$	$t < s$
(2)	$(a_2, s) : \Diamond(B \rightarrow C)$	

We proceed to create a new label r using $\Diamond E$ rule. The relevance label is carried over. We have $t < s < r$.

3. $(a_2, r) : B \rightarrow C$

Using $\Box E$ rule with relevance label carried over, we have:

4. $(a_1, s) : \Box B$

5. $(a_1, r) : B$

Using modus ponens with relevance label updated

6. $(a_1, a_2, r) : C$

Using $\Diamond I$ rule:

7. $(a_1, a_2, s) : \Diamond C$

8. $(a_1, a_2, t) : \Diamond\Diamond C$

(8) means that we got $t : \Diamond\Diamond C$ using both assumptions a_1 and a_2.

There are two serious problems in modal and temporal theorem proving. One is that of Skolem functions for $\exists x \Diamond A(x)$ and $\Diamond \exists x A(x)$ are not logically the same. If we skolemise we get $\Diamond A(c)$. Unfortunately it is not clear where c exists, in the current world $((\exists x = c)\Diamond A(x))$ or the possible world $(\Diamond(\exists x = c)A(x))$.

If we use labelled assumptions then, $t : \exists x \Diamond A(x)$ becomes $t : \Diamond A(c)$ and it is clear that c is introduced at t.

On the other hand, the assumption $t : \Diamond \exists x A(x)$ will be used by the $\Diamond E$ rule to introduce a new point $s, t < s$ and conclude $s : \exists x A(x)$. We can further skolemise at s and get $s : A(c)$, with c introduced at s. We thus need the mechanism of remembering or labelling constants as well, to indicate where they were first introduced.

Labelling systems for modal and temporal logics is studied in [22].

EXAMPLE 2.5. The following example describes the logic of modal **S4** strict implication. In this logic the labels can be read either as relevance labels or as possible worlds. **S4** strict implication $A \rightarrow B$ can be understood as a temporal connective, as follows:

"$A \rightarrow B$ is true at world t iff for all future worlds s to t and for t itself we have that if A is true at s then B is true at s". Thus $A \rightarrow B$ reads "From now on, if A then B".

Suppose we want to prove that $A \rightarrow B$ and $A \rightarrow (B \rightarrow C)$ imply $A \rightarrow C$. To show this we reason semantically and assume that at time t, the two assumptions are true. We want to show that $A \rightarrow C$ is also true at t. To prove that we take any future time s, assume that A is true at s and show that C is also true at s. We thus have the following situation:

1. $t : A \rightarrow B$

2. $t : A \rightarrow (B \rightarrow C)$

3. show $t : A \rightarrow C$
from box

3.1	Assume $s : A$ Show $s : C$
	Since s is in the future of t, we get that at s, (1) and (2) are also true.
3.2	$s : A \to B$ from (1)
3.3	$s : A \to (B \to C)$ from (2)
	We now use modus ponens, because $X \to Y$ means "from now on, if X then Y"
3.4	$s : B$ from (3.1) and (3.2)
3.5	$s : B \to C$ from (3.2) and (3.3)
3.6	$s : C$ modus ponens from (3.4) and (3.5)

exit $t : A \to C$

Notice that any $t : D$ can be brought into (reiterated) the box as $s : D$, provided it has an implicational form, $D = D_1 \to D_2$. We can thus regard the labels above as simply naming assumptions (not as possible worlds) and the logic has the reiteration rule which says that only implications can be reiterated.

Let us add a further note to sharpen our understanding. Suppose $\to$ is read as a **K4** implication (i.e., transitivity without reflexivity). Then the above proof should fail. Indeed the corresponding restriction on modus ponens is that we do perform $X, X \to Y \vdash Y$ in a box, provided $X \to Y$ is a reiteration into the box and was not itself derived in that same box. This will block line (3.6).

EXAMPLE 2.6. Another example has to do with the Barcan formula

	Assumption	*Configuration*
(1)	$t : \forall x \Box A(x)$	$t < s$

We show

$$s : \forall x A(x)$$

We proceed intuitively

1. $t : \Box A(x)$ (stripping $\forall x$, remembering x is arbitrary).
2. Since the configuration contains $s, t < s$ we get
$$s : A(x)$$
3. Since x is arbitrary we get
$$s : \forall x A(x)$$

The above intuitive proof can be restricted.

The rule

$$\frac{t : \Box A(x), t < s}{s : A(x)}$$

is allowed only if x is instantiated.

To allow the above rule for arbitrary x is equivalent to adopting the Barcan formula axiom:

$$\forall x \Box A(x) \to \Box \forall x A(x)$$

EXAMPLE 2.7. To show $\forall x \Box A(x) \to \Box \forall x A(x)$ in the modal logic where it is supposed to be true.

1. Assume $t : \forall x \Box A(x)$
   We show $\Box \forall x A(x)$ by the use of the meta-box:

	create α,	$t < \alpha$
(2)	$t : \Box A(x)$	from (1)
(3)	$\alpha : A(x)$	from (2) using a rule which allows this with x a variable.
(4)	$\alpha : \forall x A(x)$	universal generalisation.

(5) Exit: $t : \Box \forall x A(x)$.

This rule has the form:

Create α,	$t < \alpha$
Argue to get	$\alpha : B$
Exit with	$t : \Box B$

The above are just a few examples for the scope we get using labels. The exact details and correspondences are worked out in our monograph [19].

EXAMPLE 2.8. (Relevance reasoning.) The indices are α, β, and $\gamma = (\beta - \alpha)$. The reasoning structure is:
Assume $\alpha : A$
Show $\beta : B$
If $\beta \supseteq \alpha$ then exit with $(\beta - \alpha) : A \to B$.
To show $A \to (B \to C) \vdash B \to (A \to C)$

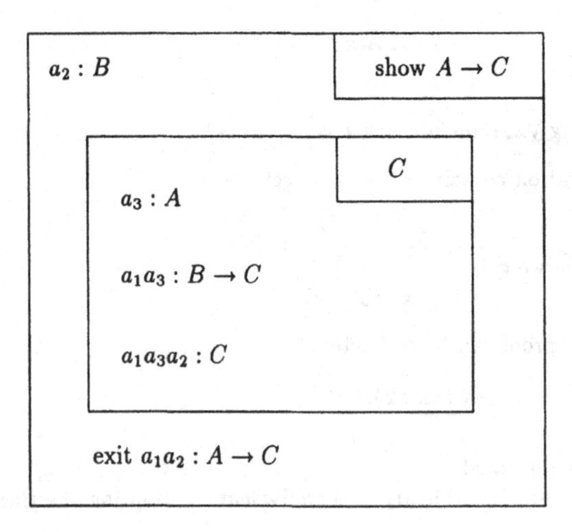

exit $a_1 : B \to (A \to C)$

Figure 2

Assume

$$a_1 : A \to (B \to C)$$

we use the meta-box to show $B \to (A \to C)$. See figure 2.

EXAMPLE 2.9. (Lukasiewicz many-valued logics.) Consider Lukasiewicz infinite-valued logic, where the values are all real numbers or rationals in [0,1]. We designate 0 as *truth* and the truth table for implication is

$$x \to y = \max(0, y - x)$$

Here the language contains atoms and implication only, assignments h give values to atoms in [0,1], $h(q) \in [0,1]$ and h is extended to arbitrary formulas via the table for $\to$ above. Define the relation

$$A_1, \ldots, A_n \vdash B$$

to mean that for all h, $h(A_1) + \cdots + h(A_n) \geqq h(B)$, where $+$ is numerical addition.

This logic can be regarded as a labelled deductive system, where the labels are values $t \in [0,1]$. $t : A$ means that $h(A) = t$, for a given background assignment h. The interesting part is that to show $t : A \to B$ (i.e., that $A \to B$ has value t) we assume $x : A$ (i.e., that A has value x) and then have to show that B has value $t + x$, i.e., show $t + x : B$.

This is according to the table of $\to$.

Thus figure 3 shows the deduction in box form:

$x : A$	assumption
$\vdots$	
$\vdots$	
$t + x : B$	

exit $t : A \to B$

Figure 3

This has the *same structure* as the case of relevance logic, where $+$ was understood as concatenation.

A full study of many valued logics from the LDS point of view is given in [19].

EXAMPLE 2.10. (Formulas as types.) Another instance of the natural use of labels is the Curry–Howard interpretation of formulas as types. This interpretation conforms exactly to our framework. In fact, our framework gives the incentive to extend the formulas as types interpretation in a natural way to other logics, such as linear and relevance logics and surprisingly, also many valued logics, modal logics, and intermediate logics. A formula is considered as a type and its label is a *definable* λ-term of the same type. Given a system for defining λ-terms, the theorems of the logic are all those types which can be shown to be non-empty.

The basic propagation mechanism corresponding to modus ponens is:

$$\frac{\begin{array}{c} t^A : A \\ t^{A\to B} : A \to B \end{array}}{t^{A\to B}(t^A) : B}$$

It is satisfied by *application.*

Thus if we read the + in $t^{A\to B} + t^A$ as application, we get the exact parallel to the general schema of propagation. Compare with relevance logic where + was concatenation, and with many valued logics where + was numerical addition!

To show $t : A \to B$ we assume $x : A$, with x arbitrary, i.e., start with a term x of type A, use the proof rules to get B. As we saw, applications of modus ponens generate more terms which contain x in them via application. If we accept that proofs generate functionals, then we get B with a label $y = t(x)$. Thus $t = \lambda x t(x)$. This again conforms with our general schema for $\to$.

In our paper [18] on the Curry–Howard interpretation we exploit this idea systematically. There are two mechanisms which allow us to restrict or expand our ability to define terms of any type. We can restrict λ-abstraction, (e.g. allow $\lambda x t(x)$ only if x actually occurs in t), this will give us logics weaker than intuitionistic logic, or we can increase our world of terms by requiring diagrams to be closed e.g., for any φ of classical logic such that

$$\vdash (A \to B) \to [\varphi(A) \to \varphi(B)]$$

in classical logic, we want the following diagram to be complete, i.e., for any term t there must exist a term t' (see figure 4).

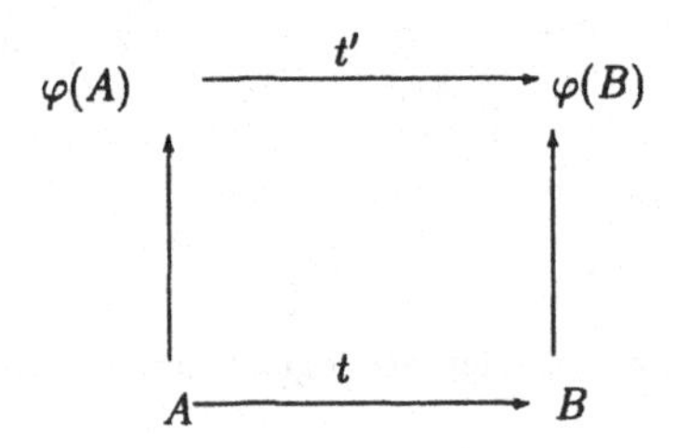

Figure 4

Take for example the formula $A \to (B \to A)$ as type. We want to show a definable term of this type, we can try and use the standard proof (see figure 5),

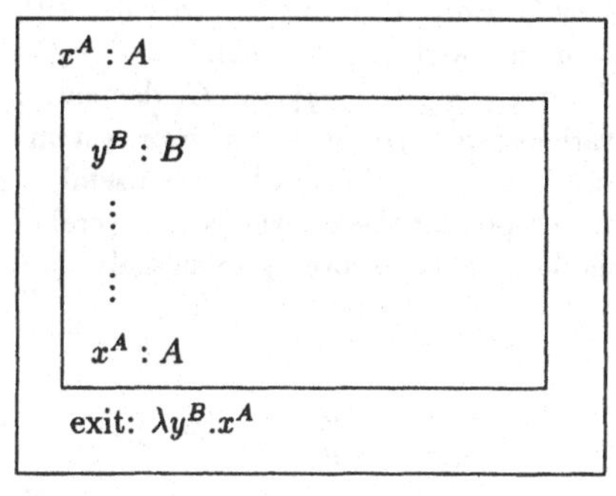

Figure 5

however, with the restriction on λ-abstraction which requires the abstracted variable to actually occur in the formula, we cannot exit the inner box. For details see [18].

EXAMPLE 2.11. (Realisability interpretation.) The well-known realisability interpretation for intuitionistic implication is another example of a functional interpretation for $\rightarrow$ which has the same universal *LDS* form. A notation for a recursive function $\{e\}$ realises an implication $A \rightarrow B$ iff for any n which realises A, $\{e\}(n)$ realises B. Thus

$$e : A \rightarrow B \text{ iff } \forall n[n : A \Rightarrow \{e\}(n) : B]$$

It is an open problem to find an axiomatic description of the set of all wffs which are realisable.

§3. Examples from non-monotonic logics.

The examples in the previous section are from the area of monotonic reasoning. This section will give examples from non-monotonic reasoning. As we have already mentioned, we hope that the idea of *LDS* will unify these two areas.

EXAMPLE 3.1. (Ordered logic.) An ordered logic database is a partially ordered set of local databases, each local database being a set of clauses. The diagram (figure 6) describes an ordered logic database.

The local databases are labelled t_1, t_2, t_3, s_1, s_2 and $\varnothing$ and are partially ordered as in the figure.

To motivate such databases, consider an ordinary logic program $C_1 = \{p \leftarrow \neg q\}$. The computation of a logic program assumes that, since q is not a head of any clause, $\neg q$ is part of the data (this is the *closed world assumption*). Suppose we relinquish this principle and adopt the principle of asking an ***advisor*** what to do with $\neg q$. The advisor might say that $\neg q$ succeeds or might say that $\neg q$ fails. The advisor might have his own program to consult. If his program is C_2, he might run the goal q (or $\neg q$), look at what he gets and then advise. To make the situation symmetrical and general we must allow for Horn programs to have rules with both q and $\neg q$ (i.e., literals) in heads and bodies and have any number of negotiating advisors. Thus we can have $C_2 = \{\neg q\}, C_1 = \{q \leftarrow \neg q\}$ and C_1 depends on C_2. Ordered logic develops and studies various aspects of such an advisor system which is modelled as a partially ordered set of theories. Such a logic is useful, e.g. for multi-expert systems where we want to represent the knowledge of several experts in a single system. Experts may then be ordered according to an "advisory" or a relative preference relation.

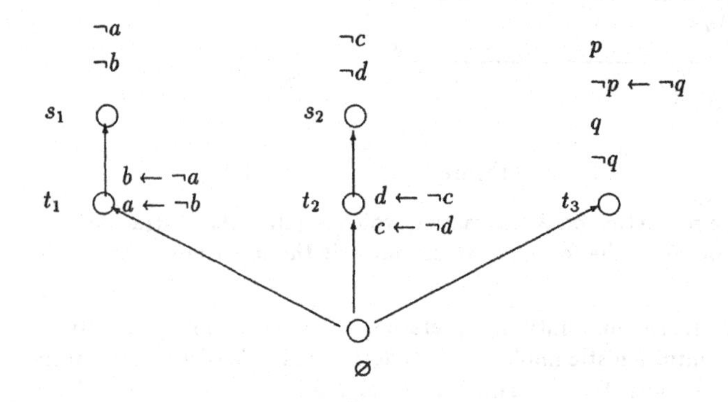

Figure 6

A problem to consider is what happens when we have several advisors that are in conflict. For example, C_1 depends on C_2 and C_1 depends on C_3. The two advisers, C_2 and C_3, may be in conflict. One may advise $\neg q$, the other q. How to decide? There are several options:

1. We can accept q if all advisors say "yes" to q.

2. We can accept q if at least one advisor says "yes" to q.

3. We can apply some non-monotonic or probabilistic mechanism to decide.

If we choose options (1) or (2) we are essentially in modal logic. To have a node t and to have $?q$ refer to advisors $t_1, \ldots, t_n$ with $t < t_i, i = 1, \ldots, n$ is like considering $?\Box q$ at t in modal logic with $t_1, \ldots, t_n$ possible worlds in option 1 and like considering $\Diamond q$ at t in option (2). Option (3) is more general, and here an *LDS* approach is most useful. We see from this advisors examples an application area where the labels arise naturally and usefully. The area of ordered logic is surveyed in [13].

EXAMPLE 3.2. (Defeasible logic.) This important approach to non-monotonic reasoning was introduced by Nute [14]. The idea is that rules can prove either an atom q or its negation $\neg q$. If two rules are in conflict, one proving q and one proving $\neg q$, the deduction that is stronger is from a rule whose antecedent is logically more specific. Thus the database:

$$\begin{aligned} t_1 &: \text{Bird }(x) \rightarrow \text{Fly }(x) \\ t_2 &: \text{Big }(x) \wedge \text{Bird }(x) \rightarrow \neg \text{ Fly }(x) \\ t_3 &: \text{Big }(a) \\ t_4 &: \text{Bird }(a) \end{aligned}$$

$$\begin{array}{l} t_1 < t_2 \\ t_3 \\ t_4 \end{array}$$

can prove:

$$\begin{aligned} t_2 t_3 t_4 &: \neg\text{Fly}(a) \\ t_1 t_4 &: \text{Fly}(a) \end{aligned}$$

The database will entail $\neg$Fly (a) because the second rule is more specific.

As an *LDS* system the labelling of rules in a database Δ is very simple. We label a rule by its antecedent. The ordering of the labels is done by logical strength relative to some background theory Θ (which can be a subtheory of Δ of some form). Deduction pays attention to strength of labels.

EXAMPLE 3.3. (Propositional circumscription.) Circumscription is defined semantically via satisfaction in minimal models. Surprisingly, results of Olivetti [26] allow one to present an *LDS* discipline for (at least) propositional circumscription.

To explain the idea let $\vdash_m$ denote consequence in minimal models. For this consequence we have, for example, $p \vee q \vdash_m \neg p \vee \neg q$, which does not follow in classical logic. Suppose we try and find a semantic tableaux counter-model for the above. In classical logic we try the tableaux construction and if all the top nodes are *closed* then there is no countermodel. For $\vdash_m$ we just change the notion of "*closed*." This can depend on labelling. A more precise study of this theme will be done later.

§4. Conclusion.

Logic is widely applied in computer science and artificial intelligence. The needs of the application areas in computing are different from those in mathematics and philosophy. In response to computer science needs, intensive research has been directed in the area of non-classical and non-monotonic logic. New logics have been developed and studied. Certain logical features, which have not received extensive attention in the pure logic community, are repeatedly being called upon in computational applications. Two features in logic seem to be of crucial importance to the needs of computer science and stand in need of further study. These are:

1. The meta-level features of logical systems

2. The "logic" of Skolem functions and unification

The meta-language properties of logical systems are usually hidden in the object language. Either in the proof theory or via some higher-order or many-sorted devices. The logic of Skolem functions is non-existent. Furthermore, the traditional presentation of classical and non-classical logics is not conducive to bringing out and developing the features needed for computer science applications. The very concept of what is a logical system seems to be in need of revision and clarification. A closer examination of classical and non-classical logics reveals the possibility of introducing a new approach to logic; the discipline of *Labelled Deductive Systems* (*LDS*) which, I believe, will not only be ideal for computer science applications but will also serve, I hope, as a new unifying logical framework of value to logic itself. What seem to be isolated local features of some known logics turn out to be, in my view, manifestations of more general logical phenomena of interest to the future development of logic itself.

LDS is part of a more general view of logic. This view is discussed elsewhere [19, 20, 21], however in brief, we claim the following. The new concept of a logical system is that of a *network* of *LDS* systems which has mechanisms for *communication* (through the labels, which code meta-information) and *evolution* or change.

Evaluation is a general concept which can embrace updating, abduction, consistency maintenance, action and planning. The above statement of position is vague but it does imply that we believe that notions like abduction and updating are logical notions of equal standing to those of provability. See [17].

References

[1] A. R. Anderson and N. D. Belnap, ***Entailment***, Princeton University Press, 1975.

[2] D. M. Gabbay, ***Semantical Investigations in Heyting's Intuitionistic Logic***, D. Reidel, 1981.

[3] D. M. Gabbay, *Theoretical Foundations for non monotonic reasoning*, in ***Expert Systems, Logics and Models of Concurrent Systems***, K. Apt (ed.), Springer-Verlag, 1985, pp. 439–459.

[4] D. M. Gabbay, *Theory of Algorithmic Proof*, in ***Handbook of Logic in Computer Science***, Volume 1, S. Abramsky, D. M. Gabbay, T. S. E. Maibaum (eds.), Oxford University Press, 1993, pp. 307–408.

[5] D. Makinson, *General Theory of Cumulative Inference*, in ***Non-monotonic Reasoning***, M. Reinfrank, J. de Kleer, M. L. Ginsberg and E. Sandewall (eds.), Lecture Notes on Artificial Intelligence, no. 346, Springer-Verlag.

[6] D. Makinson, *General Patterns in nonmonotonic Reasoning*, in ***Handbook of Logic in Artificial Intelligence and Logic Programming***, Volume 2, D. M. Gabbay, C. J. Hogger, J. A. Robinson (eds.), Oxford University Press, to appear, 1993.

[7] D. SCOTT, *Completeness and axiomatizability in many valued logics*, in ***Proceedings of the Tarski Symposium***, American Mathematical Society, 1974, pp. 411–436.

[8] A. TARSKI, *On the Concept of Logical Consequence* (in Polish), 1936. Translation in ***Logic Semantics Metamathematics***, Oxford University Press, 1956.

[9] R. WOJCICKI, *An Axiomatic treatment of non monotonic arguments*, ***Studia Logica***, to appear.

[10] R. WOJCICKI, *Heuristic Rules of Inference in non-monotonic arguments*, ***Studia Logica***, to appear.

[11] S. KRAUS, D. LEHMANN, and M. MAGIDOR, *Preferential models and cumulative logics*, ***Artificial Intelligence***, vol. 44 (1990), pp. 167–207.

[12] D. LEHMANN, *What does a conditional knowledge base entail?* in ***KR 89, Toronto, May 89***, Morgan Kaufmann Publisher, pp. 1–18.

[13] D. VERMEIR and E. LAENENS, *An overview of ordered logic*, in ***Abstracts of the Third Logical Biennial***, Varga, Bulgaria, 1990.

[14] D. NUTE, *LDR—A Logic for Defeasible Reasoning*, 1986, ACMC Research Report 01-0013.

[15] D. M. GABBAY, *Algorithmic Proof with Diminishing Resource, I*, in ***Proceedings CSL 90***, LNCS 533, Springer-Verlag, pp. 156–173,

[16] D. M. GABBAY, *The Craig Interpolation Theorem for Intuitionisic Logic I* and *II*, in ***Logic Colloquium 69***, R. O. Gandy (ed.), North-Holland Pub. Company, pp. 391–410.

[17] D. M. GABBAY, *Abduction in labelled deductive systems, a conceptual abstract*, in ***ECSQAU 91***, R. Kruse and P. Siegel (eds.), Lecture notes in Computer Science 548, Springer-Verlag, 1991, pp. 3–12.

[18] D. M. GABBAY and R. J. G. B. DE QUEIROZ, *Extending the Curry–Howard Interpretation to Linear, Relevant and other Resource Logics*, ***The Journal of Symbolic Logic***, vol. 57 (1992), pp. 1319–1365.

[19] D. M. GABBAY, *Labelled Deductive Systems*, 1st Draft September 1989, 6th draft February 1991. Published as a report by CIS, University of Münich. To appear as a book with Oxford University Press.

[20] D. M. GABBAY, *Theoretical Foundations for non monotonic reasoning Part 2: Structured non-monotonic Theories*, in ***SCAI '91***, Proceedings of the Third Scandinavian Conference on AI, IOS Press, Amsterdam, pp. 19–40.

[21] D. M. GABBAY, *A General Theory of Structured Consequence Relations*, to appear in a volume on substructured logics, P. Schröder-Heister and K. Dosen (eds.), Oxford University Press.

[22] D. M. GABBAY, *Modal and Temporal Logic Programming II*, in ***Logic Programming—Expanding the Horizon***, T. Dodd, R. P. Owens, S. Torrance (eds.), Ablex, 1991, pp. 82–123.

[23] D. M. GABBAY, *How to construct a logic for your application*, in ***Proceedings of the 16th German AI Conference, GWAI 92***, Lecture Notes on AI, vol. 671, Springer-Verlag, 1992, pp. 1–30.

[24] D. M. GABBAY, *Labelled Deductive Systems and Situation Theory*, to appear in ***Proceedings STA-III***, 1992.

[25] D. M. GABBAY, *Modal and Temporal Logic Programming III, Metalevel Features in the Object Language*, in ***Non-Classical Logic Programming***, L. Fariñas del Cerro and M. Penttonen (eds.), Oxford University Press, 1992, pp. 85–124.

[26] N. OLIVETTI, *Tableaux and Sequent Calculus for Minimal Entailment*, ***Journal of Automated Reasoning***, vol. 9 (1992), pp. 99–139.

Department of Computing
Imperial College, 180 Queen's Gate
London SW7 2BZ.
e-mail: dg@doc.ic.ac.uk
Tel: 071 589 5111

TEMPORAL EXPRESSIVE COMPLETENESS IN THE PRESENCE OF GAPS

D. M. GABBAY, I. M. HODKINSON, and M. A. REYNOLDS[1]

Abstract

It is known that the temporal connectives *until* and *since* are expressively complete for Dedekind complete flows of time but that the Stavi connectives are needed to achieve expressive completeness for general linear time which may have "gaps" in it. We present a full proof of this result.

We introduce some new unary connectives which, along with *until* and *since* are expressively complete for general linear time. We axiomatize the new connectives over general linear time, define a notion of complexity on gaps and show that *since* and *until* are themselves expressively complete for flows of time with only isolated gaps. We also introduce new unary connectives which are less expressive than the Stavi connectives but are, nevertheless, expressively complete for flows of time whose gaps are of only certain restricted types. In this connection we briefly discuss scattered flows of time.

§1. Introduction: the problem of expressive completeness.

This section will present the problem of expressive completeness of temporal connectives within the more general model theoretic concept of the existence of a finite G-basis for m-adic theories. The known results in this area will then be outlined.

We begin with the ordinary propositional temporal logic. Assume we are given a flow of time $(T, <)$, where T is the set of moments of time and $<$ is a transitive and irreflexive relation on T, thought of as the earlier-later relation. We define the notion of m-dimensional temporal logic on $(T, <)$. An m-dimensional atomic proposition q on $(T, <)$ can be associated with a subset Q of T^m, representing the set of all m-tuples of moments of time where q is true. The boolean logical operations on temporal formulas, such as $\wedge$, $\vee$, $\sim$ and $\rightarrow$ correspond naturally to operations on these subsets. It is clear that a temporal assignment h to the atoms associating with atoms q_i subsets $h(q_i) \subseteq T^m$, gives rise to an ordinary model for $(T, <, Q_i, =)$. To be able to express formally the connections between propositional temporal formulas and subsets of T^m, we need to use the m-adic language with $(T, <, Q_i, =)$, where $Q_i \subseteq T^m$ are m-place predicates and $=$ is equality.

[1]The work of M. Reynolds was supported by the U.K. Science and Engineering Research Council under the MetateM project (GR/F/28526).

DEFINITION 1.1.

1. *We define the temporal propositional language $L[C_1, \ldots, C_n]$, with connectives $C_1, \ldots, C_n$ as follows:*

   (a) *Any atom q is a wff.*

   (b) *If A and B are wffs so are $A \wedge B$, $A \vee B$, $\sim A$ and $A \rightarrow B$.*

   (c) *If C_i is n_i-place and $A_1, \ldots, A_{n_i}$ are wffs so is $C_i(A_1, \ldots, A_{n_i})$.*

2. *Let $(T, <)$ be a flow of time. Let Π be a set of m-place predicates. The m-adic theory $(T, <, \Pi, =)$ is defined as the language with $(<, =, Q_i \in \Pi)$ and wffs as follows:*

   (a) *$Q_i(x_1, \ldots, x_m), x_i = x_j$ and $x_i < x_j$ are wffs, for x_j variables and $Q_i \in \Pi$.*

   (b) *If φ and ψ are wffs so are $\varphi \wedge \psi$, $\varphi \vee \psi$, $\sim\varphi$, $\varphi \rightarrow \psi$, $\forall x\varphi$ and $\exists x\varphi$.*

3. *The temporal language and the m-adic language can be connected in the following manner.*

   (a) *Enumerate the atomic propositions of $L[C_1, \ldots, C_n]$ as $q_1, q_2, \ldots$ and enumerate the m-adic predicates of Π as $Q_1, Q_2, \ldots$ and associate q_i with Q_i.*

   (b) *Associate with the connective $C(p_1, \ldots, p_n)$, where $p_1, \ldots, p_n$ are propositional variables, a formula $\psi_C(t_1, \ldots, t_m, P_1, \ldots, P_n)$ with m free variables $t_1, \ldots, t_m$ and n m-adic variables $P_1, \ldots, P_n$. ψ_C is called a table for C.*

   (c) *Any model $(T, <, \Pi)$ of the m-adic language will now give rise to an m-dimensional temporal model as follows. Let the assignment h be*

   $$h(q_i) = \{(s_1, \ldots, s_m) \mid (T, <, \Pi) \models Q_i(s_1, \ldots, s_m)\}.$$

   Extend h to all wff by the equations:

   $$\begin{aligned} h(A \wedge B) &= h(A) \cap h(B) \\ h(\sim A) &= T^m - h(A) \\ h(C(A_1, \ldots, A_n)) &= \{(t_1, \ldots, t_m) \mid (T, <, \Pi) \models \\ &\quad \psi_C(t_1, \ldots, t_m, h(A_1), \ldots, h(A_n))\} \end{aligned}$$

   for any connective C.

It is obvious from Definition 1.1 that any formula $\psi(t_1, \ldots, t_m, Q_1, \ldots, Q_n)$ defines an n-place connective $C_\psi(q_1, \ldots, q_n)$ via the following truth table: $C_\psi(q_1, \ldots, q_n)$ holds at $t_1, \ldots, t_m$ iff $\psi(t_1, \ldots, t_m, Q_1, \ldots, Q_n)$ holds, where $Q_i = \{(s_1, \ldots, s_m) \mid q_i \text{ holds at } (s_1, \ldots, s_m)\}$.

In particular the connectives *since* (S) and *until* (U) correspond to the monadic tables:

$$\psi_S(t, Q_1, Q_2) = \exists s < t(Q_1(s) \wedge \forall u(s < u < t \rightarrow Q_2(u)))$$

and

$$\psi_U(t, Q_1, Q_2) = \exists s > t(Q_1(s) \wedge \forall u(s > u > t \rightarrow Q_2(u))).$$

Clearly we can use the connectives $S(p, q)$ and $U(p, q)$ to build arbitrary wffs $A(q_1, \ldots, q_n)$. It is easy to see that for each A, there exists a formula $\psi_A(t, Q_1, \ldots, Q_n)$ of the monadic language such that for all t and $q_1, \ldots, q_n$, $A(q_1, \ldots, q_n)$ holds at t iff $\psi_A(t, Q_1, \ldots, Q_n)$ holds, where $Q_i = \{s \mid q_i \text{ holds at } s\}$.

The family of all ψ_A can be defined inductively as follows:

DEFINITION 1.2. *Let $W_1(\{\psi_S, \psi_U\})$ be the smallest set of well formed formulas of the monadic language with one free variable satisfying the following conditions:*

1. *$Q_i(t) \in W_1$ for Q_i atomic.*
2. *If φ, $\psi \in W_1$ so are $\varphi \wedge \psi$, $\sim\varphi$, $\varphi \vee \psi$ and $\varphi \rightarrow \psi$.*
3. *$\psi_U, \psi_S \in W_1$.*
4. *If $\psi(t, Q_1, \ldots, Q_n) \in W_1$ with t the free variable and Q_i the monadic letters in ψ and if $\psi_i(t) \in W_1$, for $i = 1, \ldots, n$ then $\psi(t, \psi_1, \ldots, \psi_n)$ is also in W_1, where $\psi(t, \psi_1, \ldots, \psi_n)$ is obtained from $\psi(t, Q_1, \ldots, Q_n)$ by substituting simultaneously $\lambda t \psi_i(t)$ for $\lambda t Q_i(t)$, $i = 1, \ldots, n$.*

DEFINITION 1.3. *In general given formulas $\psi_1(t_1, \ldots, t_m), \ldots, \psi_k(t_1, \ldots, t_m)$ with m free variables the set $W_m(\{\psi_1, \ldots, \psi_k\})$ can be defined in the m-adic language as follows:*

- *$Q_i(t_1, \ldots, t_m) \in W_m$ for Q_i atomic.*
- *If φ, $\psi \in W_m$ so are $\varphi \wedge \psi$, $\sim\varphi$, $\varphi \vee \psi$ and $\varphi \rightarrow \psi$.*
- *$\psi_1, \ldots, \psi_k \in W_m$.*
- *If $\psi(t_1, \ldots, t_m, Q_1, \ldots, Q_n) \in W_m$ with $t_1, \ldots, t_m$ exactly the free variables of ψ and Q_i exactly the m-adic predicates in ψ and if $\psi_i(t_1, \ldots, t_m) \in W_m$, for $i = 1, \ldots, n$ then*
  $$\psi(t_1, \ldots, t_m, \psi_1, \ldots, \psi_n)$$
  is also in W_m, where $\psi(t_1, \ldots, t_m, \psi_1, \ldots, \psi_n)$ is obtained from $\psi(t_1, \ldots, t_m, Q_1, \ldots, Q_n)$ by substituting simultaneously $\lambda t_1, \ldots, t_m \psi_i(t_1, \ldots, t_m)$ for $\lambda t_1, \ldots, t_m Q_i(t_1, \ldots, t_m)$, $i = 1, \ldots, n$.

DEFINITION 1.4.

1. *The problem of expressive completeness for a set of m-adic wffs*
   $$\{\psi_1, \ldots, \psi_k\}$$
   over a class $\mathcal{K}$ of flows of time is the question of whether $W_m(\{\psi_1, \ldots, \psi_k\})$ is essentially the set of all m-adic wffs over $\mathcal{K}$: namely whether for any ψ there exists $\phi \in W_m$ such that $\mathcal{K} \models \psi \leftrightarrow \phi$.

2. *The problem of finite G_m-basis for the m-adic language over a class $\mathcal{K}$ of flows $(T,<)$ is whether the m-adic language can be represented as equal to a $W_m(\{\phi_1,\ldots,\phi_k\})$ for some finite set $\{\phi_1,\ldots,\phi_k\}$.*

3. *The problem of expressive completeness of since and until over a class $\mathcal{K}$ is whether $\{\psi_U,\psi_S\}$ form a finite G_1-basis for all monadic wffs over the class $\mathcal{K}$ of models $(T,<,\Pi)$.*

The problem of finite basis is a general model theoretic one. Let $\mathcal{K}$ be a class of models in some language, e.g. it might be the class $\mathcal{G}$ of all groups.

Let $(\mathcal{G},Q_1,Q_2,\ldots,=)$ be the m-adic theory of $\mathcal{G}$ where Q_i are new additional m-ary relational variables. Let $\phi_1,\ldots,\phi_k$ be m-adic formulas with m free variables. We can still define $W_m(\{\phi_1,\ldots,\phi_k\})$ and ask whether W_m essentially equals the set of all m-adic wffs over $\mathcal{G}$. We can thus ask whether the theory of groups admits a finite G_m-basis for its m-adic theory.

H. Kamp in [K] has shown that *since* and *until* form a finite G_1-basis for the monadic theory of Dedekind complete linear orderings. J. Stavi put forward two additional connectives which are shown in Theorem 3 to be a finite G_1-basis for general linear time. A first complete proof of this result is given in this paper. Schlingloff [S] has produced a finite G_1-basis for binary trees. The current paper studies finite bases for linear orderings with manageable gaps.

The problem of the existence of a finite G_m-basis for a class of models $\mathcal{K}$ is related to the notion of Gabbay's H_m-dimension.

DEFINITION 1.5.

1. *A theory T is said to have a finite H_m-dimension $\leq n$ over a class of models $\mathcal{K}$ iff every wff $\phi(t_1,\ldots,t_m,Q_1,\ldots,Q_k)$ with at most m free variables t_i and arbitrary number k of m-adic predicates is equivalent over $\mathcal{K}$ to a wff $\psi(t_1,\ldots,t_m,Q_1,\ldots,Q_k)$ where ψ uses no more than n distinct bound variable letters.*

2. *The minimal n satisfying (a) above is called the H_m-dimension of T.*

THEOREM 1 [GHR].

- *A class $\mathcal{K}$ of models has a finite H_m-dimension if it has a finite G_m-basis.*

- *Let the class $\mathcal{K}$ have H_m-dimension n, then it has a finite G_{m+n}-basis.*

- *There is a class $\mathcal{K}$ of models with H_1-dimension 3 but with no finite G_1-basis.*

Another notion of interest is that of weak m/m'-dimensional logic where $1 \leq m' < m$. This notion arises from m dimensional logic where the atoms $Q_i(t_1,\ldots,t_m)$ depend only on the first m' places. For example for $m'=1$ the weak $(m/1)$ m-dimensional temporal logic has the Q_i unary. In this case the existence of a finite G_1-basis implies the existence of a finite $G_{m/1}$-basis for any m.

Of special interest for applications are one or two dimensional temporal logics over a linear flow of time. In intuitive terms we are evaluating formulas at points or at intervals (or pairs of points). The problem of finding an expressively complete set of connectives is of special importance. Such connectives are extensively studied in [GHR]. We quote one theorem here of relevance.

DEFINITION 1.6. *Let $\mathcal{K}$ be a class of linear flows of time.*

1. *A formula A of a one-dimensional temporal logic is said to be pure future (past) iff its truth value at a point of any (T, h) for any $T \in \mathcal{K}$ and any h, depends only on the value of the atoms at the future (past) of that point.*
2. *A set of one-dimensional connectives is said to have the separation property over $\mathcal{K}$ iff every formula A can be rewritten equivalently (over K) as a boolean combination of pure past, atomic and pure future formulas.*

THEOREM 2. *A set of one-dimensional connectives $\{C_1, \ldots, C_k\}$ has the separation property over $\mathcal{K}$ iff it forms a G_1-basis over $\mathcal{K}$.*

Separation can be combinatorially checked by trying actually to rewrite any formula into a separated boolean combination. In the case of linear ordering the presence of gaps seems to be of combinatorial importance. As atoms are true or false over stretches of time, the first or last point of truth is very useful. If no such point exists we have a gap. We therefore need to study temporal behaviour around gaps. The case of Dedekind complete flows is simple. *Since* and *until* form a G_1-basis.

If the flow allows for gaps then a lot depends on the kind of gaps allowed. It is clear that in the general case new connectives are needed. It is not hard to show, and indeed our Lemma 3 below shows, that U and S are then not adequate to express some first-order connectives. However, as mentioned in [GPSS], Stavi was able to introduce two new connectives U' and S' so that the set $\{U, S, U', S'\}$ is expressively complete over all linear time. We present what we believe is the first full published proof of this result in Section 8.

For the sake of completeness, we consider the question of whether there are intermediate connectives appropriate for structures in which the gaps are in certain senses nice. In this paper we classify the gaps appearing in linear orders and are then able to introduce new connectives to talk about the behaviour of atoms in the neighbourhood of such gaps. Natural questions arise about the expressive power of sets of these connectives and we are able to present a fairly comprehensive (although by no means complete) range of answers.

The authors would like to thank Tony Hunter, Rob Hubbard and Robin Hirsch for many useful discussions during the development of this work.

§2. Gaps in the flow of time.

We identify gaps in a flow of time with supremum-less non-empty proper initial segments of the order and insert the gap in the appropriate place in the order. Dedekind complete orders then are those without gaps. The completion T^* of an order T is another order consisting of T and all the gaps in the right places and is Dedekind complete.

The simplest kind of gap imaginable is an *isolated* gap which exists in an open interval of time which is otherwise gap-free. Taking one point out of the reals or sticking two copies of the integers together are two straightforward ways of producing an isolated gap.

We are going to define a hierarchy of kinds of gaps. For any (zero, successor or limit) ordinal α, an αth order gap is a gap which is not of lesser order but lies in an open interval which contains, apart from itself, only gaps of order less than α. So a zero order gap is just an isolated gap.

Of course this hierarchy does not include all the gaps possible. For example, nowhere in the rationals is there a gap of any order at all.

We will use the game characterisation of unranked gaps. Let γ_0 be a gap of T. Players $\forall$ and $\exists$ move alternately, defining a sequence γ_i $(0 < i < \omega)$ of gaps. In each round, $\forall$ chooses an open interval I_i containing γ_i, and $\exists$ chooses $\gamma_{i+1} \in I_i$ with $\gamma_{i+1} \neq \gamma_i$. $\exists$ wins iff the game goes on for ω moves. γ_0 is unranked iff $\exists$ has a winning strategy for the game.

To see this one can employ a straightforward transfinite induction to show that if γ_0 is ranked then $\forall$ has a winning strategy. This simply involves continually choosing open intervals around gap γ_i which contain, apart from γ_i itself, only gaps of lesser ranks. Conversely it can be seen that if γ_0 is unranked then every open interval containing it also contains other unranked gaps. $\exists$ can win by always choosing unranked gaps.

It is interesting to note that if all gaps in a flow of time have ordinal order then the cardinality of the flow is at least as great as the cardinality of any of those orders and for every infinite ordinal α, there exists a flow of time of cardinality the same as α with a gap of order α. Let us prove the first of these statements.

DEFINITION 2.1. *Let T be a linear order of cardinality κ, and suppose that Γ is a set of gaps of T. A gap γ of T is said to be Γ-rich if every open interval I of T containing γ contains $\geq \kappa^+$ gaps from Γ. Here, κ^+ is the next largest cardinal after κ.*

PROPOSITION 1. *Let T be a linear order of cardinality κ. Suppose that Γ is a set of gaps of cardinality $\geq \kappa^+$. Then there is a Γ-rich gap $\gamma \in \Gamma$.*

PROOF. If not, for each $\gamma \in \Gamma$ choose an open interval I_γ with endpoints $a_\gamma < b_\gamma$ in T, such that $\gamma \in I_\gamma$ and $| I_\gamma \cap \Gamma | \leq \kappa$. As $| \Gamma | > \kappa$, there is $\Gamma' \subseteq \Gamma$ with $| \Gamma' | > \kappa$ and $I_\gamma = I$ say, for all $\gamma \in \Gamma'$. Then $\Gamma' \subseteq I$, a contradiction. □

COROLLARY 1. *Let T be a linear order of cardinality κ. Then T has at most κ ranked gaps.*

PROOF. Assume not. Let Γ be a set of κ^+ ranked gaps of T. We will show that any Γ-rich gap is unranked; this will contradict the proposition.

Let γ_0 be an Γ-rich gap. $\forall$ and $\exists$ will play the game above, starting with γ_0. $\exists$ will privately construct sets $\Gamma_i (i < \omega)$ of κ^+ ranked gaps, so that each γ_i is

Γ_i-rich. She begins by defining $\Gamma_0 = \Gamma$.

Inductively assume that $i < \omega$, and γ_i is a Γ_i-rich gap. $\forall$ chooses an interval $I_i = (a, b)$ say, around γ_i. I_i contains κ^+ gaps from Γ_i. Let Γ_{i+1} be the gaps from Γ_i contained in (a, γ_i) if this set has cardinality κ^+; otherwise let Γ_{i+1} be the gaps from Γ_i contained in (γ_i, b). So in any case, $| \Gamma_{i+1} |= \kappa^+$. By the proposition, $\exists$ can choose a Γ_{i+1}-rich gap $\gamma_{i+1} \in \Gamma_{i+1}$. If she does this, the game goes on forever and she wins. Hence γ_0 was unranked, as required. □

COROLLARY 2. *Let T be a linear order of cardinality κ and let γ be a ranked gap of T. Then $|rank\ (\gamma)| \leq \kappa$.*

PROOF. Any gap γ of rank α has gaps of rank β arbitrarily close, for all $\beta < \alpha$. So if T has a gap of rank α with $|\alpha| > \kappa$, then T has more than κ ranked gaps. The result follows from the previous corollary now. □

§3. Connectives to talk about gaps.

Recall that $U'(A, B)$ is as pictured:

B <· · · ¬B
now ——()—— A
a gap

S' is defined dually i.e., with past and future swapped. Despite involving a gap, U' is in fact a first-order connective and its table is given by:

$$
\begin{array}{llll}
U'(p,q) \equiv & & & \\
\exists s & t < s & & \\
\wedge\ \forall u & (& t < u < s \rightarrow & \\
 & ([& \exists v(u < v \wedge \forall w(t < w < v \rightarrow q(w)) &] \\
 \vee & [& \forall v(u < v < s \rightarrow p(v)) & \\
 & \wedge & \exists v(t < v < u \wedge \neg q(v)) &])) \\
\wedge & \multicolumn{3}{l}{\exists u[t < u < s \wedge \neg q(u)]} \\
\wedge & \multicolumn{3}{l}{\exists u[t < u < s \wedge \forall v(t < v < u \rightarrow q(v))]}
\end{array}
$$

By presenting our new connectives below in terms of U, S, U' and S' we thus guarantee that they are also first-order.

We start off with some new unary connectives which talk about a single gap located by the vicissitudes of a single temporal formula. First we need to know that there is a gap coming up.

$$
\begin{aligned}
\gamma^+(A) \quad = \quad & U(\neg A, \top) \wedge U(A, A) \\
& \wedge \neg U(\neg A, A) \wedge \neg U(\neg U(\top, A), A)
\end{aligned}
$$

This is true whenever A holds up until a gap but fails to hold arbitrarily soon afterwards. We call such a gap an A left gap: A is true on the left of the gap. Dually we can define γ^- and A right gaps. Notice that $\gamma^\pm$ are expressible in $\{U, S\}$.

Next we specify that the gap coming up is isolated, as far as gaps definable by the same formula and in the same direction go.

$$\gamma_0^+(A) = \gamma^+(A) \wedge U'(\neg\gamma^+(A), A)$$

Dually we can define γ_0^-. Notice that we use the Stavi connectives here.

Now we can recursively define a hierarchy of connectives. For every $n \geq 0$, define

$$\gamma_{\leq n}^+(A) = \gamma_0^+(A) \vee \cdots \vee \gamma_n^+(A)$$

and

$$\begin{aligned}\gamma_{n+1}^+(A) &= \gamma^+(A) \\ &\quad \wedge \neg\gamma_{\leq n}^+(A) \\ &\quad \wedge U'(\gamma^+(A) \rightarrow \gamma_{\leq n}^+(A), A)\end{aligned}$$

$\gamma_{\leq n}^-$ and γ_n^- are defined dually.

Notice that there is a distinction between gaps in the flow of time and gaps definable by a particular temporal formula or even by any temporal formula. Thus we need to define another hierarchy of gaps—this time within a temporal structure rather than just in a flow of time. Let A be a temporal formula. For any ordinal α, an αth order A left gap is an A left gap which is not of lesser order but begins an interval containing only A left gaps of lesser order. Dually we can define A right gaps of each order.

For $\alpha < \omega$, gap γ is an αth order A left gap if and only if $\gamma_\alpha^+(A)$ holds in an interval on the left of γ. We consider the possibility of $\gamma_\alpha^+(A)$ for $\alpha \geq \omega$ later.

We have mentioned the distinction between αth order A gaps and αth order gaps in the flow of time. Nevertheless, it is clear that there is only an A gap when there is a gap in time at the right place and that it is an A gap of order α when that gap in time is at least of order α or possibly of non-ordinal order.

Let us finish this section by demonstrating the existence of definable gaps which do not fit into our scheme of classification. The idea is Robin Hirsch's.

We create a flow of time from a certain subset of the set Q^* of finite sequences of rational numbers. Let T consist of those non-empty sequences in which every rational number but the last is a power of $1/2$ and the last number in the sequence is neither a power of $1/2$ nor zero. We order the sequences as follows: $(a_0, a_1, \ldots, a_m) < (b_0, b_1, \ldots, b_n)$ iff there is some $k \geq 0$ such that $k \leq n$, $k \leq m$, for all $i < k$, $a_i = b_i$ and $a_k < b_k$.

We turn $(T, \leq)$ into a $\{p\}$-structure by making $T \models p(t)$ if and only if the last number in the sequence t is negative.

Each p left gap in T occurs just after the segment $(a\hat{}(-1), a\hat{}0)$ where a is a sequence of powers of $1/2$ (possibly the empty sequence).

It is easy to prove that none of these gaps is isolated. Let $(a\hat{}0, t)$ be any open interval after a gap. If t is not of the form $a\hat{}q\hat{}b$ for some possibly empty sequence b and some rational q then we show that there is a left gap in $(a\hat{}0, a\hat{}1)$ and note that $a\hat{}1$ which is of that form must be less that t. So wlog t is $a\hat{}q\hat{}b$. Let r be any power of $1/2$ less than q. Clearly all elements of T of the form $a\hat{}r\hat{}s$ are in the interval $(a\hat{}0, t)$ and so is the gap at $a\hat{}r\hat{}0$.

Now a very straight forward transfinite induction proves that

LEMMA 1. *For any formula A and any non-zero ordinal α, each αth order A left gap is followed arbitrarily soon by zero order A left gaps.*

So if a flow has no isolated p left gaps then it has no p left gaps of any order at all and we have our result.

§4. Expressive power.

Before we state our new results we mention the theorem which makes our job a lot easier.

THEOREM 3 [GPSS]. *$\{U, S, U', S'\}$ is expressively complete over all linear time.*

We will prove this in Section 8. From this theorem, our first result falls out easily:

LEMMA 2. *Over flows of time with only isolated gaps, $\{U, S\}$ is expressively complete.*

This is because, over such flows,

$$U'(A, B) \equiv \gamma^+(B) \wedge U(\neg B, \gamma^+(B) \vee A).$$

Kamp's pioneering theorem is then a special case of this lemma.

Our next lemma shows that gaps don't have to get much more complicated before *until* and *since* are not sufficient.

LEMMA 3. *In general linear time, $\{U, S\}$ is not expressively complete. There are even flows of time with a single non-isolated gap on which γ_0^+ is not expressible in terms of $\{U, S\}$.*

PROOF. We take a flow of time $(T, \leq)$ with a single non-isolated gap and show that there is no temporal formula built from $\{U, S\}$which is equivalent to $\gamma_0^+(p)$ on all temporal structures over T.

T is constructed in two successive parts: the first one is got by taking a copy of $\mathbb{Z}$ for each negative integer and joining them into one long line and the second has a copy of $\mathbb{Z}$ for each integer arranged in order. There is a gap at the beginning of the whole order and a gap at the end of each copy of $\mathbb{Z}$. The only non-isolated gap is that between the two parts.

A p-structure over T will be called nice iff

- on each little copy of the integers, either p is always true or always false and
- every point has both p and $\neg p$ true in both its past and future.

An easy induction with several cases shows that for any formula ϕ constructed from p in the language $\{U, S\}$, there is a formula p, $\neg p$, $\top$ or $\bot$ which is uniformly equivalent to ϕ everywhere in all nice structures over T. It is easy to show, though, that these four formulae are all distinct in their truth conditions. For example, $U(p, p)$ is always equivalent to p.

So suppose, for contradiction, that we can express $\gamma_0^+(p)$ in $\{U, S\}$. By the above argument, we have a formula ψ always equivalent to it over nice structures.

Look now at a particular nice p-structure in which p alternates in truth on copies of $\mathbb{Z}$ but is true in the last copy of the first part. Here $\gamma_0^+(p)$ is false. Thus ψ must be either $\neg p$ or $\bot$.

Look next at a structure in which p alternates in the first part, is true in the last copy of $\mathbb{Z}$ there, is false for an initial segment of the second part and then alternates again. Here $\gamma_0^+(p)$ is true in the end of the first part. Thus ψ must be either p or $\top$ and we have our contradiction. □

A similar proof to the above readily shows that

LEMMA 4. *If for all i, $m < n_i$ then $\gamma_m^\pm$ is not expressible over all linear flows by any formula built from U,S and any (finite) number of $\gamma_{n_i}^\pm$.*

It is a bit harder to prove that any $\gamma_m^\pm$ can be expressed in terms of $\gamma_n^\pm$ (in combination with U and S) for any $n < m$.

LEMMA 5. *For any temporal formula P and any $n \geq 0$,*

$$\gamma_{n+1}^+(P) = \gamma_n^+(P \wedge \gamma^+(P) \wedge \neg\gamma_n^+(P))$$

The dual result also holds.

PROOF. This is immediate from the more informative lemma which follows the next:

LEMMA 6. *Let $n \geq 0$ and P be any temporal formula. We write Q for*

$$P \wedge \gamma^+(P) \wedge \neg\gamma_n^+(P)$$

and consider the left P gaps in a structure.

- *Every left Q gap is a left P gap.*
- *No order n left P gap is a left Q gap.*
- *All the other left P gaps are left Q gaps.*

The dual result also holds.

PROOF.

- To prove the first claim let us examine a left Q gap α say. Q is true in an interval, containing a point t say, on the left of α and false arbitrarily soon after. P, as a conjunct of Q, is thus true from t until α. If P is false arbitrarily soon after α then we have a left P gap at α as required. Suppose for contradiction that P is instead true for a while after α. Thus, like P , $\gamma^+(P)$ must stay true for a while after α. Finally look at the third conjunct, $\neg\gamma_n^+(P)$, of Q. Since it is also true at t, β can not be an order n left P gap and again the conjunct remains true after α at least as far as β. We have shown that Q remains true before and after α and we have our desired contradiction.

- The second observation is clear as $\gamma_n^+(P)$ is true arbitrarily recently before an order n left P gap.

- For the third let us look at a non-nth order left P gap. For a while, on the left, P $\gamma^+(P)$ and $\neg\gamma_n^+(P)$ are all true. Since P, and hence Q, is false arbitrarily soon after the gap, we have a left Q gap.

Now we can actually be more specific about the orders of the gaps involved:

LEMMA 7. *Let k and n be whole numbers and P be any temporal formula. We write*

$$Q = P \wedge \gamma^+(P) \wedge \neg\gamma_n^+(P)$$

and consider the left P gaps in a structure.

Any order k left P gap is

- *an order k left Q gap if $k < n$*
- *not a left Q gap at all if $k = n$ and*
- *an order $k - 1$ left Q gap if $k > n$.*

Any order k left Q gap is

- *an order k left P gap if $k < n$ and*
- *an order $k + 1$ left P gap if $k \geq n$.*

The dual result with right substituted for left also holds.

PROOF. Fix n. Now we proceed by induction on k.

First part. Suppose that we have an order k left P gap at α. If $k = n$ then the previous lemma gives us our result. So suppose not. Thus α is a left Q gap. We will show that α is an order K left Q gap where

$$K = \begin{cases} k & \text{if } k < n \\ k-1 & \text{if } k > n. \end{cases}$$

Now for a while after α any left P gaps are of order less than k. Any left Q gaps which are in this interval are then by the previous lemma, left P gaps and so of order less than k as left P gaps. If $k < n$ then these gaps are by the inductive hypothesis, left Q gaps of the same order less than $k = K$. If $k > n$ then these gaps are, also by the inductive hypothesis, left Q gaps of order one less than their order as P gaps which is less than $K = k - 1$. In either case, for a while after α all left Q gaps are of order less than K.

If $K = 0$ then we have shown than α is an isolated left Q gap.

Otherwise, since α is an order k left P gap it must have order $k-1$ left P gaps arbitrarily soon afterwards. There are three cases:

- if $k < n$ then by the inductive hypothesis these are order $K - 1 = k - 1$ left Q gaps;
- if $k > n + 1$ then these are order $k - 2 = K - 1$ left Q gaps and

- if $k = n + 1 > 1$ then the order $k - 1 = n > 0$ left P gaps are also followed arbitrarily closely by order $k - 2$ left P gaps which are by the inductive hypothesis also order $k - 2 = K - 1$ left Q gaps.
- if $k = n + 1 = 1$ then $K = k - 1 = 0$ which we have supposed to not be the case.

This proves that α is a left Q gap of order K.

Second part. Suppose that α is an order k left Q gap. It is also a left P gap. We will show that it is also an order K left P gap where

$$K = \begin{cases} k & \text{if } k < n \\ k+1 & \text{if } k \geq n. \end{cases}$$

For a while after α all left Q gaps are of orders less than k. By our inductive hypothesis they will also be left P gaps of various finite orders. Thus α must be a finite order left P gap say of order l. We know that l is not n for then α wouldn't be a left Q gap at all.

By part one, if $l < n$ then $k = l$ so $K = k = l$ as required.

If $l > n$ then $k = l - 1 \geq n$ so $K = k + 1 = l$ as required. □

Now what if we can use $\gamma_0^{\pm}$? Let us consider the new set of connectives $\{U, S, \gamma_0^{\pm}\}$ and ask about its expressive power. In fact, the connectives which talk of higher order gaps are redundant. In expressive power, the γ_i^+ hierarchy collapses: for each $n \geq 0$,

$$\begin{array}{rrl} \gamma_{n+1}^+(p) & \equiv & \gamma^+(p) \wedge \neg\gamma_{\leq n}^{\pm}(p) \\ & \wedge & \gamma_0^+(\gamma^+(p) \wedge \neg\gamma_{\leq n}^{\pm}(p)) \\ & \wedge & U(\gamma_{\leq n}^{\pm}(p), p \vee U(\gamma_{\leq n}^{\pm}(p), \neg\gamma^+(p) \vee \gamma_{\leq n}^{\pm}(p))) \end{array}$$

Thus one might think that higher order gaps hold no surprises for $\{U, S, \gamma_0^{\pm}\}$. In fact we do not even need to stop at finite orders.

LEMMA 8. *$\{U, S, \gamma_0^{\pm}\}$ is expressively complete over general linear time.*

PROOF. We will exhibit a $\{U, S, \gamma_0^{\pm}\}$ formula which is equivalent to $U'(p, q)$ in any $\{p, q\}$-structure. Because the dual formula will be equivalent to $S'(p, q)$, we will have shown that $\{U, S, \gamma_0^{\pm}\}$ is expressively complete over such structures.

Let ϕ be:

$$\begin{array}{rl} & \gamma^+(q) \wedge U(U(\neg q, p), q) \\ \\ \vee & \gamma^+(q) \\ \wedge & U(\neg q, \gamma^+(\neg U(\neg q, p)) \vee p) \\ \wedge & \neg U(\neg q, \neg U(\neg q, p)) \\ \wedge & \gamma_0^+(\neg U(\neg q, p)) \end{array}$$

Suppose that B is a $\{p, q\}$-structure. We will show that for any $b \in B$,

$$\begin{array}{rl} & B \models U'(p, q)(b) \\ \Leftrightarrow & B \models \phi(b) \end{array}$$

($\Longleftarrow$) Let us assume that ϕ holds at b. We must show that $U'(p, q)$ is true at b. There are two cases.

If the first disjunct holds then it is clear that $U'(p,q)$ does too.

Now suppose that the second disjunct of ϕ holds at b but that the first does not. The first conjunct guarantees that q is true from b up until a gap which we can call β. $\neg q$ is true arbitrarily soon after β.

For contradiction we also suppose that $U'(p,q)$ does not hold at b. Thus p is false arbitrarily soon after β. Since $U(\neg q, \gamma^+(\neg U(\neg q,p)) \vee p)$ holds at b, we have $\gamma^+(\neg U(\neg q,p))$ true arbitrarily soon after β.

Since q is true up until β but p is false arbitrarily soon afterwards, we must have $\neg U(\neg q,p)$ holding from b at least up until β. But

$$\neg U(\neg q, \neg U(\neg q,p))$$

holds at b so $U(\neg q,p)$ must be true arbitrarily soon after β.

So $\neg U(\neg q,p)$ is true up until β but false arbitrarily soon afterwards. Thus $\gamma^+(\neg U(\neg q,p))$ holds at b and β is the $\neg U(\neg q,p)$ left gap involved.

Knowing that both $U(\neg q,p)$ and $\gamma^+(\neg U(\neg q,p))$ are true arbitrarily soon after β tells us that there are $\neg U(\neg q,p)$ left gaps arbitrarily soon after β.

Thus β is not an isolated $\neg U(\neg q,p)$ left gap and this contradicts

$$\gamma_0^+(\neg U(\neg q,p))$$

holding at b. We are done.

($\Longrightarrow$) Suppose that $B \models U'(p,q)(b)$. So q is true for a while after b up until a gap, called β say. We must show that ϕ is true at b. There are two cases.

If p is true for a while before β as well as after then it is clear that the first disjunct of ϕ holds at b.

So let us assume that that p is false arbitrarily soon before β.

In this case it is not hard to see that the second disjunct of ϕ holds at b. To see that those conjuncts involving $\neg U(\neg q,p)$ hold one need only notice that $\neg U(\neg q,p)$ holds from b up until β and is false for a while afterwards. It is false after the gap, i.e., $U(\neg q,p)$ holds, because p is true for a while and $\neg q$ is true arbitrarily soon after β. □

§5. Other connectives.

The connective

$$\rho^+(q) = U'(\neg q, q)$$

and its dual ρ^- could equally well have been used in this paper instead of $\gamma_0^\pm$. We can define

$$\gamma_0^+(p) = \rho^+(\gamma^+(p))$$

so that the set $\{U, S, \rho^\pm\}$ is expressively complete.

$Ung(p,q)$ iff q holds until a gap after which it is arbitrarily soon false and after which there are no left p gaps for a while. Dually we define Sng. Note that $\gamma_0^+(p)$ equals $Ung(p,p)$. Thus $\{U, S, Ung, Sng\}$ is expressively complete.

We say that a p left gap is pure if it is not itself an isolated p left gap but there are no isolated p gaps for a while after the gap. Lemma 1 above shows that pure gaps have non-ordinal order. The non-ordinal order gaps constructed in the example above are pure. We can define a new connective using purity: $\pi^+(q)$ holds iff $\gamma^+(q) \wedge U(\neg q, \neg\gamma_0^+(q))$ does.

An argument similar to the proof of Lemma 2 shows that π^+ is not expressible in terms of U and S. We compare the truth of $\pi^+(p)$ before gaps in two different structures. In one p is true up until a gap after which p is false for a while. In the other p is true up until a gap after which open intervals of p being true, and open intervals of p being false replace rational numbers in an interval from that ordering.

The proof of Lemma 2 can also be employed to show that γ_0^+ can not be expressed in terms of U, S or $\pi^\pm$. Clearly $\pi^\pm(p)$ is always false in the structures defined there.

§6. An axiomatisation of $U, S, \gamma_0^\pm$ using the irreflexivity rule.

We first axiomatise U, S and $\gamma_0^\pm$ over arbitrary linear flows of time using the irreflexivity rule of [G1]. This rule allows simple axiomatisations of many temporal connectives over irreflexive flows of time. We derive some simple consequences and list some open questions. In the next section we will relate some of these questions to the class of scattered flows of time.

In this section, unless otherwise stated a *temporal formula* will mean one written with the connectives U, S, γ_0^+ and γ_0^-. We will use the standard abbreviations F, P, H and G: Fp abbreviates $U(p, \top)$ etc. Recall also that $K^+(q)$ abbreviates $\neg U(\top, \neg q)$ and $\gamma^+(q)$ abbreviates $F\neg q \wedge U(q,q) \wedge \neg U(\neg q \vee K^+(\neg q), q)$; and similarly for K^- and γ^-.

We adopt as axioms the following:

1. All truth functional tautologies.

2. $G(p \to q) \to (Gp \to Gq)$
   $H(p \to q) \to (Hp \to Hq)$

3. $q \to GPq, q \to HFq$

4. $FFq \to Fq$ [transitivity]

5. $G(p \wedge Gp \to q) \vee G(q \wedge Gq \to p)$
   $H(p \wedge Hp \to q) \vee H(q \wedge Hq \to p)$ [linearity]

6. $r \wedge H\neg r \to [U(p,q) \leftrightarrow F(p \wedge H(Pr \to q))]$
   $r \wedge H\neg r \to [S(p,q) \leftrightarrow P(p \wedge G(F(r \wedge H\neg r) \to q))]$

7. $r \wedge H\neg r \to [\gamma_0^+(q) \leftrightarrow (\gamma^+(q) \wedge F(\neg q \wedge H(P(\neg q \wedge Pr) \to \neg\gamma^+(q))))]$
   $r \wedge H\neg r \to$
   $[\gamma_0^-(q) \leftrightarrow (\gamma^-(q) \wedge P(\neg q \wedge G(F(\neg q \wedge F(r \wedge H\neg r)) \to \neg\gamma^-(q))))]$.

The rules of inference are:

modus ponens
substitution
generalisation: $\vdash A \Rightarrow \vdash GA \wedge HA$
irreflexivity: $\vdash r \wedge H\neg r \rightarrow A \Rightarrow \vdash A$
(for all A and atoms r not occurring in A).

These axioms and rules are valid over irreflexive linear time.

DEFINITION 6.1. *If A is a temporal formula, N a temporal structure, and t a point of the flow of time of N (for short, "$t \in N$"), we write $N \models A(t)$ if A holds at t in N.*

Take any set Σ of temporal formulas. A model *of Σ will be an irreflexive linear temporal structure N such that for some $t \in N, N \models A(t)$ for all $A \in \Sigma$.*

THEOREM 4. (Completeness.) *Given any countable consistent set Σ of formulas, there is a countable model N of Σ in which all instances of the axioms are valid at every point.*

PROOF (sketch; see e.g. [GH] for details). Using standard techniques we can obtain a countable irreflexive linear temporal structure N whose points are maximal consistent sets of temporal formulas. The irreflexivity rule allows us to assume that for each $t \in N$ there is an atom r with $r \wedge H\neg r \in t$. Further:

- there is $t_0 \in N$ with $\Sigma \subseteq t_0$.
- for all atoms q and all $t \in N, N \models q(t)$ iff $q \in t$.
- for each formula A there is an atom q such that $A \leftrightarrow q \in t$ for all $t \in N$.
- for all formulas A built using only F and P, and all $t \in N, A \in t$ iff $N \models A(t)$.

It now easily follows that for all $t \in N$ and all temporal formulas $A, N \models A(t)$ iff $A \in t$. The proof is by induction on the structure of A using axioms 6 and 7. Hence as $\Sigma \subseteq t_0$, we have constructed a model of Σ. □

QUESTION. Is there an axiomatisation of U, S and $\gamma_0^{\pm}$ without using the irreflexivity rule? Burgess axiomatises U and S over arbitrary linear time in [B], without using this rule.

Even if the answer is negative, we still obtain the following corollaries, whose statements do not mention the irreflexivity rule.

COROLLARY 3. (Compactness.) *Let Σ be a set of temporal formulas (of U, S, γ_0^+ and γ_0^-). Suppose that every finite subset of Σ has a model. Then Σ has a model.*

PROOF. With the given axioms and finitary rules, no contradiction is derivable from Σ. Hence by Theorem 4 Σ has a model as stated. □

COROLLARY 4.

1. *The connective* $\gamma^+_{\geq\omega}(-)$, *saying that there is a gap of rank at least* ω *coming up on the right, is not definable by any first order formula.*

2. *Not both of the connectives* $\gamma^+_\omega(-)$ *and* $\gamma^+_{\text{ordinal}}(-)$, *saying that there is coming up on the right a gap of rank* ω, *or (respectively) ordinal rank, are first order definable.*

PROOF.

1. Assume for contradiction that $\gamma^+_{\geq\omega}(q)$ has a first order table. Hence by expressive completeness of $\{\bar{U}, S, \gamma_0^+, \gamma_0^-\}$ (Lemma 8 above) there is already a temporal formula equivalent to $\gamma^+_{\geq\omega}(q)$. So consider $\Sigma = \{\neg\gamma^+_{\geq\omega}(q) \wedge \gamma^+_{\geq n}(q) : n < \omega\}$. Every finite subset of Σ has a model, but Σ does not. This contradicts the previous corollary.

2. We have $\gamma^+_{\geq\omega}(q) \equiv \gamma^+(q) \wedge [\neg\gamma^+_{\text{ordinal}}(q) \vee \gamma^+_\omega(q) \vee \neg U'(\neg\gamma^+_\omega(q), q)]$, so the definability of both of γ^+_ω and $\gamma^+_{\text{ordinal}}$ would contradict (1). □

QUESTIONS.

1. Is γ^+_ω definable? Note that γ^+_ω is definable from $\gamma^+_{\geq\omega}$ by $\gamma^+_\omega(q) \equiv \gamma^+_{\geq\omega}(q) \wedge U'(\neg\gamma^+_{\geq\omega}(q), q)$.

2. Is $\gamma^+_{\text{ordinal}}$ first order definable?

By Corollary 4(2), relevant to the definability of γ_ω is the fact that the flows of time in which there are essentially no unranked gaps are essentially exactly the scattered flows: those that do not embed the rationals. They are our next topic.

§7. Unranked gaps and scattered flows of time.

We will observe that any temporal logic with first order connectives over the class of all scattered flows of time is decidable. This gives a weak recursive axiomatisation of the temporal structures with scattered flows of time, though a strong axiomatisation is not possible.

Recall that a q-definable gap (one where $\gamma^+(q)$ holds on some interval to the left) is of rank ∞ ('*unranked*') if it is not of rank α for any ordinal α. An example of such gaps was given in Section 3.1. They can also be exhibited by first defining N_i $(i = 0, 1)$ to be a structure with flow of time $\mathbb{Q}$, on which q is always true $(i = 1)$ or always false $(i = 0)$, and then replacing each $i \in \mathbb{Q}$ by a copy of N_0 or N_1 in such a way that any interval of $\mathbb{Q}$ contains copies of both structures. Let Q be the resulting temporal structure. Each $i \in \mathbb{Q}$ that is given a copy of N_1 yields a pure unranked q-gap in Q corresponding to the 'right hand end' of that copy. Note that the flow of time of Q is isomorphic to $\mathbb{Q}$.

We defined unranked gaps of a flow of time in Section 2. As an example, all gaps in $\mathbb{Q}$ are unranked. Flow-of-time gaps may not be 'definable' by a temporal formula (i.e., detectable by γ). However, note that an unranked definable gap is also an unranked flow-of-time gap.

DEFINITION 7.1.

1. *If I is a linear order and $x, y \in I$ we will write $[x, y]$ for the closed interval of I with endpoints x, y. This extends the usual notation to the case where $x \geq y$.*

2. *An equivalence relation $\equiv$ on a linear ordering I is called a condensation if the $\equiv$-classes are convex (i.e., are intervals, but possibly one-point intervals or with gaps for endpoints). Note that if $\equiv$ is a condensation, the ordering of I induces a canonical linear ordering of $I/\equiv$. Strictly speaking, the condensation is this linear ordering, and not the corresponding relation $\equiv$.*

3. *Recall that I is said to be scattered if $\mathbf{Q}$ does not embed into I. See [Ro] for general information on scattered orderings.*

PROPOSITION 2 (cf. [D], Lemma 2.3). *A linear ordering I is scattered iff whenever $\equiv$ is a condensation of I, $I/\equiv$ is not dense.*

PROOF.

$\Rightarrow$ If $\equiv$ is a dense condensation of I, we can use the axiom of choice to choose a set of representatives of the $\equiv$-classes. Some subset of this will have order type $\mathbf{Q}$.

$\Leftarrow$ If $\mathbf{Q} \subseteq I$ define $\equiv$ on I by $x \equiv y$ iff $[x, y] \cap \mathbf{Q}$ is finite. Clearly $I/\equiv$ is dense.

□

THEOREM 5. *Let I be a linear ordering.*

1. *Suppose that I is scattered. Then there are no unranked flow-of-time gaps in I.*

2. *Assume that I is countable and that no temporal structure M with flow of time I has unranked definable gaps. Then I is scattered.*

PROOF.

1. Clearly $(*)$ any open interval of I containing an unranked gap contains infinitely many unranked gaps. Suppose that γ is an unranked gap of I. We define a chain of finite sets $S_n \subseteq I$ by induction on n so that for all adjacent points $i < j$ in S_n, the open interval (i, j) contains (a) an unranked gap, and (b) a point of S_{n+1}.

   Choose $i_0 < \gamma < i_1$ arbitrarily and let $S_0 = \{i_0, i_1\}$. Let $S_n = \{s_0, \ldots, s_k\}$ be given, satisfying (a) and with $s_0 < s_1 < \cdots < s_k$. By $(*)$, for each $i < k$ we can take $s_i < t_i < s_{i+1}$ such that both (s_i, t_i) and (t_i, s_{i+1}) contain unranked gaps. Define $S_{n+1} = S_n \cup \{t_i : i < k\}$. Clearly (b) holds now for S_n, and (a) holds for S_{n+1}.

Having defined the S_n, we observe that $\bigcup_{n<\omega} S_n$ has order type $\mathbf{Q} \cap [0,1]$, so that $\mathbf{Q}$ embeds into I. Hence I is not scattered.

Note that in the case where I is already a temporal structure and γ is a q-definable gap, the same argument shows that the extensions (truth sets) in I of q and of $\neg q$ both embed $\mathbf{Q}$.

2. The example $I = \mathbf{R}$ shows that the theorem can fail if the assumption of countability is discarded. Assume that I is not scattered. Let $\equiv$ be a condensation of I such that $(I/\equiv) \cong \mathbf{Q} \cap [0,1]$ (use Proposition 2, the countability of I and Cantor's theorem). Let Q^* be obtained from the structure Q made from N_0 and N_1 as above, by adding left and right endpoints at which q is false (say). Hence there is an order isomorphism $\theta : I/\equiv \to Q^*$. Define I as a q-structure M by: if $m \in I, M \models q(m)$ iff $Q^* \models q(\theta(m/\equiv))$. Then each unranked q-definable gap of Q^* gives rise to a similar gap in M. □

If the compactness theorem held for the scattered orderings, non-definability of γ_ω (even in the class of scattered orderings) would again follow. For the previous argument using compactness would show that $\gamma_{\geq\omega}$ is not definable even over the scattered orderings. But $\gamma^+_{\geq\omega}(q) \equiv \gamma^+(q) \wedge [\neg\gamma^+_{\text{ordinal}}(q) \vee \gamma^+_\omega(q) \vee \neg U'(\neg\gamma^+_\omega(q), q)]$, as above. In scattered orderings, because of Theorem 5 we have $\gamma^+_{\geq\omega}(q) \equiv \gamma^+_\omega(q) \vee \gamma^+(\neg\gamma^+_\omega(q))$, so that γ_ω's being definable would force $\gamma_{\geq\omega}$ to be definable, a contradiction.

However, we now show that this is not the case.

PROPOSITION 3. *The compactness theorem fails for the class of scattered orderings.*

PROOF. Introduce propositional atoms $q_i (i \in \mathbf{Q})$. Let $\Sigma = \{P(q_i \wedge H\neg q_i \wedge Pq_j) : j < i \text{ in } \mathbf{Q}\}$. Then any finite subset of Σ has a scattered model. But if M were a scattered model of Σ, then $\mathbf{Q}$ would embed into M via $i \mapsto m_i$ where $m_i \in M$ satisfies $M \models (q_i \wedge H\neg q_i)(m_i)$. □

Now the rules of inference are finitary, so completeness implies compactness. Hence, for the class of scattered orderings, there is no completeness theorem of the form: Σ is consistent iff Σ has a scattered model. However, there is a weak completeness theorem that deals with the case where Σ is finite. That is, there is a recursive set of axioms such that $\vdash A$ iff $\models A$ for all temporal formulas A. This follows trivially from the following decidability result.

PROPOSITION 4.

1. *The monadic second order theory of the class of countable scattered linear orders is decidable.*

2. *Over scattered flows of time, any temporal logic using connectives with first order tables is decidable.*

PROOF.

1. Let σ be a monadic second order sentence in the signature $\{=,<\}$, where quantification over elements and subsets is allowed. Let Q be a new unary relation symbol and let σ^Q denote the relativisation of σ to Q. (I.e., the first order quantifiers $\exists x, \forall x$ are replaced by $\exists x \in Q, \forall x \in Q$ respectively, and the second order quantifiers $\exists X, \forall X$ by $\exists X \subseteq Q$ and $\forall X \subseteq Q$ respectively. Later we give a formal definition of relativisation in the first order case.) Let $\xi(Q)$ be the formula

$$\forall R \subseteq Q([\exists xy(R(x) \land R(y) \land x < y)] \rightarrow$$
$$\exists xy(R(x) \land R(y) \land x < y \land \neg\exists z(x < z < y \land R(z))).$$

So $\xi(Q)$ says that the set of points where Q holds is a scattered ordering. Now any countable linear ordering embeds into $(\mathbf{Q}, <)$. So $\mathbf{Q} \models \exists Q(\xi(Q) \land \sigma^Q)$ iff σ has a countable scattered model.

It follows from the celebrated result of Rabin [R] that the monadic second order theory of $\mathbf{Q}$ is decidable: cf. [BG, Theorem 2.6]. Hence there is an algorithm to decide whether $\mathbf{Q} \models \exists Q(\xi(Q) \land \sigma^Q)$. This completes the proof.

2. It follows from the downward Löwenheim–Skolem theorem (see [CK]) that if A is a temporal formula with a first order table, then A has a scattered model iff A has a countable scattered model. Let A use atoms $p_1, \ldots, p_n$ and have table $\alpha(x, P_1, \ldots, P_n)$, where the P_i are unary relation symbols corresponding to the atoms. Then A has a scattered model iff the monadic second order sentence

$$\exists P_1 \cdots P_n \exists x \alpha(x, P_1, \ldots, P_n)$$

holds in some countable scattered linear order. By (1) there is an algorithm to decide this question. □

REMARKS.

1. It follows trivially that given any set of connectives with first order tables, there is a recursive axiomatisation of the class K of temporal structures with scattered flow of time. We simply take as axioms $\{A : A$ is valid in every structure in $K\}$; this set is recursive by Proposition 4. The only proof rule required is substitution.

2. In [GH] a finite (not merely recursive) axiomatisation of the temporal logic with Until and Since over the real numbers $\mathbf{R}$ was given. In that proof a certain condensation $\sim_r$ $(r < \omega)$ was defined, and the irreflexivity rule used to show that every $\sim_r$-class was a closed interval of the flow of time. (Reynolds [Re] has since eliminated the use of the IRR rule.) The temporal translation B of $\neg\exists y < x(y \sim_r x)$ was then true exactly

at the left-hand endpoint of each $\sim$-class, so a single axiom could be used to specify properties of the condensation $M/\sim_r$, uniformly in r. In our case the relevant axiom would be $\Diamond(B \wedge FB) \rightarrow \Diamond(B \wedge U(B, \neg B))$ (cf. Proposition 2), but we have not found a formula true exactly once in each $\sim_r$-class (our proof of Proposition 2 uses the axiom of choice). So this method does not appear to be applicable in the scattered case.

3. [BG, Theorem 2.9] proves the decidability of the temporal logic with Until and Since over the real numbers. Their argument is a variant of the 'finite model property' approach to decidability, and goes back to [LL] and [Ra]. Also see [D]. This technique can be used to give another proof of our Proposition 4(2).

§8. Expressive completeness of U, S & Stavi connectives over linear time.

In this section we will prove Theorem 3. That is, we establish expressive completeness of U, S and the Stavi connectives for arbitrary linear flows of time. The formal statement follows after some initial definitions. Our argument was sketched in [GPSS] for the case of U and S over natural numbers time; the generalisation to arbitrary linear time was indicated but not proved.

DEFINITION 8.1.

1. *We fix an arbitrary finite set L of propositional atoms. We will consider first order formulas $\varphi(\overline{x})$ in the 'monadic' language with $=, <$ and a unary relation symbol Q for every atom $q \in L$. We also consider temporal formulas. Unless otherwise stated, a temporal formula will be one built from the atoms of L using the Boolean connectives and the binary temporal connectives U, S, U' and S' (standing for* Until and Since *and the Stavi connectives).*

2. *A* temporal *(L-)* structure *is formally a triple $N = (T, <, h)$, where $(T, <)$ is an irreflexive poset (the flow of time of N) and $h : L \rightarrow \mathcal{P}(T)$ is the assignment map. We will often abuse notation by identifying N with its flow of time T. Moreover, as every temporal formula A defines a subset of a structure—the set of time points $h(A)$ (cf. Definition 1.1(3)) where A is true—we will regard A as an extra atom and use it in monadic first order formulas as a monadic relation symbol. This simplifies the notation a little. So for example, $N \models \forall x U(A, B)(x)$ iff $U(A, B)$ is true at every point of N.*

3. *We will usually use Roman letters for temporal formulas and Greek for classical first order ones.*

In this setting, Theorem 3 becomes:

For all L-formulas $\varphi(x)$ there is a temporal formula A such that if N is a linear temporal structure (i.e., one with linear flow of time) in which the atoms of

φ have interpretations then for all $t \in N, N \models \varphi(t)$ iff $N \models A(t)$. Moreover, A is effectively obtainable from φ (i.e., by an algorithm).

This says that the temporal logic with Until, Since and the Stavi connectives is *expressively (functionally) complete* over linear time. Our proof here is based on the sketch in [GPSS]; an alternative proof using separation (cf. Theorem 2, and [G2]) will appear in [GHR]. The algorithm resulting from separation is probably more efficient than ours.

We begin with some definitions.

DEFINITION 8.2. (Rank.) *The rank of a temporal formula A is defined to be the maximum depth of nesting of temporal connectives in A. Example: if p, q are atoms then rank$(p \wedge q) = 0$, and rank$(\neg U(p, \neg S'(\neg q, q))) = 2$. Since L is finite, it is easy to show by induction on r that for each $r < \omega$ there is a finite set of temporal formulas of rank r such that every rank r formula is logically equivalent to one of them.*

DEFINITION 8.3. (Gaps.) *We will use the definition of a gap in a linear order discussed in Section 2. We need a few extra notions. Let $M = (M, <, h)$ be any linear temporal structure. If γ is a gap and $S \subseteq M$, we say that γ =sup(S) if for all $t \in M, t \geq s$ for all $s \in S$ iff $t \geq \gamma$. We also say that γ =inf(S) if for all $t \in M, t \leq s$ for all $s \in S$ iff $t \leq \gamma$.*

Let γ be a gap and let D be a temporal formula. We say that γ is definable on the left *by D if D is true at all points of M in some non-empty interval (t, γ) on the left of γ, and not true throughout any non-empty interval (γ, u) on the right. The definition of a gap's being* definable *(by D)* on the right *is made in a similar way. If $r < \omega$, an* r-definable gap *is one that is definable (on the left or right) by a formula D of rank at most r. For $r < \omega$ we let $M_r = M \cup \{r$−definable gaps of $M\}$, with the induced ordering $<$. So in general $M \subset M_0 \subset M_1 \subset \cdots$. For example, if M has no last element then $+\infty$ is a gap of M definable on the left by $\top$, so that $\infty \in M_0 \backslash M$. The situation for $-\infty$ is similar.*

We will refer to the elements of M as *points.*

DEFINITION 8.4. (Relativised connectives.) *There is a natural way of evaluating temporal formulas of the form $\sharp(A, B)$ for $\sharp \in \{U, S, U', S'\}$ at gaps. For example, $U(A, B)$ holds at a gap γ (i.e., $\gamma \in M_n$ for some n) iff there is a* point *$t > \gamma$ (so $t \in M$) where A holds, with B holding at all* points *$u \in (\gamma, t)$. To formalise this we relativise our connectives to points.*

Fix $r < \omega$ and let $\mu \notin L$ be a new propositional atom. We define M_r as a temporal $L \cup \{\mu\}$−structure $(M_r, <, h')$ by:

$$h'(q) = h(q) \subseteq M \text{ for all } q \in L$$
$$h'(\mu) = M.$$

We will relativise U, S, U' and S' to μ.

Let $\varphi(\overline{x})$ be any first order formula in the signature consisting of $=, <$ and a unary relation symbol for each atom of $L \cup \{\mu\}$. We define the relativisation *φ^μ of φ to μ by induction on φ:*

if φ is quantifier free then $\varphi^\mu = \varphi$;

$$(\neg\varphi)^\mu = \neg(\varphi^\mu);$$
$$(\varphi \wedge \psi)^\mu = \varphi^\mu \wedge \psi^\mu;$$
$$[\exists y\varphi(\overline{x}, y)]^\mu = \exists y(\mu(y) \wedge \varphi^\mu(\overline{x}, y)).$$

We introduce connectives U^μ, S^μ, U'^μ and S'^μ whose tables are the relativisations to μ of the tables of U, S, U' and S' respectively. We can write formulas using these connectives that are meaningful in any $L \cup \{\mu\}$-structure. In particular we can interpret them in M_r. If A is any formula of $USU'S'$, we let A^μ be the formula obtained by replacing each U in A by U^μ, and similarly for S, U' and S'.

REMARK.

1. Let $\alpha(x)$ be the canonical *first order table* of the temporal formula A, as defined in Definition 1.2: in any temporal structure T, $\{t \in T : T \models A(t)\} = \{t \in T : T \models \alpha(t)\}$. Then it is easily seen that the table of A^μ is just α^μ: i.e., for all $t \in M_r$, $M_r \models A^\mu(t)$ iff $M_r \models \alpha^\mu(t)$ (this holds for any $L \cup \{\mu\}$-structure).
2. If $t \in M$ then $M \models A(t)$ iff $M_r \models A^\mu(t)$.
3. Let $A = S'(B, C)$ where C has rank $\leq r$. If $t \in M_r$ and $M_r \models A^\mu(t)$, then the gap that A asserts the existence of actually lies in M_r (as C defines it on the right).

The Stavi connectives can express existence of gaps, but cannot talk directly about what formulas are 'true' at them. So we need to transform properties of a gap into properties of 'real' points. This is done in the following definition and lemma.

DEFINITION 8.5. *Let D be any temporal L-formula. We define a temporal L-formula* left(A, D) *by induction on A:*

- *left$(p, D) = \bot$ for atomic p*
- *left$(\neg A, D) = U'(\top, D) \wedge \neg left(A, D)$*
- *left$(A \wedge B, D) =$ left$(A, D) \wedge$ left(B, D)*
- *left$(U(A, B), D) = U'(B \wedge U(A, B), D)$*
- *left$(U'(A, B), D) = U'(B \wedge U'(A, B), D)$*
- *left$(S(A, B), D) = U(D \wedge B \wedge S(A, B) \wedge U'(\top, B \wedge D) \wedge \neg U'(D, B \wedge D), D)$*
- *left$(S'(A, B), D) = U(D \wedge B \wedge S'(A, B) \wedge U'(\top, B \wedge D) \wedge \neg U'(D, B \wedge D), D)$.*

So rank(left$(A, D)) \leq$ max$(rk(A), rk(D)) + 2$. We define right(A, D) *similarly by swapping each U with S and U' with S' in the definition above.*

The point of this definition is given by the following lemma.

LEMMA 9. *Let A, D be temporal formulas with D of rank at most r. Let $m \in M_r$. Then the following are equivalent:*

1. $M_r \models \text{left}(A, D)^\mu(m)$;

2. *There is $\gamma \in M_r - (M \cup \{\pm\infty\}), \gamma$ a gap of M defined by D to the left, with (a) $\gamma > m$, (b) D holds in M on (m, γ), and (c) $M_r \models A^\mu(\gamma)$.*

PROOF. Clear. A corresponding result holds for right(A, D). □

DEFINITION 8.6. (Games.) *We will need some results on Ehrenfeucht–Fraïssé games. Let Σ be any finite first order signature without function symbols. Let M, N be Σ-structures. If $n < \omega$ we define a game $G^n(M, N)$ between two players, $\forall$ (male) and $\exists$ (female). The game has n rounds. In each round, $\forall$ chooses an element from whichever of M, N he wishes. Then $\exists$ responds by choosing an element of the other structure. After n rounds, two n-tuples $\bar{a}, \bar{b}$ of elements have been chosen from M, N respectively; the order of the elements in each tuple is the order in which they were chosen as the game was played. $\exists$ wins this 'play' $(\bar{a}, \bar{b})$ of the game iff for all quantifier-free formulas $\varphi(\bar{x})$ of Σ, $M \models \varphi(\bar{a})$ iff $N \models \varphi(\bar{b})$. This is slightly stronger than saying that the map $\bar{a} \mapsto \bar{b}$ is a partial isomorphism, since Σ may have constant symbols.*

A strategy for $\exists$ in a game is a set of rules (not necessarily deterministic) telling her what to do—this can be formalised as a family of functions. The strategy is said to be winning if whenever she uses it she wins.

The following is a well-known result of Ehrenfeucht–Fraïssé game theory.

PROPOSITION 5. *Let Σ be any signature as above. Let M, N be Σ-structures and let $n < \omega$. The following are equivalent:*

1. *$\exists$ has a winning strategy for $G^n(M, N)$*

2. *$M \models \sigma$ iff $N \models \sigma$ for all Σ-sentences σ of quantifier depth of nesting at most n.*

PROOF. See [E]. As is well known, (2) $\rightarrow$ (1) can fail if Σ is assumed infinite or to have function symbols. □

NOTATION. If $x < y$ in M_r we write (x, y) for $\{t \in M : x < t < y\}$, and if $n \leq r, (x, y)_n$ for $\{t \in M_n : x < t < y\}$. We write $[x, y]_n$ for $\{t \in M_n : x \leq t \leq y\}$, etc. We do not require that $x, y \in M_n$.

DEFINITION 8.7. (Special games on temporal structures.) *We now introduce a modified version of the game above. Let M and N be linear temporal structures. The game $G_{n;r}(M, xy; N, x'y')$ for $n, r < \omega, x < y$ in M_r, and $x' < y'$ in N_r, is played as follows. There are only two rounds. $\forall$ begins by choosing n elements $a_1, \ldots, a_n \in [x, y]_r$; $\exists$ responds with elements $a'_1, \ldots, a'_n \in [x', y']_r$. Then $\forall$ chooses one more element $b' \in [x', y']$—so b' must not be a gap—and $\exists$ replies with $b \in [x, y]$.*

$\exists$ wins iff:

1. *the tuples $xy\bar{a}b$ and $x'y'\overline{a'}b'$ have the same order type; and if $t \in xy\bar{a}b$ and t' is the corresponding element of $x'y'\overline{a'}b'$, then:*

2. *t is a gap of M iff t' is a gap of N*

3. *for each temporal L-formula A of rank at most r, $M_r \models A^\mu(t)$ iff $N_r \models A^\mu(t')$.*

LEMMA 10. *Let M, N etc. be as above. Suppose that $\exists$ has a winning strategy σ for $G_{n;r}(M, xy; N, x'y')$ for some $n, r < \omega$. Let $n' \leq n, r' \leq r$. Then σ gives in the natural way a winning strategy for $G_{n';r'}(M, xy; N, x'y')$ provided that $x, y \in M_{r'}$ and $x', y' \in N_{r'}$.*

PROOF. Recall that K^+X abbreviates the formula $\neg U(\top, \neg X)$, and $K^-X = \neg S(\top, \neg X)$.

Suppose in a play of $G_{n';r'}(M, xy; N, x'y')$, $\forall$ chooses $a_1, \ldots, a_{n'} \in [x, y]_{r'}$. Then $\exists$ defines $a_{n'+1}, \ldots, a_n$ to be x, say. So $a_1, \ldots, a_n \in [x, y]_r$. She applies σ to $\bar{a}$ to obtain $\bar{e} \in [x', y']_r$.

We claim that each $e_i \in [x', y']_{r'}$. This is clear if $r' = r$, so assume that $r' < r$. Take i; certainly if $a_i \in M$ then $e_i \in N$. Otherwise a_i is defined by some formula $\neg D$ of rank $\leq r'$. So letting $D' = (K^+D \wedge \neg K^-D) \vee (K^-D \wedge \neg K^+D)$, a formula of rank $\leq r' + 1 \leq r$, we have $M_r \models D'^\mu(a_i)$. As σ is winning, $N_r \models D'^\mu(e_i)$. Hence e_i is also a gap defined by $\neg D$; so $e_i \in N_{r'}$.

If $\forall$ now chooses $a' \in [x', y']$ then $\exists$ simply uses σ to respond with $e' \in [x, y]$. Then $\bar{a}e'$ and $\bar{e}a'$ satisfy the same order relations and rank r temporal formulas, hence also the same temporal formulas of rank r'. Hence $\exists$ has won the play. □

We want to characterise the formulas associated with these games.

DEFINITION 8.8.

1. *Let $r < \omega$ and $t \in M_r$ be given. Define X_t to be the conjunction of all temporal L-formulas X of rank $\leq r$ with $M_r \models X^\mu(t)$. This conjunction is effectively finite, as because L is finite there are up to logical equivalence only finitely many distinct formulas of any rank. Hence X_t can be taken to be a temporal formula of rank r.*

   If $t < u$ in M_r, define $X_{(t,u)}$ to be $\bigvee_{v \in (t,u)} X_v$. Again the disjunction is effectively finite, so that $X_{(t,u)}$ can be taken to be a formula of rank r. Note that only points (non-gaps) contribute to the disjunction.

2. *(This definition is from [GPSS].) An $n;r$-decomposition formula is a first order formula of the form:*

   $$\psi(x_1, x_2) = \exists y_1, \ldots, y_n[x_1 < y_1 < \cdots < y_n < x_2 \wedge \forall z \chi(x_1, x_2, \bar{y}, z)],$$

   where χ is a conjunction of formulas of the following kinds:

   (a) *$\theta(t)$, where t is an element of $x_1x_2\bar{y}$ and θ is either $\mu, \neg\mu$, or A^μ for some temporal L-formula A of rank $\leq r$;*

(b) $\mu(z) \wedge a < z < b \rightarrow B^{\mu}(z)$, where $a < b$ are adjacent elements of the sequence $x_1 y_1 \cdots y_n x_2$, and B is a temporal formula of rank $\leq r$.

LEMMA 11. *Let $M, N, x, y, x,', y'$ be as above. Let $n, r < \omega$. Then the following are equivalent:*

1. *$\exists$ has a winning strategy for $G_{n;r}(M, xy; N, x'y')$.*

2. *for all $n;r$-decomposition formulas $\varphi(x_1, x_2)$, $M_r \models \varphi(x, y) \Rightarrow N_r \models \varphi(x', y')$.*

PROOF. (1) $\Rightarrow$ (2)—clear.
(2) $\Rightarrow$ (1) Let $\forall$ choose $a_1, \ldots, a_n \in [x, y]$ in his first move. Assume without loss that $x < a_1 < \cdots < a_n < y$. Write a_0 for x and a_{n+1} for y. Let $\psi(y_0, y_{n+1}) = \exists y_1 \cdots y_n [y_0 < y_1 < \cdots < y_{n+1} \wedge \forall z (\bigwedge_{a_i \in M} \mu(y_i) \wedge \bigwedge_{a_i \notin M} \neg\mu(y_i) \wedge \bigwedge_{i \leq n+1} X_{a_i}(y_i) \wedge \bigwedge_{i \leq n} (\mu(z) \wedge y_i < z < y_{i+1} \rightarrow X_{(a_i, a_{i+1})}(z))]$. Then ψ is an $n;r$-decomposition formula and $M_r \models \psi(x, y)$. Hence by assumption $N_r \models \psi(x', y')$, and so there are $e_i \in (x', y')$ witnessing the $\exists$'s in ψ. If $\exists$ chooses the e_i she can easily win the game. □

The main step in our proof is

THEOREM 6. *Suppose that M, N are linear temporal structures. Then $(*)_n$ holds for all $n < \omega$:*

$(*)_n$ *For all $r < \omega$, if $x < y$ in M_r, $x' < y'$ in N_r, and $\exists$ has a winning strategy for*

$$G_{1+3n;r+4n}(M, xy; N, x'y'),$$

then $\exists$ has a winning strategy for $G_{n;r}(N, x'y'; M, xy)$.

This says that if $\exists$ possesses winning strategies for enough 'forward' games $G(M, xy; N, x'y')$ then she has a winning strategy for a given 'backward' game $G(N, x'y'; M, xy)$. The proof does not use compactness and is really a syntactic result—we could equally prove that a certain class of formulas is closed under negation up to equivalence, which is what is done (for the case of U and S over $\mathbb{N}$ and without full proof) in [GPSS]. However, the game approach, though still complicated, seems rather simpler to present.

Before we prove Theorem 6 we finish our result on expressive completeness.

PROPOSITION 6. *Let M, N be linear temporal structures and let $x \in M$, $y \in N$. Suppose $n, r < \omega$ and that x and y satisfy the same temporal formulas of rank $r + 4n + 1$ in their respective structures. Then $\exists$ has a winning strategy for $G_{n;r}(M, -\infty x; N, -\infty y)$ and $G_{n;r}(M, x\infty; N, y\infty)$.*

PROOF. (Sketch.) Suppose for simplicity that $\forall$ chooses n points $x < a_1 < \cdots < a_n$ in M in the future of x. Let $a_0 = x$. Define C_n to be $X_{a_n} \wedge \neg U(\neg X_{(a_n, \infty)}, \top)$, and for $i < n$, C_i to be $X_{a_i} \wedge U(C_{i+1}, X_{(a_i, a_{i+1})})$. So rank $(C_i) = r + n + 1 - i$. Then $M \models C_0(x)$, so that $N \models C_0(y)$. $\exists$ can use the form of C_0 to choose points $y = e_0 < e_1 < \cdots < e_n$ in N such that $N \models X_{a_i}(e_i)$ and $N \models X_{(a_i,a_{i+1})}(t)$ for all (non-gaps) $t \in (e_i, e_{i+1})$. If $\forall$ now chooses $t \in (e_i, e_{i+1})$ then $N \models X_u(t)$ for some $u \in (a_i, a_{i+1})$. If $\exists$ responds with such a u, she wins the game. The argument for the 'past' game is similar. If some of the a_i are gaps, the idea is the same but the formulas C are more complicated and involve formulas D defining the gaps, together with the formulas left(X_{a_i}, D) or right(X_{a_i}, D)—cf. the proof of Cases III, IV of Theorem 6. In all cases we have rank$(C_0) \leq r+4n+1$. □

DEFINITION 8.9. *Let f, g be any functions on ω satisfying $f(0) = g(0) = 0$, $f(n+1) \geq (1+3f(n)).(2k_n)+1$, and $g(n+1) \geq g(n)+4f(n)$, where k_n is the number of inequivalent $(1+3f(n)); (g(n)+4f(n))$-decomposition formulas.*

PROPOSITION 7. *For all $n < \omega$ the following holds. Let M, N be linear temporal structures and let $x_1 < \cdots < x_m, y_1 < \cdots < y_m$ be increasing m-tuples of elements of M, N respectively, for arbitrary $m < \omega$. Define $x_0 = -\infty$ and $x_{m+1} = \infty$ in M_0. Define y_0, y_{m+1} similarly.*

Suppose that $\exists$ has winning strategies for

$$G_{f(n);g(n)}(M, x_i, x_{i+1}; N, y_i, y_{i+1})$$

and

$$G_{f(n);g(n)}(N, y_i, y_{i+1}; M, x_i, x_{i+1})$$

for all $0 \leq i \leq m$. Then $\exists$ has a winning strategy for the Ehrenfeucht–Fraïssé game $G^n((M, \overline{x}), (N, \overline{y}))$.

PROOF. By induction on n. If $n = 0$ the result is trivial. Assume it true for n, let $r = g(n)+4f(n) \leq g(n+1)$, and suppose that $\exists$ has winning strategies for the games $G_{f(n+1);r}(M, x_i, x_{i+1}; N, y_i, y_{i+1})$ and $G_{f(n+1);r}(N, y_i, y_{i+1}; M, x_i, x_{i+1})$.

Let $\forall$ begin $G^{n+1}((M, \overline{x}), (N, \overline{y}))$ by choosing without loss $a \in M$. (If $\forall$ chooses in N the proof is the same as we have complete symmetry.) If $a \in \{x_1, \ldots, x_m\}$ then $\exists$ chooses the corresponding element of $\overline{y}$, and the result then follows using the induction hypothesis and Lemma 10. So let $i \leq m$ be such that $x_i < a < x_{i+1}$. List as $\varphi_1, \ldots, \varphi_j$ the $[1+3f(n)];r$-decomposition formulas $\varphi(u, v)$ such that $M_r \models \varphi(x_i, a)$, and as $\psi_1, \ldots, \psi_k$, the $[1+3f(n)];r$-decomposition formulas $\psi(u, v)$ with $M_r \models \psi(a, x_{i+1})$.

Let $\exists$ choose witnesses for the existential quantifiers of each φ, ψ, together with a, making at most $n' = (1+3f(n)).(j+k)+1 \leq f(n+1)$ elements of $(x_i, x_{i+1})_r$ in all. She now applies her winning strategy for $G_{f(n+1);r}(M, x_i x_{i+1}; N, y_i y_{i+1})$. Let e be the point she chooses corresponding to a. Clearly (cf. Lemma 11) we have $N_r \models \varphi_s(y_i, e)$ for all $s \leq j$ and $N_r \models \psi_s(e, y_{i+1})$ for $s \leq k$. By Lemma 11, $\exists$ has a winning strategy for $G_{1+3f(n);r}(M, x_i a; N, y_i e)$ and for $G_{1+3f(n);r}(M, a x_{i+1}; N, e y_{i+1})$. Crucially, by Theorem 6, she also has winning strategies for

$$G_{f(n);g(n)}(N, y_i e; M, x_i a)$$

and

$$G_{f(n);g(n)}(N, e y_{i+1}; M, a x_{i+1}).$$

By the induction hypothesis, $\exists$ has a winning strategy σ for

$$G^n((M, \bar{x}a), (N, \bar{y}e)).$$

So in $G^{n+1}(M, \bar{x}), (N, \bar{y}))$, $\exists$ can choose e in response to $\forall$'s choice of a and then follow σ. This strategy wins the game for her. □

COROLLARY 5. *Let M, N be linear temporal structures and let $x \in M, y \in N$. Suppose that x and y satisfy the same temporal formulas of rank $g(n+1)+1$ in their respective structures. Then for all monadic first order formulas φ (of L) of quantifier depth $\leq n$, $M \models \varphi(x)$ iff $N \models \varphi(y)$.*

PROOF. By Propositions 5, 6, 7. □

Expressive completeness now follows easily. For given $\varphi(x)$ of quantifier depth n, we may choose a finite L with atoms corresponding to the monadic predicates of φ. Now take a finite set Ψ of temporal formulas of rank $1+g(n+1)$ such that (1) if $A, B \in \Psi$ and $A \wedge B$ is consistent then $A = B$; (2) each temporal formula C of rank $1+g(n+1)$ is equivalent to a disjunction of formulas in Ψ. Let $\Psi' = \{B \in \Psi :$ for some linear M and $t \in M, M \models B(t)$ and $M \models \varphi(t)\}$. Then by Corollary 5, φ is equivalent over linear time to the rank $1+g(n+1)$-formula $\bigvee \Psi'$.

Note that by a result of Gurevich [BG, 2.7(a)], the universal monadic second order theory of linear order is decidable. Hence Ψ' is computable by an algorithm, so that the translation of first order formulas into temporal ones is effective.

PROOF (of Theorem 6). We must prove

$(*)_n$ For all $r < \omega$, if $x < y$ in $M_r, x' < y'$ in N_r, and $\exists$ has a winning strategy for $G_{1+3n;r+4n}(M, xy; N, x'y')$, then $\exists$ has a winning strategy for $G_{n;r}(N, x'y'; M, xy)$.

We prove $(*)_n$ by induction on n. For the case $n = 0$ (r is arbitrary) assume that $\exists$ has a winning strategy σ for $G_{1;r}(M, xy; N, x'y')$ and that $\forall$ chooses $a \in (x, y)$ in the second round of $G_{0;r}(N, x'y'; M, xy)$ (as $n = 0$ the first round is 'empty'). a is not a gap. $\exists$ simply applies σ to choose a response $e \in (x', y')$. Clearly $\exists$ has won.

Assume $(*)_n$ for $n < \omega$; we prove $(*)_{n+1}$. Fix $r < \omega, x < y$ in M_r and $x' < y'$ in N_r. Assume that $\exists$ has a winning strategy for

$$G_{4+3n;r+4(n+1)}(M, xy; N, x'y').$$

We will construct a winning strategy for $\exists$ in $G_{n+1;r}(N, x'y'; M, xy)$.

Suppose $\forall$ chooses $n+1$ points $x' < a_0 < \cdots < a_n < y'$ in N_r (we may assume that they are all distinct, for otherwise the result follows by the inductive hypothesis and Lemma 10). Define the following rank r temporal formulas:

$$A = X_{(a_{n-1}, a_n)}$$
$$C = X_{(a_n, y')}$$

where if $n = 0$ we take a_{n-1} in A to be x'. Clearly in N, A holds on (a_{n-1}, a_n) and C on (a_n, y'). Let

- $c = \inf\{t \in [x, y] : M \models C(u) \text{ for all } u \in (t, y)\}$.

If $c \notin M$ then either $c = x \in M_r$ already, or c is a gap definable on the right by C. Hence $c \in M_r$. Define $c \in N_r$ similarly.

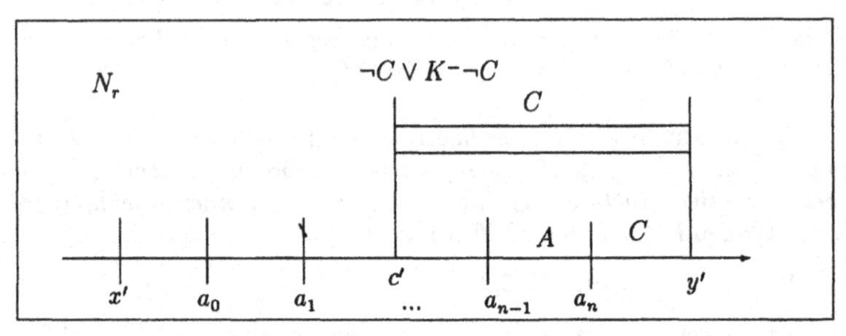

Claim 1.
Consider a play of the game $G_{m;r'}(M, xy; N, x'y')$ for arbitrary $r' > r, m \geq 1$ in which $\exists$ uses a winning strategy. Let $\forall$ begin by choosing c plus $m-1$ other points, and let $\exists$'s response to c be d (plus $m - 1$ other points). Then $d = c'$.

Proof of Claim.
As the strategy is winning, any rank r' temporal formula satisfied by one of $\forall$'s choices must also be satisfied by the corresponding choice of $\exists$. Now the rank $r+1$ formula $C' = \neg C \vee K^{-}\neg C$ satisfies $M_r \models C'^{\mu}(c)$. Hence also $N_r \models C'^{\mu}(d)$, so $d \leq c'$.

If $d < c'$ then $\forall$ can choose $d' \in (d, y')$ with $N \models \neg C(d')$. $\exists$ now has no winning response, a contradiction. Hence $d = c'$. This proves the claim.

Claim 2.
$\exists$ has a winning strategy for

$$G_{1+3n;r+4(n+1)}(M, xc; N, x'c')$$

and for

$$G_{1+3n;r+4(n+1)}(M, cy; N, c'y').$$

Proof of Claim.
Let $r' = r+4(n+1)$. Suppose that $\forall$ chooses $1+3n$ elements in the interval $[x, c]_{r'}$. By assumption $\exists$ has a winning strategy σ for the game $G_{4+3n;r'}(M, xy; N, x'y')$. $\exists$ adds c to $\forall$'s choices and applies σ (cf. Lemma 10). As the order of $\exists$'s element choices from σ matches the order of $\forall$'s, Claim 1 ensures that her responses to $\forall$'s choices all lie in $[x', c']_{r'}$. If $\forall$ then chooses in $[x', c']$ then again $\exists$'s strategy will yield an answer in $[x, c]$. The strategy is clearly winning. To sum up, the restriction of σ to games in which $\forall$ always chooses in $[x, c]_{r'}$ and then in $[x', c']$ can yield a winning strategy for $G_{1+3n;r+4(n+1)}(M, xc; N, x'c')$. Similarly for the intervals $[c, y], [c', y']$. This establishes the claim. We will use this argument repeatedly.

Hence by inductive hypothesis $(*)_n$, $\exists$ has winning strategies σ, τ for the backward games $G_{n;r+4}(N, x'c'; M, xc)$ and $G_{n;r+4}(N, c'y'; M, cy)$.

Now clearly $c' \leq a_n$, so $(x', c')_r$ contains at most n points from $\{a_0, \ldots, a_n\}$. The proof will divide into cases, mainly according to whether a_n is a point of N, a left- or a right-definable gap.

Case I: $a_0 \leq c'$.

Then $(c', y')_r$ also contains at most n points from $\{a_0, \ldots, a_n\}$. So as $\exists$ is trying to win

$$G_{n+1;r}(N, x'y'; Mxy),$$

she can use σ and τ to choose points $e_0, \ldots, e_n \in M_r$. She applies σ to those a_i in $(x', c')_r$ and τ to the rest using the method of Lemma 10; if an a_i happens to be c' it can be dealt with by either strategy. If $\forall$ then responds in $[x, c)$ she uses σ, and if in $[c, y]$, τ. If she does this then by Lemma 10 she will win the game.

Case II: All the points $a_0, \ldots, a_n$ lie in (c', y'), and $a_n \in N$ is not a gap.

Recall that $\exists$ is trying to win $G_{n+1;r}(N, x'y'; M, xy)$—i.e., to preserve all rank r formulas. Define $B = X_{a_n}$, and $b = \sup\{t \in (x, y) : M \models B(t)\}$. As before, either $b \in M, b = y$ or b is an r-definable gap, defined on the right by $\neg B$, so that $b \in M_r$. Define $b' \in N_r$ similarly. Then clearly $b' \geq a_n$.

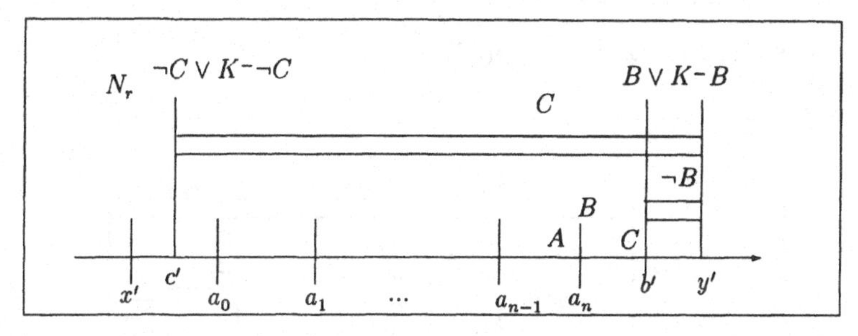

As in Claim 1, in any play of $G_{4+3n;r+4(n+1)}(M, xy; N, x'y')$ in which $\exists$ is using her winning strategy and $\forall$ chooses b, c amongst other points, $\exists$ will respond with b', c' amongst others. Hence again $\exists$ has a winning strategy for $G_{1+3n;r+4(n+1)}(M, cb; N, c'b')$. So by the induction hypothesis $(*)_n$ she has a winning strategy τ for $G_{n;r+4}(N, c'b'; M, cb)$. She already has a winning strategy σ for $G_{n;r+4}(N, x'c'; M, xc)$.

Let her first use τ in response to $a_0, \ldots, a_{n-1}$. It delivers n points $e_0, \ldots, e_{n-1} \in (c, b)_r$ (cf. Lemma 10). Now clearly $N_r \models U(B, A)^\mu(a_{n-1}) : a_n$ is a witness to this. (This holds even if a_{n-1} is a gap; if $n = 0$ we take $a_{-1}$to be c' and (see below) e_{-1} to be c.) $U(B, A)$ has rank $r + 1$, so as τ preserves formulas up to rank $r + 4, M_r \models U(B, A)^\mu(e_{n-1})$. Hence there is $z > e_{n-1}$ in M with $M \models B(z)$ and $M \models A(t)$ for all $t \in (e_{n-1}, z)$. But $e_{n-1} < b$. Hence we can assume that $z \leq b$. $\exists$ defines e_n to be such a z, completing her move. Clearly e_n and a_n satisfy the same temporal formulas of rank r, as they both satisfy B.

Suppose that $\forall$ continues by choosing $t \in [x, y]$. Recall that by the game rules, t is not a gap. If $t < c$ then $\exists$ uses σ to respond, and if $c \leq t \leq e_{n-1}$ she uses τ. If $t \in (e_{n-1}, e_n)$ then $M \models A(t)$. By definition of A there is $t' \in (a_{n-1}, a_n)$ with $M \models X_{t'}(t)$. $\exists$ can then choose any such t' as her response. It follows that t and t' agree on all rank r temporal formulas, as required. If $t = e_n$ then $\exists$ responds with

a_n. Finally, if $y > t > e_n$ then certainly $t > c$, so $M \models C(t)$. By definition of C there is $t' > a_n$ with $M \models X_{t'}(t)$, and $\exists$ can choose such a t' in response to t. If $\exists$ follows these directions she will win.

The remaining cases are similar to Case II, which gave a response e_n to a_n by letting B describe a_n and making $U(B, A)$ true at e_{n-1}. But a_n will now be a gap, so we must use the Stavi U'—and $U'(B, A)$ does not say that B^μ is true at the gap. So we use the formulas left(-,-) and right(-,-) instead.

Case III: All the points $a_0, \ldots, a_n$ lie in $(c', y')_r$, and a_n is a gap defined on the left by some formula D of rank $\leq r$. Clearly a_n is also defined by $A \wedge D$, so we can assume that $D \vdash A$.

Write B for X_{a_n}, and δ for $A \wedge$ left(B, D). δ is a formula of rank $\leq r+2$, and $N_r \models U(\delta, A)^\mu(a_{n-1})$ (again we set a_{n-1} to be c' if $n = 0$). Define d', g' by:

- $d' = \sup\{t \in (x', y') : N \models \neg D(t)\}$
- $g' = \sup\{t \in (x', d') : N \models \delta(t)\}$.

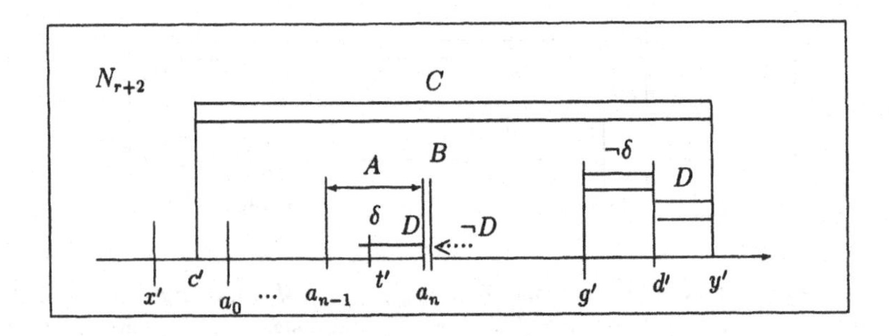

Define d, g similarly. Note that as before, all these points lie in M_{r+2}, N_{r+2}. Clearly, $a_n < d'$ and the fact that $N_r \models U(\delta, A)^\mu(a_{n-1})$ is witnessed at a point $t' \in N$ where δ holds, with $t' < g'$.

Now if $\exists$ uses a winning strategy for $G_{4+3n;r+4(n+1)}(M, xy; N, x'y')$ and adds c, g and d to $\forall$'s choices, then as before, her strategy delivers *inter alia* c', g' and d' in response. So again, $\exists$ has a winning strategy for $G_{1+3n;r+4(n+1)}(M, cg; N, c'g')$ for all m, r'. By $(*)_n$, $\exists$ has a winning strategy for $G_{n;r+4}(N, c'g'; M, cg)$. Let her use it to choose $e_0, \ldots, e_{n-1}$ in response to $a_0, \ldots, a_{n-1}$. Then as in Lemma 10, $e_0, \ldots, e_{n-1} \in (c, g)_r$, and as rank $r+4$ formulas are preserved, $M_r \models U(\delta, A)^\mu(e_{n-1})$. As $e_{n-1} < g$ we can choose $t \leq g$ in M with $M \models \delta(t)$ and such that A holds at all $u \in (e_{n-1}, t]$.

By definition of δ and Lemma 9, there is a gap $e_n \in (t, d)_r$ defined by D on the left, and such that A holds between t and e_n. Moreover, any rank r formula holds at e_n iff it holds at a_n, as they both satisfy B. $\exists$ chooses e_n in response to a_n, so completing her move. The same argument as in Case I allows $\exists$ to complete the remainder of the game, winning it.

Case IV: $a_0, \ldots, a_n \in (c', y'), a_n \in N_r - N$, and a_n is not definable on the left by any formula of rank $\leq r$.

It follows from the case assumption that A holds throughout some interval containing a_n. Choose D of rank $\leq r$ defining a_n on the right. Define $B = X_{a_n}$ and $\delta = A \wedge \neg D \wedge U(\text{right}(B, D), A)$ (rank $r + 3$). Let $d' = \sup\{t \in (x', y') : N \models \text{right}(B, D)(t)\}$, and then $g' = \sup\{t \in (x', d') : N \models \delta(t)\}$. Define $d, g \in M_{r+3}$ similarly.

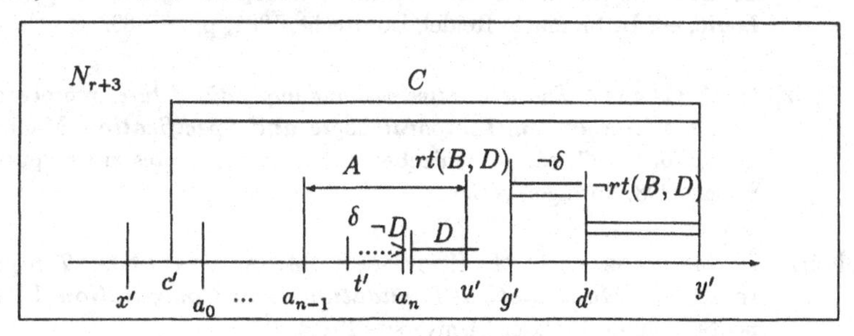

Clearly there are $a_{n-1} < t' < a_n < u' < y'$, with $t', u' \in N$, $N \models \delta(t')$, $N \models \text{right}(B, D)(u')$, and A holding on (t', u') (if $n = 0$ we take a_{n-1} to be c' as usual). Hence $t' \leq g'$ and $u' \leq d'$. As usual, if $\exists$ uses a winning strategy for $G_{4+3n;r+4(n+1)}(M, xy; N, x'y')$ and adds c, g and d to $\forall$'s choices she can derive a winning strategy for $G_{1+3n;r+4(n+1)}(M, cg; N, c'g')$. So by $(*)_n$ she has a winning strategy for $G_{n,r+4}(N, c'g'; M, cg)$. Let her use it to respond to $a_0, \ldots, a_{n-1}$ with $e_0, \ldots, e_{n-1}$. So as $U(\delta, A)$ has rank $\leq r + 4$, $M_r \models U(\delta, A)^{\mu}(e_{n-1})$. We can choose $e_{n-1} < t \leq g$ with $t \in M$, $M \models \delta(t)$, and A holding on (e_{n-1}, t). Then we can choose $u \in M$ with $t < u \leq d$, $M \models \text{right}(B, D)(u)$ and such that A holds in (e_{n-1}, u).

By Lemma 9 there is a gap $e_n \in (t, u)$ defined by D and at which the same relativised rank r formulas hold as at a_n in N_r. (We have $e_n > t$ because $M \models \neg D(t)$.) Then $\exists$ adds e_n to her choices to complete the move. The remainder of the game is as before.

This ends the proof of the theorem. □

REFERENCES

[B] J. P. BURGESS, *Axioms for tense logic I: "Since" and "Until"*, ***Notre Dame J. Formal Logic***, vol. 23 no. 2 (1982), pp. 367–374.

[BG] J. P. BURGESS, Y. GUREVICH, *The decision problem for linear temporal logic*, ***Notre Dame J. Formal Logic*** vol. 26 no. 2 (1985), pp. 115–128.

[CK] C. C. CHANG, H. J. KEISLER, ***Model Theory***, North-Holland, Amsterdam, 3rd edn., 1990.

[D] KEES DOETS, *Monadic Π^1_1-theories of Π^1_1-properties*, **Notre Dame J. Formal Logic**, vol. 30 no. 2 (1989), pp. 224–240.

[E] A. EHRENFEUCHT, *An application of games to the completeness problem for formalized theories*, **Fund. Math.**, vol. 49 (1961), pp. 128–141.

[G1] D. M. GABBAY, *An irreflexivity lemma*, in **Aspects of Philosophical Logic**, ed. U. Monnich, Reidel, Dordrecht, 1981, pp. 67–89.

[G2] D. M. GABBAY, *The declarative past and imperative future*, in proceedings, **Colloquium on Temporal Logic and Specification**, Manchester, April 1987, ed. B. Banieqbal et. al., Lecture Notes in Computer Science 398, Springer-Verlag.

[GHR] D. M. GABBAY, I. M. HODKINSON, M. A. REYNOLDS, **Temporal Logic: Mathematical Foundations and Computational Aspects**, Volume 1, Oxford University Press, 1993.

[GH] D. M. GABBAY, I. M. HODKINSON, *An axiomatisation of the temporal logic with Until and Since over the real numbers*, **J. Logic Computat.**, vol. 1 no. 2 (1990), pp. 229–259.

[GPSS] D. M. GABBAY, A. PNUELI, S. SHELAH, J. STAVI, *On the temporal analysis of fairness*, **7th ACM Symposium on Principles of Programming Languages**, Las Vegas, 1980, pp. 163–173.

[K] J. A. W. KAMP, *Tense logic and the theory of linear order*, Ph.D. dissertation, University of California, Los Angeles, 1968.

[LL] H. LAÜCHLI, J. LEONARD, *On the elementary theory of linear order*, **Fund. Math.**, vol. 59 (1966), pp. 109–116.

[R] M. O. RABIN, *Decidability of second order theories and automata on infinite trees*, **Trans. Amer. Math. Soc.**, vol. 141 (1969), pp. 1–35.

[Ra] FRANK P. RAMSEY, *On a problem of formal logic*, **Proc. London Math. Soc.**, vol. 30 (1930), pp. 264–286.

[Re] MARK A. REYNOLDS, *An axiomatization for Until and Since over the reals without the IRR rule*, **Studia Logica**, vol. 51 (1992), pp. 165–193.

[Ro] JOSEPH G. ROSENSTEIN, **Linear Orderings**, Academic Press, New York, 1982.

[S] B.-H. SCHLINGLOFF, *Expressive completeness of temporal logic over trees*, ***J. Applied Non-Classical Logics***, vol. 2 (1992), pp. 157–180.

Department of Computing
Imperial College
London SW7 2BZ
United Kingdom

NEW FOUNDATIONS FOR MATHEMATICAL THEORIES

JAAKKO HINTIKKA

§1. The motivation. In this paper, I shall outline a new approach to the logical foundations of mathematical theories. One way of looking at its motivation is as follows (I am following here Hintikka, 1989):

In the foundational work around 1900, e.g. in Hilbert's ***Foundations of geometry***, a crucial role was played by assumptions of *extremality* (i.e., ***minimality*** and *maximality*). For instance, Hilbert's so-called axiom of completeness is a maximality assumption. The Archimedean axiom can be thought of as a minimality assumption, the principle of induction likewise as a minimality axiom, and Dedekind's assumption of the existence of a real for each cut as a maximality assumption. Slowly, it has become clear to everybody that such extremality assumptions cannot normally be expressed as ordinary first-order axioms. To what extent they can or cannot be expressed in other ways, e.g. as higher-order axioms or set-theoretical axioms, and to what extent we should try to express them in such ways, will not be discussed here. In any case, in spite of the tremendous *prima facie* interest and power of extremality assumptions, they have not attracted much interest lately.

The approach proposed and outlined here relies crucially on extremality assumptions but seeks to implement them in a new way on a first-order level. Instead of introducing extremality assumptions on the top of a ready-made logic as explicit axioms, I propose to build them into the very logic we are employing, thus by-passing the difficulties the earlier uses of extremality assumptions encountered.

A logic is in effect specified by a space Ω of models together with a definition of what it means for a statement (closed formula) to be true in a model $M \in \Omega$ (and for a formula to be satisfied with in M). I shall not modify the latter ingredient. Instead, I propose to modify the usual space of models (of a given first-order language L) in the simplest possible way, viz. by omitting some of its members.

Even though this kind of modification looks innocuous, it facilitates a radical new look at the prospects of mathematical and logical theories. Most importantly, the possibility of reaching completeness can be profoundly affected.

What are the different kinds of completeness relevant here? Here are four candidates, which have not always been distinguished from each other sufficiently clearly:

(1) *Descriptive completeness.* It is an attribute of a ***non-logical theory.*** It means that the theory has as its models only the intended (standard) ones, i.e., that it has no non-standard ones. If there is only one standard

model, a descriptively complete theory must be categorical. Here the notion of standardness has to be characterized independently.

(2) *Semantical completeness.* It is a property of an *axiomatization of* (some branch of) *logic*. It means that the theorems of the axiomatization exhaust all valid formulas, where validity means truth in all the models of the space Ω. Thus semantical completeness amounts essentially to the recursive enumerability of the set of valid formulas.

(3) *Deductive completeness.* It is a property of a *non-logical axiom system together with an axiomatization of logic*. It means that, for each statement C, either C or $\sim C$ can be derived from the (non-logical) axioms by means of the given logic.

(4) *Hilbert's* so-called *axiom of completeness* is in effect a maximality assumption in a non-logical axiom system. While Hilbert's own intentions are not clear, this axiom can be taken to say that one cannot add new individuals to a (standard or intended) model without violating other axioms.

What happens when the space of models Ω is replaced by some $\Omega^* \subset \Omega$?

(1) Descriptive completeness becomes *ceteris paribus* easier to reach since some (or all) of non-standard models in Ω may belong to $\Omega - \Omega^*$.

(2) Since there are fewer models, there are *ceteris paribus* more valid formulas (i.e., formulas true in all of them). Hence semantical completeness can become more difficult to achieve.

In other words, by moving from Ω to a suitable Ω^* ($\Omega^* \subset \Omega$) we can trade in the semantical completeness of our underlying logic in order to achieve the descriptive completeness of suitable mathematical theories. I have argued elsewhere that this would represent a major gain in philosophical and conceptual clarity. (See Hintikka 1989.)

For example, Gödel showed elementary arithmetic to be incomplete in the sense (3) (deductive incompleteness). From this it does not by itself follow that elementary arithmetic is descriptively incomplete. This does follow if the underlying logic is complete, which Gödel had proved (for first-order logic) prior to proving the the incompleteness of elementary arithmetic. However, if we are willing to change this logic (strengthen it) so as to render it semantically incomplete, we can very well hope to reach a descriptively complete first-order theory of arithmetic. This in fact turns out to be possible.

More generally, by means of suitable extremality restrictions on models, it will turn out to be possible to formulate categorical first-order axiom systems *inter alia* for elementary number theory, the theory of reals, Euclidean geometry, and the second number class (countable ordinals). (See §5 below.)

(3) Deductive completeness, being a kind of combination of descriptive and semantical completeness, is not necessarily affected by any trade-off between the other two kinds of completeness (1)–(2).

(4) The restriction of the space of models to a suitable subset serves the same purpose as the axiom of completeness, and is supposed to replace any such explicit axiom.

§2. Model theory via constituents. The crucial question obviously is how the ideas of minimality and maximality can be implemented in precise and general terms. Since we are speaking of the minimality and maximality of models (or parts thereof), the obvious resource here is the theory of models. I shall review some of the basic ideas of model theory, but for the sake of certain further developments I shall do so in an unfamiliar way. I shall use in the review the technique of constituents and distributive normal forms. Even though this technique may in the last analysis be dispensable in favor of more commonly employed conceptualizations (e.g. back-and-forth techniques), it offers heuristic advantages by allowing an almost geometrical (tree-theoretical) visualization of the logical relationships under scrutiny in this paper.

An approach to model theory via constituents leads us straight to the field of stability theory. However, it is in fact quicker and much more perspicuous to develop the necessary theory directly here without going by way of stability theory. (For stability theory, see e.g. Baldwin, 1988.)

In what follows, it is assumed that we are dealing with a given first-order language L with a finite list of predicate constants, no function symbols, and an unspecified supply of individual constants. I shall deal only with languages without identity. It turns out that the presence or absence of identity does not matter very much for the central purposes of this work.

A constituent L with the free variables $x_1, x_2, \ldots, x_k$ will be expressed as follows:

$$C_i^{(d)}[x_1, x_2, \ldots, x_k] \tag{2.1}$$

It is a well-formed formula which has a number k of arguments $x_1, x_2, \ldots, x_k$ and it is also characterized by its *depth d*. The subscript i serves to distinguish different constituents with the same arguments and with the same depth from each other.

Constituents can be defined recursively as follows:

$$C^{(0)}[x_1, x_2, \ldots, x_k] \tag{2.2}$$

is of the form

$$\bigwedge_{j \in J} (\pm_j) A_j[x_1, x_2, \ldots, x_k] \tag{2.3}$$

where the $A_j[x_1, x_2, \ldots, x_k]$, $j \in J$, are all the different atomic formulas that can be formed from the predicate constants of L and of $x_1, x_2, \ldots, x_k$, and where each $(\pm_j)$ is either $\sim$ or nothing, depending on j.

$$C_i^{(d+1)}[x_1, x_2, \ldots, x_k] \tag{2.4}$$

is of the form

(2.5)

$$\bigwedge_{j \in J} (\exists y) C_j^{(d)}[y, x_1, x_2, \ldots, x_k] \;\&\; (\forall y) \bigvee_{j \in J} C_j^{(d)}[y, x_1, x_2, \ldots, x_k]$$
$$\& \;\; C_l^{(0)}[x_1, x_2, \ldots, x_k].$$

Here the last conjunct is simply some one constituent without quantifiers with $x_1, x_2, \ldots, x_k$ as its only free individual symbols. The index set J is a subset of the set of the subscripts of all the different constituents

$$C^{(d)}[y, x_1, x_2, \ldots, x_k]. \tag{2.6}$$

Intuitively, a constituent like (2.5) of depth $d+1$ tells us what kinds of individuals there exist (in relation to $x_1, x_2, \ldots, x_k$) and do not exist. The latter is accomplished in the universally quantified disjunction of (2.5) by saying that each individual must be of one of the kinds listed in the first of the three conjuncts in (2.5). Here the "kinds of individuals" are in turn specified by constituents (2.6) of a lesser depth d.

Each constituent (2.5) thus has a tree structure where the nodes of this labeled tree are constituents of increasingly smaller depth each occurring in its predecessor. Intuitively, each branch of such a tree describes a sequence of $d+1$ individuals that you can find in a model of L in which (2.5) is true. The tree structure show how the initial segments of such sequences limit their possible continuations.

In a sense, a constituent thus presents an explicit description of certain salient structural features of a model M in which it is true. The constituent tells you which (ramified) sequences of individuals (up to the length $d+1$) you can hope to find in a model in which it is true.

In this work, the term "constituent" will also be applied to substitution-instances of (2.4) with respect to the individual constants of L, i.e., to formulas like

$$C^{(d+1)}[a_1, a_2, \ldots, a_k] \tag{2.7}$$

or

$$C^{(d)}[y, a_1, a_2, \ldots, a_k]. \tag{2.8}$$

If identity is present in L, the definition of a constituent can be changed as follows: (2.2) is now of the form

$$\bigwedge_{j \in J} (\pm_j) A_j[x_1, x_2, \ldots, x_k] \;\&\; \bigwedge_{m,n \leq k}^{m \neq n} (x_m \neq x_n) \tag{2.9}$$

and (2.5) is now of the form

$$\bigwedge_{j \in J} (\exists y) C_j^{(d)}[y, x_1, x_2, \ldots, x_k] \;\& \tag{2.10}$$

$$(\forall y)\Big(\bigwedge_{m=1}^{m=k} (y \neq x_m) \supset \bigvee_{j \in J} C^{(d)}[y, x_1, x_2, \ldots, x_k]\Big) \;\&\; C_l^{(0)}[x_1, x_2, \ldots, x_k].$$

What this means is that in the presence of identity constituents can be written precisely in the same way as in the identity-free case provided that an exclusive interpretation of quantifiers and free individual variables is adopted.

In the rest of this paper, I shall assume that identity is not present.

The concept of (quantificational) depth of $d(S)$ of a formula S can be defined for arbitrary formulas as follows:

(d.i) If there are no quantifiers in S, $d(S) = 0$.
(d.ii) $d(S_1 \ \& \ S_2) = d(S_1 \vee S_2) = \max[d(S_1), d(S_2)]$
(d.iii) $d(\sim S) = d(S)$
(d.iv) $d((\exists x)S[x]) = d((\forall x)S[x]) = d(S[x]) + 1$

It is easily seen that this definition agrees with the way the notion of depth was used in connection with constituents.

In discussing the identity of constituents we shall consider (i) the order of disjuncts and conjuncts, (ii) the choice of bound variables, and (iii) possible repetitions of identical (*modulo* (i)–(ii)) members as inessential. If this idea is used in the numbering (indexing) constituents we can prove the following:

LEMMA 2.1: *If $i \neq j$,*

$$C_i^{(d)}[x_1, x_2, \ldots, x_k] \vdash \sim C_j^{(d)}[x_1, x_2, \ldots, x_k]. \tag{2.11}$$

This is easily proved by induction on d. It is also obvious on the basis of the intuitive meaning of constituent.

We can also prove

LEMMA 2.2: *If S is a closed formula of L of depth d, then for each i either*

$$C_i^{(d)} \vdash S \tag{2.12}$$

or

$$C_i^{(d)} \vdash \sim S. \tag{2.13}$$

This, too, can be proved by induction on d.

The same can be proved for formulas $S[x_1, x_2, \ldots, x_k]$ and constituents $C^{(d)}[x_1, x_2, \ldots, x_k]$ having the same free variables $x_1, x_2, \ldots, x_k$. We shall call this result Lemma 2.3.

In particular,

LEMMA 2.4: *For each constituent of the form*

$$C_i^{(d+1)}[x_1, x_2, \ldots, x_k] \tag{2.14}$$

there is precisely one constituent

$$C_j^{(d)}[x_1, x_2, \ldots, x_k] \tag{2.15}$$

such that (2.14) logically implies (2.15). For other values of j, (2.14) logically implies the negation of (2.15).

In the former case (2.15) can be obtained from (2.14) by omitting it from all constituents of depth 1, together with connectives that thereby become idle, and all repetitions. It is obvious that the result is implied by (2.14).

What Lemma 2.4 says is in effect that you can omit the last layer of quantifiers from any constituent and obtain a shallower one which is implied by the original. In fact you can omit any one layer of quantifiers in a given constituent.

Together Lemmas 2.1–2.4 entail

LEMMA 2.5: *Each consistent formula $S^{(d)}[x_1, x_2, \ldots, x_k]$ of depth d with the free variables $x_1, x_2, \ldots, x_k$ is logically equivalent with a disjunction of constituents of the form*

$$C_j^{(d)}[x_1, x_2, \ldots, x_k].$$

Not only can we omit layers of quantifiers from a constituent; we can likewise omit arguments from it.

LEMMA 2.6: *Given a consistent constituent*

$$C_i^{(d)}[x_1, x_2, \ldots, x_k], \tag{2.16}$$

consider the constituent

$$C_i^{(d)}[x_1, x_2, \ldots, x_{k-1}, \{x_k\}] \tag{2.17}$$

obtained from (2.16) by omitting from it all atomic formulas containing x_k, all connectives which thereby become vacant, and all repetitions. Then (2.17) is the only constituent of the form

$$C^{(d)}[x_1, x_2, \ldots, x_{k-1}] \tag{2.18}$$

which is implied by (2.16).

Proof: It is again obvious that (2.16) implies (2.17). If it implied any other constituent of form (2.18), it would be inconsistent by Lemma 2.1.

Several of the basic concepts of model theory are easily and naturally defined by reference to constituents.

A consistent sequence of constituents

$$C_{i(d)}^{(d)}[x_1, x_2, \ldots, x_k] \tag{2.19}$$

with a fixed k ($k > 0$), but with an ever increasing $d = 1, 2, 3, \ldots$ defines a k-type. It is easily shown that this definition is equivalent with the usual one, according to which a k-type is the maximal consistent set of formulas with $x_1, x_2, \ldots, x_k$ as their only free variables.

When $k = 0$, we have a sequence of closed constituents

$$C_{i(d)}^{(d)} \qquad (d = 1, 2, \ldots). \tag{2.20}$$

From Lemma 2.5, it is seen that (2.20) defines a complete theory, and that each complete theory can be represented in this way.

Notice that each k-type (2.19) implies a unique complete theory (2.20). For each member of the sequence (2.19) implies a unique constituent without any individual constants in virtue of Lemma 2.6. Those types (2.19) which so imply (2.20) are the only ones consistent with (2.20).

The k-type (2.19) is compatible with the complete theory iff constituents (2.19) all occur in the successive members of (2.20).

The types compatible with (2.20) will be called the types of the complete theory (2.20). Each type satisfied in a model of (2.20) is a type of (2.20), but all the types of (2.20) need not be satisfied in a given model of (2.20). The question as to which of them are satisfied is one of the central ones in model theory. Different kinds of models are distinguished from each other by the types that are satisfied in them.

One particularly useful result concerning constituents is the following:

LEMMA 2.7: *Given a complete theory (2.20), a model M of (2.20), and a constituent C (or a substitution instance of a constituent containing names of members of dom(M)), if C is compatible with the set of sentences true in M, C is satisfied in M. For constituents without names, it suffices to assume that they are compatible with Th(M).*

We can here perhaps see some of the advantages of the use of constituents. All the lemmas of this section can be seen to be valid directly on the basis of the import of a constituent. (Cf. the explanation of the meaning of the tree structure of a constituent given above.) Lemma 2.7 is a case in point, though perhaps slightly less obvious at first than the earlier lemmas.

Other results can likewise be read off from the intuitive meaning of a constituent, albeit not equally directly. As an example of such a result, we can mention the following result off almost immediately from the intuitive meaning of a constituent in the following:

LEMMA 2.8: *Assume that*

$$C_i^{(d)}[a_1, a_2, \ldots, a_k] \tag{2.21}$$

is compatible with

$$C_j^{(d+k)}. \tag{2.22}$$

Then (2.22) implies

$$(\exists x_1)(\exists x_2)\cdots(\exists x_k)C_i^{(d)}[x_1, x_2, \ldots, x_k]. \tag{2.23}$$

Proof (informal): In exploring a world in which (2.22) is true, we can come upon $x_1 = a_1, x_2 = a_2, \ldots, x_k = a_k$ in this order. If (2.21) is likewise true in the same world, as it can be if (2.21) and (2.22) are compatible, the rest of the world is described by $C_i^{(d)}[a_1, a_2, \ldots, a_k]$.

This lemma holds by the same token if there are additional free variables or constant parameters in (2.21) and (2.22).

Many of the well-known results in model theory are proved easily and in a perspicuous way by means of constituents. As an example of the use of constituents to systematize old results and to obtain new ones, Rantala's monograph ***Aspects of definability*** (1977) can be mentioned.

More illustrations of the use of constituents are offered in the next few sections.

§3. A wrong implementation of the extremality idea. At this point, it might seem to be easy to implement extremality conditions on models. The natural way to interpret our extremality requirements is to say the following: A model is minimal iff as few *kinds of individuals* as possible are instantiated in it; a model is maximal iff as many *kinds of individuals* as possible are instantiated in it. Then the *prima facie* plausible idea is to take the concept of type defined above as the explication of the pre-theoretical idea of a kind of individuals (or a kind of k-tuples of individuals). Then the question raised at the end of the preceding chapter (Which types are satisfied in a model?) would become highly relevant to the extremality project.

What can we say by way of a response to this suggestion? In order to answer the question, we need a few further concepts. It may happen that the k-type

$$C^{(d)}_{i(d)}[x_1, x_2, \ldots, x_k] \qquad (d = 1, 2, \ldots) \tag{3.1}$$

compatible with the complete theory T_j

$$C^{(d)}_{j(d)} \qquad (d = 1, 2, \ldots) \tag{3.2}$$

stops branching from some point on. Then there is an initial segment of (3.1) such that only one continuation of it is compatible with (3.2). Such a type is called *atomic* in T_j = (3.2).

It may happen that each initial segment of each k-type (for each k) compatible with (3.2) is consistent with an atomic k-type. Then the entire complete theory (3.2) will be called *atomic.*

Clearly, each atomic k-type compatible with (3.2) must be satisfied in each model of (3.2). The interesting question is whether any other types need to be satisfied. This question turns out to be more complicated than one might first suspect. A model M is called *atomic* iff each k-tuple of the elements of the domain dom(M) of M satisfies an atomic k-type. A partial answer to the question just posed is given by

LEMMA 3.1: *A complete theory has a countable atomic model iff it is atomic.*

In the other direction, there are models M such that each k-type, for each k, compatible with the complete theory Th(M) true in M, is satisfied. Such models are called *weakly saturated.*

A related requirement is the following: Suppose a k-type $t_1[x_1, x_2, \ldots, x_k]$ is compatible with a $(k+1)$-type $t_2[x_1, x_2, \ldots, x_k, x_{k+1}]$ in Th(M) and that $a_1, a_2, \ldots, a_k \in$ dom(M) satisfy $t_1[a_1, a_2, \ldots, a_k]$. If there always exists $a_{k+1} \in$ dom(M) such that $a_1, a_2, \ldots, a_k, a_{k+1}$ satisfy $t_2[a_1, a_2, \ldots, a_k, a_{k+1}]$, then a weakly saturated model M is called *saturated.*

Saturated models are interesting "special models." It is not difficult to prove that any consistent complete theory has such a model.

Atomic models are—or seem to be—minimal models in some reasonable sense, and saturated models seem to be maximal models in an equally clear sense. The idea is this: It seems that 1-types constitute the finest partition of individuals into different "kinds" that can be affected by first-order means; and *mutatis mutandis* for k-types with $k > 1$. Hence it seems that the poorest models one can characterize by first-order means are the ones in which only those "kinds" (i.e., types) are exemplified which must be satisfied in any case, i.e., atomic models. Likewise, it appears that the richest models that can be dealt with on the first-order level are the ones in which all the different "kinds" (i.e., types) are instantiated, perhaps with the proviso that these types are instantiated so as to allow all possible steps from k-types to $(k+1)$-types. In other words, the richest models seem to be weakly saturated or perhaps saturated models.

In brief, the concepts of atomicity and saturation seem to be the natural explications of the ideas of minimality and maximality that are guiding my thinking. Yet they do not do this job well at all. Extremality requirements so interpreted do not allow us to capture the intended (standard) models in the interesting cases.

For instance, we cannot in this way capture naturally the intended "standard" model of Peano arithmetic. On the contrary, it is known (see, e.g., Chang and Keisler, Example 3.4.5) that any consistent complete extension of Peano arithmetic is an atomic theory and hence has atomic models. This holds also for complete theories not true in the intended structure of natural numbers. Hence the atomicity requirement does not do the job of capturing the structure of natural numbers.

Another example is offered by the (first-order) theory of dense linear order. It has a model which has the structure of the rationals. This model is at the same time an atomic model and a saturated one. But it is not really a minimal model in some intuitive sense, for you can omit elements from it and yet preserve its status as a model. It is not really a maximal model, either, in some striking sense, because it can be embedded in a richer one, viz. the structure of the reals, which is not isomorphic with it.

The notions of atomicity and saturation of course do not exhaust the resources of contemporary model theory. For instance, there is the notion of prime model.

DEFINITION: *A model M_0 of a theory T is a **prime model** iff it can be elementarily embedded in every model M of T.*

Prime models might look like plausible candidates for the role of a minimal model. However, on a closer look even the notion of prime model is not an adequate explication of the idea of a minimal model. For one thing, the way this notion is usually introduced is not useful to us as such. What we are looking for are some intuitive structural characterizations of minimality and maximality, and the notion of primeness does not give us such a characterization.

When I say this I mean the following: What made the idea of atomicity so appealing is that there is a clean syntactically definable notion of a kind of individual which enabled us to speak of a model in which a minimum of such "kinds" were instantiated. More generally, what we are looking for are characterizations of a minimal model M in terms of the constituent representation of the complete theory Th(M) true in M. For the constituent representation is in some obvious sense an explicit description of the most easily understandable features of the structure of M. In a sense, therefore, we want to have a characterization of minimality whose applicability can so to speak seem directly from the theory Th(M). Now it surely cannot be seen directly from model M itself or from the theory Th(M) whether M can be embedded elementarily in certain other models.

Of course, another way of defining primeness might do the trick. But the most prominent alternative characterization of a prime model, viz. to characterize it as a countable atomic model, does not fare much better. From a theory it is very hard to see directly what the cardinality of its several models might be.

Moreover, the notion of prime model is subject to most of the same objections as were marshaled above against atomic models as implementations of minimality. For instance, even though the structure of natural numbers N is the unique prime model of the Peano arithmetic, it is not the prime model of all the consistent extensions of this arithmetic, viz. of those which are not true in N.

The most flagrant source of dissatisfaction is the fact that a prime model of a complete theory might be elementary equivalent with a proper submodel of itself. An example is offered by the theory of dense linear order, where a prime model, for instance, the structure of the rationals, could obviously be elementarily equivalent with its proper submodel. Hence prime models are not always minimal models in any intuitive sense of the word.

§4. **Super models.** The explanation of the failure of special models to implement the extremality idea is not very hard to see. Types are not the right explication of the idea of "kinds of individuals" existing in a model M. A type, say a one-type, characterizes a kind of individual in so far as this individual is considered alone. In order to catch full the idea of a kind of individual, we have to consider them also in relation to the other individuals in the model.

This refined idea of a "kind of individual" can be captured by means of the following definition:

Let M be a model and let $a_1, a_2, \ldots$ be a sequence of members of the domain dom(M) of M. Let

$$C^{(d)}_{i(d,k)}[x, a_1, a_2, \ldots, a_k] \qquad (d = 1, 2, \ldots, k = 0, 1, 2, \ldots) \tag{4.1}$$

be a (double) sequence of mutually consistent constituents compatible with the complete theory Th(M) true in M. Assume also that the constituents $C^{(d)}_{i(d,k)}[\{x\}, a_1, a_2, \ldots, a_k]$ are all true in M. Then (4.1) is said to define *supertype* in $A = \{a_1, a_2, \ldots\}$ relative to M.

The justification of formulating the definition of a supertype in this way is that the theory defined by (4.1) clearly does not depend on the order of the $a_1, a_2, \ldots$

The corresponding sequence with individual variables instead of constants, i.e.,

$$C^{(d)}_{i(d,k)}[x, y_1, y_2, \ldots y_k] \tag{4.2}$$

can be called the *structure of the supertype* (4.1), alias a *supertype structure.*

Many of the same things can be said *mutatis mutandis* of supertypes as can be said of types. For instance, the supertype (4.1) is said to be *atomic* iff it stops branching after a certain point. More explicitly, (4.1) is atomic iff it has a member

$$C^{(d)}_{i(d,k)}[x, a_1, a_2, \ldots, a_k] \tag{4.3}$$

such that for any e and any $c_1, c_2, \dots c_l \in \mathrm{dom}(M)$ there is only one constituent of the form

$$C_j^{(d+e)}[x, a_1, a_2, \dots, a_k, c_1, c_2, \dots c_l] \tag{4.4}$$

compatible with (4.3).

One of the basic properties of supertypes is the following

LEMMA 4.1: *Given M and a supertype (4.1) compatible with the complete theory $Th(M)$ true in M, each member of the sequence (4.1) (and hence each initial segment of (4.1)) is satisfied in M by some individual b, i.e., there is $b \in dom(M)$ such that*

$$M \models C_{i(d,k)}^{(d)}[b, a_1, a_2, \dots, a_k]. \tag{4.5}$$

This follows clearly from Lemma 2.7.

Hence in a sense each initial segment of each supertype compatible with $\mathrm{Th}(M)$ is satisfied in M. The only open question is whether the entire supertype is.

From what has been said it follows that each atomic supertype is satisfied.

Assume now that (4.1) defines an atomic supertype in M and that

$$C_{i(e,l)}^{(e)}[x, a_1, a_2, \dots, a_l] \tag{4.6}$$

is the member of (4.1) after which (4.1) no longer branches. We shall say that (4.6) *determines* the atomic supertype. Let us also assume that $\mathrm{Th}(M)$, represented in the form (3.2), is the complete theory true in M.

One the assumptions just stated, we have

LEMMA 4.2: *Let*

$$C_j^{(e+f)}[x, a_1, a_2, \dots, a_l, b_1, b_2, \dots, b_m] \tag{4.7}$$

be any constituent compatible with (4.6) and $Th(M)$. Then we must have $j = i(e+f, l+m)$, i.e., (4.7) must be a member of (4.1).

Proof: This is what it means for (4.1) to stop branching at (4.6).

LEMMA 4.3: *On the same assumptions, each member of (4.1) later than (4.6) is equivalent with all of its successors, given $Th(M)$.*

Proof: Each member of (4.1) is implied by its successors by Lemmas 2.4 and 2.6. Hence what we have to prove is that it implies them, given $\mathrm{Th}(M)$. For this purpose, it suffices to show that (4.6) is not compatible (together with $\mathrm{Th}(M)$) with any other constituent of the form

$$C_j^{(e+f)}[x, a_1, a_2, \dots, a_l, y_1, y_2, \dots, y_m]. \tag{4.8}$$

In order to see this, let (4.8) be compatible with $\mathrm{Th}(M)$ and (4.6). Then by the same reasoning as in Lemma 2.7, there are $b_1, b_2, \dots, b_m \in \mathrm{dom}(M)$ such that (4.7) occurs in some supertype (4.1) of M. But if so, by Lemma 4.2, $j = i(e+f, l+m)$, in other words, there is only one constituent of form (4.8) compatible with $\mathrm{Th}(M)$ and (4.6).

LEMMA 4.4: *If Th(M) implies that, in a sequence like (4.1), each member is equivalent with its successors from (4.6) on, then (4.6) determines an atomic supertype.*

Proof: If (4.6) is equivalent with each if its successors, say (4.7) with $j < i(e+f, l+m)$, then by Lemma 2.1 it is incompatible with (4.7) with any other j. In other words, (4.1) can be continued from (4.6) in only one way, i.e., (4.6) determines an atomic supertype.

THEOREM 4.1: *On the same assumptions as in Lemma 4.2,*

(4.9)

$$Th(M) \vdash (4.6) \supset (\forall y_1)(\forall y_2)\cdots(\forall y_m)C^{(e+f)}_{i(e+f,l+m)}[x, a_1, a_2, \ldots, a_l, y_1, y_2, \ldots, y_m]$$

Moreover, if (4.9) holds for all F, M, (4.6) determines an atomic supertype.

Proof: From Lemma 4.3 we have

(4.10) $$\text{Th}(M) \vdash ((4.6) \supset C^{(e+f)}_{(e+f,l+m)}[x, a_1, a_2, \ldots, a_l, b_1, b_2, \ldots, b_m])$$

From this (4.9) follows by first-order logic.

Conversely, if (4.9), then (4.6) implies (given Th(M)) all its successors. By Lemma 4.4, it determines an atomic supertype.

THEOREM 4.2: *On the same assumptions*

(4.11) $$Th(M) \vdash (\forall z_1)(\forall z_2)\cdots(\forall z_l)(C^{(e)}_{i(e,l)}[x, z_1, z_2, \ldots, z_l] \supset (\forall y_1)(\forall y_2)\cdots(\forall y_m)C^{(e+f)}_{i(e+f,l+m)}[x, z_1, z_2, \ldots, z_l, y_1, y_2, \ldots, y_m])$$

Conversely, if (4.11) holds, (4.6) determines an atomic supertype, provided that (4.6) is true in M.

Proof: In the same way as in Theorem 4.1.

THEOREM 4.3: *Assume that (4.6) determines an atomic supertype. Then the same supertype structure is determined by a constituent of the form*

(4.12) $$C^{(e+l)}_j[x].$$

Proof: Consider

(4.13) $$C^{(e+l)}_{i(e+l,l)}[x, a_1, a_2, \ldots, a_l].$$

By Lemma 4.3, (4.13) is equivalent with (4.6), given Th(M). The formula

(4.14) $$C^{(e+l)}_{i(e+l,l)}[x, \{a_1\}, \{a_2\}, \ldots, \{a_l\}]$$

is of the form (4.12), and in fact can serve as (4.12).

In order to show that this is what we want, it suffices in virtue of Lemma 4.4 that for each f and m there is a constituent of the form

(4.15) $$C^{(e+l+f)}[x, y_1, y_2, \ldots, y_m]$$

implied by (4.14), given Th(M).

Now, in virtue of Lemma 2.8, (4.14) implies

(4.16) $$(\exists z_1)(\exists z_2)\cdots(\exists z_l)c^{(e)}_{i(e,l)}[x, z_1, z_2, \ldots z_l].$$

In virtue of Lemma 4.3, there is a unique constituent of the form (4.15) implied by Th(M) and

$$C^{(e)}_{i(e,l)}[x, z_1, z_2, \ldots, z_l]. \tag{4.17}$$

But there obviously is such a unique constituent implied by Th(M) and (4.14).

Results like Theorem 4.3 are interesting in a wider perspective. At first sight, it might seem that the switch from special models to superspecial models destroys the strategic advantages offered by concepts like type and atomicity. The crucial thing about them is how they help us to read off (as it were) the structural (model-theoretical) properties of the models of a theory, especially a complete theory, from the syntactical structure of this theory. The way in which the atomicity of a model of a complete theory hangs together with the structure of types in the constituent representation of this theory is a typical example of this strategy.

It might seem that the way supertypes are defined deprives us of the use of this syntax-to-models strategy. For supertypes are defined by reference to some given model. Hence it seems circular to study supertypes and their interrelations for the purpose of gaining insights into the structure of models.

What the theorems just proved show is that this impression is mistaken. Even though supertypes are defined by relation to one particular model, some of their most crucial properties depend only on the structure of the complete theory Th(M) true in M. In particular, what the atomic supertypes of M are is in a certain sense completely determined by Th(M). For instance, each atomic supertype is determined by a constituent of form (4.12) or $C^{(d)}_i[x]$ occurring in the constituent representation of the given theory. Here (4.12) does not depend on any particular member of dom(M).

One application of these observations is that we can define the notion of *superatomicity* for a complete theory in analogy with the definition of atomicity of complete theories. the complete theory Th(M) true in M is superatomic iff each initial segment of each *supertype* structure compatible with Th(M) is compatible with (the structure of) an atomic supertype.

Then we can also define a superatomic model. A ***superatomic model*** M is one in which only atomic supertypes are satisfied, and each different supertype by precisely one individual. Then we can also prove easily the following:

THEOREM 4.4: *Each superatomic complete theory has a superatomic model.*

It is also easily seen that this superatomic model is uniquely determined (up to isomorphism).

THEOREM 4.5: *A model M of a complete theory T = Th(M) is superatomic iff it is a prime model but none of its proper elementary submodels is prime.*

In order to prove Theorem 4.5, we can first prove

LEMMA 4.5: *A superatomic model M of a complete theory T = Th(M) is prime.*

In order to show that M is prime, we have to show that it can be elementarily embedded in an arbitrary model M^* of T. Now each member b_i of dom(M) satisfies a supertype determined by a constituent of the form

(4.18) $$C_i^{e(i)}[x].$$

But, by Lemma 4.1 (4.18) must be satisfied by some individual, call it b_i^* in M^*. The mapping of each b_i on b_i^* defines an embedding of M into M^*. It is easily seen that this is an elementary embedding.

To return to the proof of Theorem 4.5 it is clear that if we try to map M elementarily into itself, each $b_i \in$ dom(M) must be mapped on itself. For the image b_i^* must satisfy the same constituent (4.18) as b_i, the mapping being an elementary embedding. But the only member of dom(M) to satisfy this constituent is b_i itself. This proves Theorem 4.5.

In this way we can also see that one of the main shortcomings of prime models as explications of minimality is overcome by superatomic models.

This is *prima facie* a major difference between atomic and superatomic models in that in an atomic model, every k-type, for $k = 1, 2, \ldots$, satisfied in it must be atomic, whereas the definition of a supertype involves directly only formulas with one free individual variable. These apparently correspond to one-types only.

To reassure the reader, we can prove

LEMMA 4.6: *A superatomic model is atomic.*

In order to show this, assume first that M is a superatomic model of Th(M). Then each element of dom(M) satisfies an atomic supertype. If $\langle a_1, a_2, \ldots, a_k\rangle$ is a k-tuple of such elements, one can see by the same line of argument as was given fro Theorem 4.1 that there is a consistent structure

(4.19) $$C_i^{(d)}[x_1, x_2, \ldots, x_k]$$

satisfied by $\langle a_1, a_2, \ldots, a_k\rangle$ (and hence compatible with Th(M)) such that each formula

(4.20) $$C_i^{(d)}[a_1, a_2, \ldots, a_{j-i}, x, a_{j+i}, \ldots, a_k]$$

defines a supertype. Then for each e there is only one constituent of the form

(4.21) $$C^{(d+e)}[x_1, x_2, \ldots, x_k]$$

compatible with (4.18) and with Th(M). In other words, (4.18) determines an atomic type.

This proof illustrates how we can get along in our "supertheory" by means of supertypes with only one free variable, i.e., with what *prima facie* should be called one-supertypes. Just because in supertypes we heed the relation of the kinds of individuals characterized by them to other individuals in the models, we do not need k-supertypes with $k \geq 2$.

In order to extend one horizon to arbitrary theories instead of just complete ones, we must first extend our main concepts.

A supertype is said to be *strongly atomic* if it stops branching in all models compatible with one of its initial segments.

More explicitly, a constituent

$$C_i^{(e)}[x, a_1, a_2, \ldots, a_l] \tag{4.22}$$

determines a strongly atomic supertype iff it determines an atomic supertype in any model of (4.22).

A theory T is ***strongly superatomic*** iff each initial segment of each supertype structure compatible with T is compatible with (the structure of) a strongly atomic supertype.

A strongly superatomic model M is one in which only strongly atomic supertypes are satisfied, and each of them by precisely one individual.

It is now easy to prove suitable extensions of our earlier results, for instance

THEOREM 4.6: *Each complete theory compatible with a strongly superatomic theory has a strongly superatomic model.*

The most natural generalization of the notion of saturation (of a model M) is not equally directly connected with the structure of the complete theory Th(M).

DEFINITION: *A model M is **absolutely supersaturated** iff each supertype compatible with Th(M) is satisfied.*

A model M is supersaturated relative to a set of individuals $A = \{a_i\}(i \in I), A \subseteq \text{dom}(M)$ iff each supertype (4.1) with $a_1, a_2, \cdots \in A$ compatible with Th(M) is satisfied in M.

These notions still rely fairly heavily on the particular model M. However, the insights so far reached enable us to define a somewhat less demanding characteristic of a model which will turn out to be most useful.

DEFINITION: *A model M is **atomically supersaturated** iff*

(i) Each atomic supertype structure is satisfied in M by precisely one individual a_1; and

(ii) M is supersaturated with respect to the set $A = \{a_i\}$ of the individuals satisfying the different atomic supertype structures.

It is easily seen that the following theorem holds:

THEOREM 4.7: *Each complete theory has an atomically saturated model.*

Many results familiar from the traditional model theory have related results that can be proved for supertypes. Here I shall mention only one as an example.

THEOREM 4.8: *From a model M_1 of a complete theory T one can omit a countable number of individuals, each satisfying only nonatomic supertypes, and obtain a model M_2 of T which is an elementary submodel of M_1.*

§5. Super special models as implementations of extremality requirements. The concepts defined in the preceding section serve as excellent explications of the notions of minimality and maximality which are the focal ideas of this study. The minimality requirement (principle of paucity) is naturally captured by the idea of strong superatomicity, which literally amounts to imposing on models the least possible qualitative variety in so far as the relevant qualitative differences are understood by references to supertypes. This procedure is vindicated by the fact that superatomic models of complete theories turn out to be the minimal prime models, i.e., models elementarily embeddable into any model of the given complete theory, but not into any of their own proper submodels.

Strongly superatomic models turn out to be capable of doing the kind of job they were cast to do. For one thing, a Peano-type axiomatization of elementary arithmetic turns out to be categorical and have the structure N of natural numbers as its sole model, if the space of models is restricted to strongly superatomic ones. The only strongly superatomic model of the Peano axiomatization of elementary number theory can be shown to be the structure N of natural numbers, if the space of models is restricted to strongly superatomic models. This does not conflict with Gödel's incompleteness result, because the new "paucity logic" is not axiomatizable. Hence the new perspective on elementary arithmetic does not automatically create new avenues of actually establishing new number-theoretical results. What it nevertheless can in principle do is to facilitate the discovery of stronger and stronger proof methods. For the search for such methods can now be guided by clear-cut semantical considerations.

But how do I know that the structure of natural numbers N is the only superatomic model of the Peano axioms? These axioms are compatible with a number of different complete theories, only one of which is true in N. We can call it Th(N). It is easily seen to be strongly superatomic. It has different non-isomorphic models, of which N is one. It is easily seen that N is in fact the only superatomic model of Th(N).

But what about the other complete theories compatible with Peano arithmetic? How do we know that they do not have strongly superatomic models, too?

Perhaps the quickest way of seeing that they do not is to note a trivial-looking property of strongly superatomic models. Let each member of such a model M, say b, be correlated one-to-one with one of the constituents $C_1^{(d)}[x]$ which determines the strongly atomic supertype that b_1 satisfies. Given two such individuals b_1 and b_2, the second one being correlated with $C_2^{(e)}[x]$, one can construct effectively the formula that determines the strongly atomic supertypes satisfied by their sum, likewise for their product. This means that sum and product are recursive relations in a strongly superatomic model.

The details of this argument are given in an appendix below.

Now Tennenbaum has shown (see Tennenbaum 1959; Feferman 1958; Scott 1959; Kaye 1991, p. 153) that sum and product are not recursive in any non-standard model of Peano arithmetic (in the sense of relative recursivity just explained). From this it follows, together with the observations just made, that non-standard models of Peano arithmetic cannot be superatomic, just as was claimed.

Things are somewhat more complicated with respect to the notion of maximality (principle of plenitude). Hilbert's completeness axiom amounted to requiring that any attempted adjunction of a new individual to the intended model must lead to a violation of the other axioms. But such requirements cannot be satisfied in first-order theories in view of the upwards Skolem–Löwenheim theorem. In Hilbert's axiomatization of geometry, his completeness axiom has the intended effect only because he had also assumed the Archimedean axiom and also tacitly interpreted the notion of natural number involved in the Archimedean axiom in the standard sense. Fortunately, the intended maximality conditions can typically be interpreted so as to require only maximal qualitative richness, not necessarily the presence of the maximal selection of individuals in the intended model or models. Hence the natural course for us here is, if we want to keep our conceptualizations generally applicable, to require only maximal qualitative richness but not completeness in Hilbert's strong sense. But this does not really mean giving up Hilbert's original ideas. For even geometry, the function of the completeness axiom is to enforce continuity, not to restrict the "size" of the universe of discourse. Indeed, Hilbert's completeness axiom can be replaced by a pair of assumptions that can be roughly expressed as follows:

(H.1) If two points have the same relations to all other points and lines, they are identical.

(H.2) If M is a model of the other axioms and if there is a set of relations between an unspecified individual x and the members of M which is compatible with the other axioms and with the diagram of M, there exists in M an individual with these relations.

As you can easily see, (H.1) follows from other axioms. (Axioms of incidence and order suffice for the purpose.) Hence the import of the completeness axiom is essentially (H.2), which is an assumption of maximal qualitative richness rather than of maximal size as far as individuals are concerned. In fact, the force (H.2) is easily seen to amount to requiring that maximal number of supertypes be instantiated, compatible with the other assumptions.

Hence we can safely think of the maximality idea as being captured by a requirement of maximal qualitative richness. But in the preceding section we found that we have a genuine choice here. We can require either absolute supersaturation or atomic supersaturation of our models. The difference between the two appears nevertheless to be relatively unimportant. For one thing, the former implies the latter. Furthermore it will turn out that even atomic supersaturation is quite a strong assumption.

More has to be done here, however, than to explicate the twin notions of minimality and maximality. In the most interesting mathematical theories beyond elementary number theory, such as the theory of reals, axiomatic geometry, and set theory, the crucial thing turns out to be neither minimality assumptions nor maximality assumptions, but their interaction. Typically, we can assume that we are dealing with a theory which contains a one-place predicate, say $N(x)$, for natural numbers, and suitable axioms for natural numbers. (That is, when the axioms of the theory are relativized to $N(x)$, they must yield as consequences a reasonable axiomatization of natural numbers.) Notice that $N(x)$ does not necessarily have to be a primitive predicate. Then, we obviously have to assume that part of a model of the theory which corresponds to natural numbers is superatomic but that the rest of the model is maximal. But maximal in what sense? The crucial fact here is that we cannot simply assume that the individuals satisfying $\sim N(x)$ form a atomically supersaturated model, for that may be incompatible with the requirement that $\{x : M \models N(x)\}$ is superatomic. We can only require that the model realizes a maximal number of supertypes (either absolutely or reelative to the set of individuals satisfying superatomic types) compatible with the requirement that $\{x : M \models N(x)\}$ be superatomic.

We have thus motivated the following definitions:

Let us assume that we are given a model M of a complete theory Th(M) which contains a one-place predicate $N(x)$ for natural numbers. Let us assume further that the theory Th(M) as restricted to $\{x : M \models N(x)\}$ is superatomic. Then the model M is *absolutely Hilbertian* iff the following requirements are also satisfied:

(i) M restricted to $\{x : M \models N(x)\}$ is superatomic.

(ii) A maximal subset of supertypes compatible with (i) are instantiated in M.

M is an *atomically Hilbertian* model iff the following conditions are satisfied:

(i) As before.

(ii)* A maximal number of supertypes relative to the set of individuals satisfying a superatomic type are instantiated in M.

For instance, consider a set of axioms for real numbers which includes a predicate $N(x)$ for natural numbers. Then (i) becomes essentially the Archimedean axiom. By the usual Dedekind-type line of thought, one can then show that the structure of the (actual) reals is the only one which is also atomically Hilbertian.

Essentially, the same also happens in Hilbert's axiomatization of geometry. Hilbert needs the Archimedean axiom (utilizing the standard concept of natural number) to force as it were the multiples of the unit line to match the structure of natural numbers, and the axiom of completeness to ensure continuity. The latter point is especially clear in Hilbert (1900), where the axiom of completeness first made its appearance (as an axiom for the theory of reals rather than as an axiom of geometry). Hilbert's formulation of his axiom there also makes it clear that he thought of it as a maximality assumption.

If we give up requirement (i), we can obtain sundry non-standard models of reals. If we give up (ii), we need not any longer have all "real reals" in our model. Depending on the axiomatization, it may be sufficient, e.g., for the model to contain only all algebraic numbers.

Thus, we can again reach one of our main objectives. If we restrict the models of (the first-order language of) a theory of reals to atomically Hilbertian ones, then any reasonable theory of reals is categorical and yields the intended structure of reals as its only model (up to isomorphism). This descriptive completeness is not due to the requirement of minimality (superatomicity) alone, nor to the requirement of maximality (atomic supersaturation), but to the combination of the two in the requirement of the (atomically) Hilbertian character of the models.

An interesting pitfall here is that a complete theory need not have a unique richest model of a given cardinality, either in the sense of being absolutely Hilbertian or atomically Hilbertian. If we think of the supertypes compatible with the given theory satisfied one by one, then the ultimate outcome can so to speak depend on the order in which they are satisfied.

An interesting situation arises when the same ideas are extended to a suitable axiomatization of set theory. In order to apply the ideas sketched here, we have to assume that a predicate $N(x)$ for natural numbers is included in the language of the set theory or can be defined as the basis of the axiomatization. Then we can again stipulate that the models be restricted to atomically or absolutely Hilbertian ones, and see what happens.

I cannot here try to answer this question in general. Certain things are nevertheless relatively easy to see. Perhaps the most interesting perspective offered by our observation, is that in set theory, too, the greatest subtlety is due to the interplay of minimality and maximality requirements. On the one hand, one can construct poor (small) models of, say, ZF set theory which have a clear-cut structure but which clearly are not what is intended. On the other hand, attempts to enlarge the universe of set theory have not yielded any ultimate clarity either. It seems to me that the real source of difficulties in set theory is that the requirements of poverty and plenitude have to be balanced against each other. For another example, we can construct a theory of finite types as a many-sorted first-order theory. We might, e.g., assume that there is a primitive predicate $N(x)$ in the language for natural numbers which are among the individuals. If we then require that the models of a suitable axiomatization of such a type theory are automatically Hilbertian, the resulting theory has all sorts of nice features. For instance, the Denumerable Axiom of Choice is valid and so is the Principle of Dependent Choices for subsets of natural numbers. Furthermore, it will be easy to give a descriptively complete and indeed categorical axiomatization for a theory of the second number class (countable ordinals).

In this kind of many-sorted first-order reconstruction of type theory we can even start from an axiomatization of a discrete linear order (with an initial element) for natural numbers. Its only superatomic model is clearly $\{0, 1, 2, \dots\}$ with successor as the only relation. It is easily seen, however, that functions for addition, multiplication, etc., all necessarily exist in all the models of the full axiomatization, as indeed do all recursive functions. Thus the existence of the usual arithmetical functions does not even have to be assumed; it follows logically from the axioms. This would speak for a partial reducibility of mathematics to logic, if it were not for the fact that certain mathematical assumptions were built right into our concept of model and hence into our concept of logic.

A more sweeping philosophical perspective which opens here takes up an issue which was mooted by Hilbert and Kronecker. For Kronecker, natural numbers were the be-all if not the end-all of mathematics. In contrast, one of Hilbert's acknowledged aims in his ***Grundlagen der Geometrie*** was to show that there can be important mathematical theories which do not involve the concept of natural numbers at all. (See here Blumenthal 1922, p. 68.)

If the approach advocated here is right, there is more to be said for Kronecker and less for Hilbert than has been generally acknowledged. If the subtlety of advanced mathematical theories lies in the interplay of superatomicity requirement for the natural numbers with suitable maximality assumptions, then the concept of natural numbers is after all essentially involved in these mathematical theories, via the requirement of superatomicity. Mathematics looks more like a science of (natural) numbers than it has in a long time.

On a more technical level, there does not seem to be any obstacles in principle to use the time-honored strategy of using set theory to speak of its own semantics. In this way, e.g., Gödel captured his own metalogical construction of a constructible model in an explicit axiom. If this strategy works here, the requirement that only atomically Hilbertian models are considered would be expressible by an explicit set-theoretical axiom of the old style (without restrictions on the usual set of models). This axiom would be eminently acceptable, for (i) merely spells out the nature of natural numbers while (ii) follows from the idea that set theory is the theory of *all* sets, that in the world of set theory what *can* exist *does* exist.

Whether or not we can along these lines solve the outstanding problems of set theory remains to be seen. I do not seem to be the only one who thinks that they can be so solved. Gödel once wrote to Ulam, *apropos* John von Neumann's axiomatization of set theory:

> The great interest which this axiom [in von Neumann's axiomatization of set theory] has lies in the fact that it is a maximum principle somewhat similar to Hilbert's axiom of completeness in geometry. For, roughly speaking, it says that any set which does not, in a certain defined way, imply an inconsistency exists. Its being a maximum principle also explains the fact that this axiom implies the axiom of choice. I believe that the basic problems of abstract set theory, such as Cantor's continuum problem, will be solved satisfactorily only with the help of stronger axioms of *this* kind, which in a sense are opposite or complementary to the constructivistic interpretation of mathematics. (See Ulam 1958.)

What Gödel misses here is the crucial interplay between maximality *and* minimality assumptions, though his remark on von Neumann type axioms being complementary to constructivistic ideas perhaps suggests some degree of awareness of this fact.

Appendix. Let us assume that a strongly superatomic model M of Peano arithmetic has been given and that a one-to-one correlation has been established between all natural numbers n and all the strongly atomic supertypes of M. More explicitly, let the correlate $\varphi(n)$ of each n be one of the constituents with one free individual variable that determine a strongly atomic supertype in M, different supertypes for different values of n. (Equivalently, the correlate of n could be the Gödel number of this constituent.) This correlation establishes an isomorphism between M and a certain relational structure. (Cf. here Kaye 1991, especially sec. 11.3.) In this isomorphism, a certain numerical relation will correspond to the relation of being the sum of in M, i.e., the relation which holds between three individuals in dom(M) say a, b, c, when $S(a, b, c)$ is true in M, where S is the expression of the sum of in Peano arithmetic. The question is whether this relation is recursive.

In order to show that it is, let us suppose that we are given three constituents each of which is correlated with some natural number by φ and each of which therefore determines a strongly atomic supertype. Let these constituents be

(1) $$C_1^{(d)}[z]$$

(2) $$C_2^{(e)}[y]$$

(3) $$C_3^{(f)}[x].$$

From (1) and (2) we can form the formulas:

(4) $$C_1^{(d)}[z] \ \& \ C_2^{(e)}[y] \ \& \ S(z, y, x)$$

(5) $$(\exists z)(\exists y)(C_1^{(d)}[z] \ \& \ C_2^{(e)}[y] \ \& \ S(z, y, x))$$

By assumption, (1) and (2) each determines a strongly atomic supertype. Clearly (5) is satisfied by the sum of the two individuals in dom(M) which satisfy (1) and (2). What we are interested in here is whether it is possible to determine recursively (effectively) whether (3) is also satisfied by this sum.

For the purpose, we shall first show that (5) determines a strongly atomic supertype. In order to prove this, assume that M^* is a model in which (5) is satisfied. Hence (4) is also satisfied in M^*. What has to be shown is that, given $a_1, a_2, \ldots, a_k \in \text{dom}(M^*)$ and $g \geq \max(d, e) + 2$, there is only one constituent of the form

(6) $$C_i^{(g)}[x, a_1, a_2, \ldots, a_k]$$

compatible with (5) and Th(M^*).

In order for (6) to be compatible with (5), it must be compatible with (1) and (2). Since (1) and (2) both determine strongly atomic supertypes, there is a unique constituent of the form

(7) $$C^{(g+1)}[y, z, a_1, a_2, \ldots, a_k]$$

compatible with (5) and Th(M^*). But since it follows from the axioms of Peano arithmetic that the sum of two numbers is uniquely determined, there is in (7) one and only one constituent of the form

(8) $$C_j^{(g)}[x, y, z, a_1, a_2, \ldots, a_k]$$

compatible with (4). Hence the constituent

(9) $$C_j^{(g)}[x, \{y\}, \{z\}, a_1, a_2, \ldots, a_k]$$

is the only constituent with the parameters $a_1, a_2, \ldots, a_k \in \text{dom}(M^*)$ and g compatible with (5) and $\text{Th}(M^*)$. In other words, it is the only constituent which can serve as (6), which is therefore uniquely determined. This is just what was to be proved.

It is important to realize that this part of the overall proof is not supposed to be effective.

Consider now the unique constituent

$$C_m^{(\max(d,e))} \tag{10}$$

compatible with (4). It is true in M, and it can be obtained effectively from (4).

Assume first that $f < \max(d,e)$. Because (3) determines a strongly atomic supertype in M, there is in (9) a unique constituent of the form

$$C_n^{(\max(d,e)-1)}[x] \tag{11}$$

compatible with (3). It can be found effectively, given (3) and (14), and it clearly determines a strongly atomic supertype.

Assume then that $f \geq \max(d,e)$. Then there is a unique constituent of the form (11) in (10) compatible with (3). In this case it can be found simply by omitting layers of quantifiers from (3), hence effectively.

In either case, since (1) and (2) both determine a strongly atomic supertype in M, there is a unique constituent of the form

$$C_0^{(\max(d,e))}[z,y] \tag{12}$$

compatible with (4). It can be found effectively as follows: First we convert (4) into its distributive normal form

$$\bigvee_i C_i^{(\max(d,e))}[z,y] \tag{13}$$

This can be done effectively. However, we cannot in general know which disjuncts in (13) are consistent and which ones are not. This uncertainty can be eliminated simply by grinding out the logical consequences of (13) jointly with (4) one after the other until only one survives undisproved. But it follows from the axioms of Peano arithmetic that the sum of two individuals is uniquely determined. Hence there is in (12) a unique constituent of the form

$$C_p^{(\max(d,e)-1)}[x,z,y] \tag{14}$$

But since (14) is uniquely determined, then so is

$$C_p^{(\max(d,e)-1)}[x,\{y\},\{z\}] \tag{15}$$

Now this constituent can be compared with (11) effectively for identity. If the two are identical, (3) and (5) determine the same strongly atomic supertype, if not, they do not.

By reviewing the argument, it is easily seen that this determination can be made effectively. By Church's thesis, sum will therefore be a recursive relation in M, which was to be proved.

Acknowledgments. In working on this paper, I have profited from the comments, suggestions and criticisms by Professor Jouko Väänänen, by several members of his research group in Helsinki, and by Professors David McCarty and Philip Ehrlich. They are not responsible for any errors, however. My participation in the ASL European Summer Conference in 1990 was facilitated by a travel grant from the Academy of Finland, and my work by research support from Boston University.

REFERENCES

JOHN D. BALDWIN, ***Fundamentals of Stability Theory***, Springer-Verlag, Berlin, 1988.

OTTO BLUMENTHAL, *David Hilbert*, ***Die Naturwissenschaften***, vol. 10 (1922), pp. 67–72.

C. C. CHANG and H. J. KEISLER, ***Model Theory***, North-Holland, Amsterdam, 1973.

SOLOMON FEFERMAN, *Arithmetically Definable Models of Formalized Arithmetic*, ***Notices of the American Mathematical Society***, vol. 5 (1958), pp. 679–680.

DAVID HILBERT, ***Foundations of Geometry***, tr. by Leo Unger, tenth ed., Open Court, La Salle, 1971. German original ***Grundlagen der Geometrie***, 1899.

DAVID HILBERT, *Über den Zahlbegriff*, ***Jahresberichte der Deutschen Mathematiker-Vereinigung***, vol. 8 (1900), pp. 180–184.

JAAKKO HINTIKKA, *Is there Completeness in Mathematics after Gödel?* ***Philosophical Topics***, vol. 17, no. 2 (1989) pp. 69–90.

RICHARD KAYE, ***Models of Peano Arithmetic***, Clarendon Press, Oxford, 1991.

VEIKKO RANTALA, ***Aspects of Definability*** (Acta Philosophica Fennica, vol. 29, nos. 2–3), North-Holland, Amsterdam, 1977.

HARTLEY ROGERS, Jr., ***Theory of Recursive Functions and Effective Computability***, McGraw-Hill, New York, 1967.

DANA SCOTT, *On Constructing Models for Arithmetic*, in ***Infinitistic Methods***, Pergamon Press, Oxford, 1959, pp. 235–255.

S. TENNENBAUM, *Non-archimedean Models for Arithmetic*, ***Notices of the American Mathematical Society***, vol. 6 (1959), p. 270.

STANISLAW ULAM, *John von Neumann, 1903–1957*, ***Bulletin of the American Mathematical Society***, vol. 64 (1958, May Supplement), pp. 1–49.

Department of Philosophy
Boston University
Boston, MA 02215, USA

ABSOLUTENESS FOR PROJECTIVE SETS

HAIM JUDAH[1]

Abstract. We study an absoluteness argument for simple forcing in order to get the measurability of low projective sets of reals, such as Σ^1_3-sets.

§0. Introduction. In this note we will present absoluteness properties of the universe and show that they imply measurability (Baire property) of the low projective sets of reals. We will write $\Sigma^1_n(L)$ ($\Sigma^1_n(B)$) for the statement "Every Σ^1_n-set is Lebesgue measurable" (has the Baire property). By the construction of the Lebesgue measure we have that the Borel sets are measurable. The Δ^1_1-sets are exactly the Borel sets. So the first natural question is: Are the Σ^1_1-sets measurable? Sierpinski proved $\Sigma^1_1(L)$ and Luzin proved $\Sigma^1_1(B)$. Let us present the proof of this theorem by using absoluteness arguments.

THEOREM. $\Sigma^1_1(L)$ & $\Sigma^1_1(B)$.

Proof. Let $\varphi(x)$ be a Σ^1_1-formula. Let P be an Amoeba forcing. We know that in V^P we have that $\text{Ran}(V) = \{r : r \text{ is Random over } V\}$ is a measure one set.

Working in V^P, we let $A = \{x : \varphi(x)\}$. If $A \cap \text{Ran}(V) = \emptyset$, then A has measure zero ($\mu(A) = 0$). If $A \cap \text{Ran}(V) \neq \emptyset$ then let $r \in \text{Ran}(V)$ such that $V^P \models \varphi(r)$. But φ is a Σ^1_1-formula; therefore we have that $V[r] \models \varphi(r)$. Let $[\![\varphi(\underset{\sim}{r})]\!]_R$ be the Boolean value of $\varphi(\underset{\sim}{r})$, where $\underset{\sim}{r}$ is the canonical name for the random real. $[\![\varphi(\underset{\sim}{r})]\!]$ is an equivalence class of Borel sets. Let p be a representative.

Now $\mu(p) > 0$. Let $r_1 \in p \cap \text{Ran}(V)$. Then $V[r_1] \models \varphi(r_1)$, and by absoluteness of the Σ^1_1-formula we have that

$$V^P \models \varphi(r_1).$$

Similarly, if $r_1 \in \text{Ran}(V) - p$,

$$V^P \models \neg\varphi(r_1).$$

We actually proved that in V^P

$$\mu(A \,\triangle\, [\![\varphi(\underset{\sim}{r})]\!]) = 0.$$

Therefore we proved that A is a measurable set in V^P. But this is a Σ^1_2-sentence, and by Shoenfield's Lemma we have that this should be true in V. Thus A is measurable in V. To show $\Sigma^1_1(B)$ use $\text{Cohen}(V) = \{x : x \text{ is Cohen over } V\}$. ∎

[1] The author would like to thank Mrs. Miriam Beller for TEXing the manuscript and improving the presentation of this paper.

Throughout this paper we will generalize this idea to get models for $\Sigma^1_3(L)$ and $\Delta^1_3(L)$.

All the forcing notions in this work are defined in [S]. In [JS1] the following was proved.

THEOREM.

(i) $\Delta^1_2(L)$ *iff* $Ran(L[a]) \neq \emptyset, \forall a \in \mathbb{R}$.
(ii) $\Delta^1_2(B)$ *iff* $Cohen(L[a]) \neq \emptyset, \forall a \in \mathbb{R}$.

R. Solovay proved the following

THEOREM (see [JS1]).

(i) $\Sigma^1_2(L)$ *iff* $\mu(Ran(L[a])) = 1, \forall a \in \mathbb{R}$.
(ii) $\Sigma^1_2(B)$ *iff* $Cohen(L[a])$ *is comeager*, $\forall a \in \mathbb{R}$.

For the definition of Souslin forcing the reader should refer to [JS2].

§1. Δ^1_3-Stability. The main objective of this section is to find a characterization of the statement $\Delta^1_3(L)$ $(\Delta^1_3(B))$. We saw in the introduction that the measurability (like the Baire property) of the Δ^1_2-sets, as well as the Σ^1_2-sets, is intrinsically connected with properties of Random real forcing for the Δ^1_2-sets and Amoeba forcing for the Σ^1_2-sets, respectively.

It was proved by Gödel that if $V = L$ then the real numbers have a Δ^1_2-well order. Also, from this result, we can see that $V = L \models \neg\Delta^1_2(L)$ & $\neg\Delta^1_2(B)$. If we add a real to L then the old Δ^1_2-well order of the constructible reals is no longer Δ^1_2, but will be a Σ^1_2-well order in the big model. This kind of instability leads to the following

Definition. Let P be a forcing notion. We say that V is Δ^1_n-P-Stable if for every pair $(\varphi(x), \psi(x))$ of Σ^1_n-formulas, with parameters in V, we have

$$V \models \forall x(\varphi(x) \leftrightarrow \neg\psi(x)) \text{ iff } V^P \models \forall x(\varphi(x) \leftrightarrow \neg\psi(x))$$

This definition says that V is Δ^1_n-P-Stable when we cannot change the Δ^1_n-sets.

EXAMPLE 1. *V is always Δ^1_1-stable.*

EXAMPLE 2. *Let B be the measure product of $\aleph_1$-many random reals. Then $W = V^B$ is always Δ^1_n-Random-Stable.*

Proof. Let R be the random real forcing, and let $(\varphi(x), \psi(x))$ be a pair of Σ^1_n-formulas with parameters in W. It is well known that $R * B \cong B * R \cong B$. Therefore we may assume without loss of generality (w.l.o.g.) that the parameters in $(\varphi(x), \psi(x))$ are in V. We know, by assumption, that

$$W \models \forall x(\varphi(x) \leftrightarrow \neg\psi(x))$$

iff

$$[\![\forall x(\varphi(x) \leftrightarrow \neg\psi(x))]\!]_B = \mathbf{1} \text{ (by homogeneity of B)}$$

iff

$$[\![\forall x(\varphi(x) \leftrightarrow \neg\psi(x))]\!]_{B*R} = \mathbf{1}$$

iff

$$W^R \models \forall x(\varphi(x) \leftrightarrow \neg\psi(x)),$$

finishing the proof. ∎

The same example is true when Random is replaced by Cohen and Measure by Category, respectively.

FACT.

(i) *V is Δ^1_2-Random-Stable iff $\Delta^1_2(L)$.*
(ii) *V is Δ^1_2-Cohen-Stable iff $\Delta^1_2(B)$.*

Proof. (i) Assume V is Δ^1_2-Random-Stable, and $\neg\Delta^1_2(L)$. By §0, there is $a \in \mathbb{R}$ such that $\mathrm{Ran}(L[a]) = \emptyset$.

We define the following order on $\mathbb{R}$:

$x \prec y$ iff the first Borel measure zero set $A_x \in L[a]$ such that $x \in A_x$ has constructible order less than the first Borel measure zero set $A_y \in L[a]$ such that $y \in A_y$.

It is not hard to see that " $\prec$ " is a Σ^1_2-relation (see [JS1]).

Let us now define

$$x \preceq y \text{ iff } x \prec y \text{ or } A_x = A_y.$$

Also " $\preceq$ " is a Σ^1_2-relation.

By assumption ($\mathrm{Ran}(L[a]) = \emptyset$), we have that

$$(\forall x \forall y)(x \prec y \leftrightarrow \neg(y \preceq x)).$$

By Δ^1_2-Random-Stability we have that in $V[r]$

$$r \prec r \leftrightarrow \neg(r \preceq r),$$

when r is a Random real over V. But $V[r] \models \neg(r \prec r)$ & $\neg(r \preceq r)$, a contradiction.

Let us assume $\Delta^1_2(L)$, and assume also that $(\varphi(x), \psi(x))$ are two Σ^1_2-formulas satisfying

$$\forall x(\varphi(x) \leftrightarrow \neg\psi(x)).$$

Let a be such that the parameters of $(\varphi(x), \psi(x))$ are in $L[a]$.

Assume that

$$V^R \models \neg\forall x(\varphi(x) \leftrightarrow \neg\psi(x)).$$

Then we have two cases:

(a) $V^R \models \varphi(\tau(r))$ & $\psi(\tau(r))$, for some $\tau \in V^R$. Then

$L[a][\tau][r] \models \varphi(\tau(r))$ & $\psi(\tau(r))$ (by absoluteness of the Σ^1_2-formulas).

Therefore there is $p \in R$ such that $L[a][\tau] \models p \Vdash_R \varphi(\tau)$ & $\psi(\tau)$. But $\mathrm{Ran}(L[a][\tau]) \neq \emptyset$, therefore for $s \in \mathrm{Ran}(L[a][\tau]) \cap p$, we have

$$L[a][\tau][s] \models \varphi(\tau(s)) \text{ \& } \psi(\tau(s)).$$

Hence

$$V \models \varphi(\tau(s)) \;\&\; \psi(\tau(s)),$$

a contradiction.

(b) $V^P \models \neg\psi(\tau(s)) \;\&\; \neg\psi(\tau(s))$, we get an analogous contradiction. This finishes the proof of (i).

(ii) is proved similarly using Cohen instead of Random reals. ∎

FACT.

(i) $V \models \Delta^1_2(L) \rightarrow V[r] \models \Delta^1_2(L)$ *for* $r \in$ *Ran*(V).
(ii) $V \models \Delta^1_2(B) \rightarrow V[\dot{\tau}] \models \Delta^1_2(B)$ *for* $\tau \in$ *Cohen*(V).

Proof. (i) Assume $\Delta^1_2(L)$ and let τ be an R-name for a real. Let r be Random over V. In V we have that there is a random real over $L[\tau]$. Let s be such a real. Then r is Random over $L[\tau][s]$. But two random reals commute. Therefore s is Random over $L[\tau][r]$. Therefore s is Random over $L[\tau(r)]$. This proves $\Delta^1_2(L)$ in $V[r]$.

(ii) is similar. ∎

In the model of example 2, we have that Δ^1_n-Random-Stable holds. But also in this model $\neg\Sigma^1_2(L)$. Thus if we are looking for a statement equivalent to $\Delta^1_3(L)$ we need to strengthen "Δ^1_3-Random-Stable." We don't have yet an exact equivalent statement for $\Delta^1_3(L)$ but the following is our best candidate.

THEOREM. *If* $V \models \Sigma^1_2(L)$ & Δ^1_3*-Random-Stable then* $V \models \Delta^1_3(L)$.

Proof. Let $(\varphi(x), \psi(x))$ be a pair of Σ^1_3-formulas satisfying

$$\forall x(\varphi(x) \leftrightarrow \neg\psi(x)).$$

Therefore, by Δ^1_3-Random-Stability, we have that

$$[\![\forall x(\varphi(x) \leftrightarrow \neg\psi(x))]\!]_R = \mathbf{1}.$$

Therefore, if $\underset{\sim}{r}$ is the canonical name for the random real, we have $\varphi(\underset{\sim}{r})$ or $\psi(\underset{\sim}{r})$ holds in the generic extension by random real forcing. Let us assume, w.l.o.g., that

$$p \Vdash_R \varphi(\underset{\sim}{r}),$$

for $p \in R$.

Let $\varphi_1(x, y)$ be a Π^1_2-formula such that

$$\varphi(x) = \exists y \varphi_1(x, y),$$

and let τ be an R-name such that

$$p \Vdash_R \varphi_1(\underset{\sim}{r}, \tau)$$

is true in V. But, by using absoluteness of Π^1_2-formulas, we have that

$$L[a][p][\tau] \models p \Vdash_R \varphi_1(\underset{\sim}{r}, \tau)$$

(when a encodes the parameters of φ_1). Let b encode a, p, τ. Thus

$$L[b] \models p \Vdash_R \varphi_1(\underset{\sim}{r}, \tau).$$

Now, by $\Sigma^1_2(L)$ we have $\mu(\mathrm{Ran}(L[b])) = 1$. If $x \in \mathrm{Ran}(L[b]) \cap p$, then $L[b][x] \models \varphi_1(x, \tau(x))$ and by absoluteness

$$\varphi_1(x, \tau(x)).$$

Therefore,

$$\exists y \varphi_1(x, y) = \varphi(x).$$

And this shows that

$$\mu(\{x : \varphi(x)\}) = \mu(\llbracket\varphi\rrbracket_R). \quad \blacksquare$$

COROLLARY.

(i) *If $V \models \Sigma^1_2(L)$ then $V[\aleph_1\text{-Random}] \models \Delta^1_3(L)$.*
(ii) *If $V \models \Sigma^1_2(B)$ then $V[\aleph_1\text{-Cohen}] \models \Delta^1_3(B)$.*

Proof. (i) By example 2 we have that $V[\aleph_1\text{-Random}] \models \Delta^1_3$-$R$-Stable. We must prove that this model satisfies $\Sigma^1_2(L)$. Let a be a real. We must prove that $\mu(\mathrm{Ran}(L[a])) = 1$ in $V[\aleph_1\text{-Random}]$. It is well known that there is a random real $r \in \mathrm{Ran}(V)$ and an R-name τ such that $a = \tau(r)$. It is enough to show that $\mu(\mathrm{Ran}(L[a])) = 1$ in $V[r]$. Each measure zero set A in $L[a]$ is covered by a Borel measure zero set. Each Borel measure zero set in $L[a]$ has an R-name $\underset{\sim}{B}$ in $L[\tau]$ and also $\underset{\sim}{B}$ can be seen as a Borel set in $\mathbb{R}^2$, $\mu(\underset{\sim}{B}) = 0$. By $\Sigma^1_2(L)$ we have that there is a Borel set B of measure zero such that $\underset{\sim}{B} \subseteq B$ for every Borel set $\underset{\sim}{B} \in L[\tau]$, $\mu(B) = 0$. If r is random real over V, then $\underset{\sim}{B}(r) \subseteq B(r)$ for every $\underset{\sim}{B} \in L[\tau]$, $\mu(\underset{\sim}{B}) = 0$ and $B(r) = \{x : \langle r, x\rangle \in B\}$. Now, $\mathbb{R} - B(r) \subseteq \mathrm{Ran}(L[a])$.

(ii) is similar. $\blacksquare$

The following are open problems.

(1) $V \models \Delta^1_3(L) \to V \models \Delta^1_3$-Random-Stable?
(2) $V \models \Delta^1_3(L) \to V[r] \models \Delta^1_3(L)$, r Random?

§2. Σ^1_4-Absoluteness.

In this section we are interested in the statement $\Sigma^1_3(L)$ ($\Sigma^1_3(B)$). Again we are looking for a forcing characterization of such statements; this means that we want to find a forcing statement which is equivalent to $\Sigma^1_3(L)$. Until now we only have sufficient conditions and we conjecture that they are also necessary.

In the study of measure and category on the projective sets, the first theorem about the Σ^1_3-sets was proved by S. Shelah. This results says that the measurability of the projective sets is intrinsically related to the existency of inaccessible cardinals. The theorem says

THEOREM (Shelah). $\Sigma^1_3(L) \to (\forall r \in \mathbb{R})(\omega_1^{L[r]} < \omega_1)$.

In other words, the theorem says that $\aleph_1$ is an inaccessible cardinal in $L[r]$ for every real r. Therefore if we want to understand the statement $\Sigma^1_3(L)$, we are forced to connect it with inner models with an inaccessible cardinal. Larger cardinal assumptions are artificial in this context.

Another interesting phenomenon is that at this stage of the development of the subject we are not able to see the difference, in ZFC, between "$\Sigma^1_3(L)$" and "$\Sigma^1_4(L)$." In other words, at present all the models for $\Sigma^1_3(L)$ are models for $\Sigma^1_4(L)$.

There is a program for building inner models for larger cardinals where $\Sigma^1_3(L)$ holds and $\Sigma^1_4(L)$ fails, but this answer, if it works, will only partially solve the question. We insist that in this context only inaccessible cardinals can be accepted. We can see this inner model development as evidence that eventually an answer in ZFC will be obtained.

It is also interesting to remark that there is a very deep asymmetry between $\Sigma^1_3(L)$ and $\Sigma^1_3(B)$, namely

THEOREM (Shelah). *$Con(ZF) \to Con(ZFC + \forall n \Sigma^1_n(B))$.*

Shelah started from L and in a ω_1-stage iteration with finite support, he got a model for $\forall n \Sigma^1_n(B)$. This is one of the most sophisticated forcing constructions and it is not really clear if further development can be done using these techniques. Anyway, if we assume $\Sigma^1_2(L)$, then there is no asymmetry between $\Sigma^1_3(L)$ and $\Sigma^1_3(B)$; both imply inner models for inaccessible cardinals. The most interesting open question in the study of the asymmetry between measure and category is the following:

Does $\Sigma^1_3(L) \to \Sigma^1_3(B)$?

Using Jensen's work on the core model it is possible to give a positive answer if we assume the existence of #'s and the non-existence of a proper class of measurable cardinals. But these are very restrictive assumptions. We are sure that this question has an answer in ZFC.

We will start by giving our main definition, and we will study it on the Σ^1_2-sets. After this, we will see what the connection is to the Σ^1_3-sets.

Definition. Let P be a forcing notion. We say that V is Σ^1_n-P-Absolute if for every Σ^1_n-sentence φ with parameters in V we have

$$V \models \varphi \text{ iff } V^P \models \varphi.$$

EXAMPLE. *V is always Σ^1_2-P-Absolute.*

Proof. By Shoenfield's Absoluteness Lemma. ∎

FACT. *The following are equivalent:*

(i) $\Sigma^1_2(L)$ $(\Sigma^1_2(B))$
(ii) *Σ^1_3-Amoeba-Absolute (Σ^1_3-Amoeba-Meager-Absolute)*

Proof. (ii) $\to$ (i) Let a be a real. Then

$$V \models \mu(\mathrm{Ran}(L[a])) = 1$$

is a Π^1_3-sentence in V.

Let P be Amoeba forcing. Then $V^P \models \mu(\mathrm{Ran}(L[a])) = 1$. We have by Σ^1_3-Absoluteness that $V \models \mu(\mathrm{Ran}(L[a])) = 1$.

(i) $\rightarrow$ (ii) Let $\varphi = \exists x \psi(x)$ be a Σ^1_3-sentence (so $\psi(x)$ is a Π^1_2-formula). Assume

$$V \models \exists x \psi(x).$$

Let $a \in V \cap \mathbb{R}$ such that

$$V \models \psi(a).$$

By Shoenfield's Absoluteness Lemma, $V^P \models \exists x \psi(x)$.

Assume $V^P \models \exists x \varphi(x)$. Let τ be a P-name for a real such that

$$V^P \models \psi(\tau).$$

Let a be a real encoding the parameters used in ψ. By Shoenfield's Absoluteness Lemma, we have

$$L[a][\tau]^P \models \psi(\tau).$$

Therefore we really have

$$L[a][\tau] \models \Vdash_P \text{“}\psi[\tau]\text{”}.$$

Now, in V we have $\mu(\mathrm{Ran}(L[a][\tau])) = 1$. Therefore there is $G \subseteq P$, an $L[a][\tau]$-generic filter, G in V. Therefore

$$L[a][\tau][G] \models \psi(\tau[G]).$$

By Shoenfield's Absoluteness Lemma

$$V \models \psi(\tau[G]).$$

Thus, $V \models \varphi$, finishing the proof. ∎

FACT. *The following are equivalent:*

(i) $\Delta^1_2(L)$ $(\Delta^1_2(B))$

(ii) Σ^1_3-*Random-Absolute* (Σ^1_3-*Cohen-Absolute*)

Proof. Similar. ∎

There is a very useful fact proved by Martin–Solovay about the Σ^1_3-formulas, namely

THEOREM (Martin–Solovay). *If $\varphi(x)$ is a Σ^1_3-formula and κ is a measurable cardinal and $|P| < \kappa$ and $\underset{\sim}{r}$ is a P-name for a real, then*

$$V[\underset{\sim}{r}] \models \varphi[\underset{\sim}{r}] \text{ iff } V^P \models \varphi(\underset{\sim}{r}).$$

We can improve this theorem in our context (remember that only inaccessible cardinals are accepted) to get

THEOREM. *Assume that $(\forall r \in \mathbb{R})(\omega_1^{L[r]} < \omega_1)$ and P is a Souslin forcing and τ is a P-name for a real, and $\varphi(x)$ is a Σ^1_3-formula. Then*

$$V[\tau] \models \varphi(\tau) \text{ iff } V^P \models \varphi(\tau).$$

Proof. Clearly $V[\tau] \models \varphi(\tau)$ implies $V^P \models \varphi(\tau)$, by using Shoenfield's Absoluteness Lemma. Let us assume $V^P \models \varphi(\tau)$. Let $\varphi(\tau) = \exists x \psi(x, \tau)$. And let $\theta \in V^P$ such that

$$V^P \models \psi(\theta, \tau).$$

Let $a \in \mathbb{R}$ encode the parameters of φ, and the parameters of the definition of P. Then by absoluteness of the maximal antichain of P we have θ, τ are P-names in $L[a][\tau][\theta]$, and also

$$L[a][\tau][\theta]^P \models \psi(\theta, \tau).$$

Now let $G \subseteq P$ be V-generic. Then we have that

$$L[a][\tau[G]][\theta[G]] \models \psi[\theta(G), \tau(G)].$$

Now, $L[a][\tau[G]][\theta[G]]$ is a forcing extension of $L[a][\tau[G]]$, and call this forcing Q. Then

$$L[a][\tau[G]] \models \text{``}(\exists q \in Q)(q \Vdash_Q \psi(\theta, \tau[G]))\text{''}.$$

Now we use the following

FACT. $\omega_1^{L[a][\tau[G]]} < \omega_1$ *in* V^P.

Proof. By [JS2].

But $L[a][\tau[G]] \subseteq V[\tau[G]]$. Therefore in $V[\tau[G]]$ we have that

$$\omega_1^{L[a][\tau[G]]} < \omega_1.$$

Thus $2^{|Q|} \cap L[a][\tau[G]]$ is countable in $V[\tau[G]]$, therefore we can get $H \subseteq Q$, in $V[\tau[G]]$, a generic filter over $L[a][\tau[G]]$ such that

$$L[a][\tau[G]][H] \models \psi(\theta(H), \tau[G]).$$

Now by using Shoenfield's Absoluteness Lemma we have

$$V[\tau[G]] \models \exists x \psi(\tau[G]),$$

finishing the proof of the theorem. ∎

Next we will study the connection between Σ^1_4-Amoeba-Absoluteness and Σ^1_3-Measurability.

Let us start with the following

FACT.

(i) Σ^1_4-*Random-Absoluteness* + $\Sigma^1_2(L) \rightarrow \Delta^1_3(L)$.
(ii) Σ^1_4-*Cohen-Absoluteness* + $\Sigma^1_2(B) \rightarrow \Delta^1_3(B)$.

Proof. (i) Let $(\varphi(x), \varphi(x))$ be a pair of Σ^1_3-formulas such that

$$\forall x(\varphi(x) \leftrightarrow \neg\psi(x)).$$

But this is a Π^1_4-statement, thus is true after forcing with Random real forcing.

(ii) is similar. ∎

FACT. Σ^1_4-*Random-Absoluteness* + $\Sigma^1_2(L)$ *does not imply* $\Sigma^1_3(L)$.

Proof. Let $M \models MA + \omega_1^L = \omega_1$. By adding $\aleph_1$-many random reals to M we get a model for $\Sigma^1_2(L) + \Sigma^1_4$-Random-Absoluteness. In this model $\omega_1^L = \omega_1$, therefore $\neg\Sigma^1_3(L)$. ∎

THEOREM.

(i) Σ^1_4-*Amoeba-Absoluteness* + $(\forall r \in \mathbb{R})(\omega_1^{L[r]} < \omega_1) \to \Sigma^1_3(L)$.

(ii) Σ^1_4-*Amoeba-Meager-Absoluteness* + $(\forall r \in \mathbb{R})(\omega_1^{L[r]} < \omega_1)$ $\to \Sigma^1_3(B)$.

Proof. Let $\varphi(x)$ be a Σ^1_3-formula. We want to show that $A = \{x : \varphi(x)\}$ is a measurable set. This is a Σ^1_4-statement. Therefore it is enough to show that A is measurable in V^P, when P is Amoeba forcing. In V^P we have that $\mu(\mathrm{Ran}(V)) = 1$. If $\mathrm{Ran}(V) \cap A = \emptyset$, then

$$V^P \models \mu(A) = 0,$$

and we finish. Therefore we may assume $A \cap \mathrm{Ran}(V) \neq \emptyset$. Let $r \in A \cap \mathrm{Ran}(V)$. Then

$$V^P \models \varphi(r).$$

By the previous theorem

$$V[r] \models \varphi(r).$$

Now in V there is $p \in R$ such that

$$V \models p \Vdash_B \varphi(\underset{\sim}{r}),$$

where $\underset{\sim}{r}$ is the canonical R-name for the random real. Let $r_1 \in V^P \cap \mathrm{Ran}(V) \cap p$. Then we have

$$V[r_1] \models \varphi(r_1),$$

and thus $V^P \models \varphi(r_1)$, proving that $\mu(A) = \mu([\![\varphi(\underset{\sim}{r})]\!]_R)$.

(ii) is similar. ∎

Let us introduce a stronger principle, namely

Definition. We say that V is Σ^1_3-P-Correct if for every P-name τ for a real and $\varphi(x)$ a Σ^1_3-formula, we have $V[\tau] \models \varphi(\tau)$ iff $V^P \models \varphi(\tau)$.

COROLLARY. *If P is Souslin and $(\forall r \in \mathbb{R})(\omega_1^{L[r]} < \omega_1)$, the V is Σ^1_3-P-correct.*

FACT. *If V is Σ^1_3-Amoeba-correct and Σ^1_4-Amoeba-Absolute, then $\Sigma^1_3(L)$.*

Proof. The proof is similar to that of the previous theorem. ∎

CONJECTURE.

(i) Σ^1_4-*Amoeba-Absolute* $\to$ Σ^1_3-*Amoeba-Correctness*

(ii) Σ^1_{n+1}-*Amoeba-Meager-Absoluteness* + Σ^1_n-*Amoeba-Meager-Correctness* $\to \Sigma^1_n(B)$.

FACT.

(i) Σ^1_{n+1}-*Amoeba-Absolute* + Σ^1_n-*Amoeba-Correctness* $\to$ $\Sigma^1_n(L) + \Sigma^1_n(B)$.

(ii) Σ^1_{n+1}-*Amoeba-Meager-Absoluteness* + Σ^1_n-*Amoeba-Meager-Correctness* $\to \Sigma^1_n(B)$.

Proof. We will prove only $\Sigma^1_n(B)$ of (i). Let $\varphi(x)$ be a Σ^1_n-formula. Forcing with P-Amoeba, we have that Cohen(V) is a comeager set in V^P. We leave the rest of the details to the reader. ∎

We suggest that the reader check that Solovay's model satisfies Σ^1_n-Amoeba-Correctness and Σ^1_{n+1}-Amoeba-Absoluteness.

The last result connects Δ^1_n-Stability with Σ^1_{n+1}-Absoluteness. Really they are equivalent.

FACT (Bagaria). *V is Δ^1_n-P-Stable iff V is Σ^1_{n+1}-P-Absolute.*

Proof. One direction is trivial. We will show only the nontrivial part. Let $\exists x\varphi(x)$ be a Σ^1_{n+1}-formula. If

$$V \models \exists x\varphi(x) \text{ then } V \models \varphi(r) \text{ for some } r \in \mathbb{R},$$

by the induction hypothesis $V^P \models \varphi(r)$ then $V^P \models \exists x\varphi(x)$. Assume now $V^P \models \exists x\varphi(x)$, and $V \models \neg\exists x\varphi(x)$. Then $V \models \forall x(\neg\varphi(x) \leftrightarrow x = x)$ by Δ^1_n-Stability $V^P \models \forall x(\neg\varphi(x) \rightarrow x = x)$. Therefore $V^P \models \neg\exists x\varphi(x)$, a contradiction. ∎

REFERENCES

[BJ] J. BAGARIA and H. JUDAH. *Amoeba forcing, Suslin absoluteness and additivity of measure.* ***Set theory of the continuum***, edited by H. Judah, W. Just, and H. Woodin, MSRI Publications, vol. 26, Springer-Verlag, 1992, pp. 155–173.

[JS1] H. JUDAH and S. SHELAH. *Δ^1_2-sets of reals.* ***Annals of pure and applied logic***, vol. 42 (1989), pp. 207–233.

[JS2] H. JUDAH and S. SHELAH. *Martin's axioms, measurability and equiconsistency results.* ***The journal of symbolic logic***, vol. 54 (1989), pp. 78–94.

[JS3] H. JUDAH and S. SHELAH. *Souslin forcing.* ***The journal of symbolic logic***, vol. 53, (1988), pp. 1188-1207.

[S] S. SHELAH. *Can you take Solovay's inaccessible away?* ***Israel journal of mathematics***, vol. 48 (1984), pp. 1–47.

Department of Mathematics
Bar-Ilan University

U.C. Chile

A DIVISION ALGORITHM FOR THE FREE LEFT DISTRIBUTIVE ALGEBRA

Richard Laver[1]

In this paper we extend the normal form theorem, for the free algebra $\mathcal{A}$ on one generator x satisfying the left distributive law $a(bc) = (ab)(ac)$, which was proved in [5]. As part of the proof that an algebra of elementary embeddings from set theory is isomorphic to $\mathcal{A}$, facts about $\mathcal{A}$ itself were established. Theorem 1 summarizes some known facts about $\mathcal{A}$, including P. Dehornoy's independent work on the subject. After that the main theorem, about putting members of $\mathcal{A}$ into "division form," will be proved with the help of versions of lemmas of [5] and one of the normal forms of [5].

Let $\cdot$ denote the operation of $\mathcal{A}$. These forms take place not in $\mathcal{A}$ but in a larger algebra $\mathcal{P}$ which involves additionally a composition operation $\circ$. Let Σ be the set of laws $\{a \circ (b \circ c) = (a \circ b) \circ c,\ (a \circ b)c = a(bc),\ a(b \circ c) = ab \circ ac,\ a \circ b = ab \circ a\}$. $\mathcal{P}$ is the free algebra on the generator x satisfying Σ. Σ implies the left distributive law, and Σ is a conservative extension of the left distributive law (if two terms in the language of $\mathcal{A}$ can be proved equal using Σ, then they can be proved equal using just the left distributive law). So we may identify $\mathcal{A}$ as a subalgebra of $\mathcal{P}$ restricted to $\cdot$. If $p_0, p_1, \ldots, p_n \in \mathcal{P}$, write $p_0p_1 \cdots p_n$ (respectively, $p_0p_1 \cdots p_{n-1} \circ p_n$) for $(((p_0p_1)p_2) \cdots p_{n-1})p_n$ (respectively, $(((p_0p_1)p_2) \cdots p_{n-1}) \circ p_n$). Write $w = p_0p_1 \cdots p_{n-1} * p_n$ to mean that either $w = p_0p_1 \cdots p_n$ or $w = p_0p_1 \cdots p_{n-1} \circ p_n$. Make these conventions also for other algebras on operations $\cdot$ and $\circ$.

For $p \in \mathcal{P}$ let $p^1 = p$, $p^{n+1} = p \circ p^n$; let $p^{(0)} = p$, $p^{(n+1)} = pp^{(n)}$. Then $p^{(n+1)} = p^{(i)}p^{(n)}$ for all $i \leq n$, by induction using the left distributive law.

For $p, q \in \mathcal{P}$ let $p < q$ if q can be written as a term of length greater than one in the operations $\cdot$ and $\circ$, involving members of $\mathcal{P}$ at least one of which is p. Write $p <_L q$ if p occurs on the left of such a product: $q = pa_0a_1 \cdots a_{n-1} * a_n$ for some $n \geq 0$. Then $<_L$ and $<$ are transitive. If $a, b \in \mathcal{A}$ and $a <_L b$ in the sense of $\mathcal{P}$, then $a <_L b$ in the sense of $\mathcal{A}$; and similarly for $<$.

In [5] it was shown, via the existence of normal forms for the members of $\mathcal{P}$, that $<_L$ linearly orders $\mathcal{P}$ and $\mathcal{A}$. The proof of part of that theorem, that $<_L$ is irreflexive, used a large cardinal axiom (the existence, for each n, of an n-huge cardinal). Dehornoy ([1], [2]) by a different method independently proved in ZFC that for all $a, b \in \mathcal{A}$ at least one of $a <_L b$, $a = b$, $b <_L a$ holds. Recently ([3]) he has found a proof of the irreflexivity of $<_L$ in ZFC. This theorem has the consequence that facts about $\mathcal{P}$ (Theorem 1 below (parts (v)–(viii)), and the normal and division forms in [5] and this paper) which have previously been

[1]Supported by NSF Grant DMS 9102703.

known from a large cardinal assumption (that is, from irreflexivity), are provable in ZFC. A shorter proof of Dehornoy's theorem was found by Larue ([8]).

For u, v terms in the language of $\cdot$ in the variable x, let $u \rightarrow v$ ([1]) mean that u can be transformed into v by a finite number of substitutions, each consisting of replacing a term of the form $a(bc)$ by $(ab)(ac)$.

For λ a limit ordinal, let $\mathcal{E}_\lambda$ be the set of elementary embeddings $j : (V_\lambda, \epsilon) \rightarrow (V_\lambda, \epsilon)$, j not the identity. For $j, k \in \mathcal{E}_\lambda$, let $jk = \bigcup_{\alpha<\lambda} j(k \cap V_\alpha)$ and let $j \circ k$ be the composition of j and k. Then the existence of a λ such that $\mathcal{E}_\lambda \neq \emptyset$ is a large cardinal axiom. If $j, k \in \mathcal{E}_\lambda$, then jk, $j \circ k \in \mathcal{E}_\lambda$, and $(\mathcal{E}_\lambda, \cdot, \circ)$ satisfies Σ. For $j \in \mathcal{E}_\lambda$ let $\mathcal{A}_j$ be the closure of $\{j\}$ under $\cdot$ and let $\mathcal{P}_j$ be the closure of $\{j\}$ under $\cdot$ and $\circ$.

Some facts relating $\mathcal{P}$ to $\mathcal{A}$, such as the conservativeness of Σ over the left distributive law, may be found in [5].

THEOREM 1. *(i) If $r <_L s$, then $pr <_L p \circ r <_L ps$.*
(ii) $x \leq_L p$ for all $p \in \mathcal{P}$, $<_L$ is not well founded.
(iii) For all $p, q \in \mathcal{P}$ there is an n with $p^{(n)} > q$.
(iv) The rewriting rules for $\mathcal{A}$ are confluent, i.e., if u, v are terms in the language of $\cdot$ in the variable x, and $u \equiv v$ via the left distributive law, then for some w, $u \rightarrow w$ and $v \rightarrow w$.
(v) $<_L$ is a linear ordering of $\mathcal{A}, \mathcal{P}$.
(vi) For $p, q, r \in \mathcal{P}$, $pq = pr \Leftrightarrow q = r$, $pq <_L pr \Leftrightarrow q <_L r$.
(vii) The word problems for $\mathcal{A}$ and $\mathcal{P}$ are decidable.
(viii) $<_L = <$ on $\mathcal{A}, \mathcal{P}$.
(ix) For no $k_0, k_1, \ldots, k_n \in \mathcal{E}_\lambda$ $(n > 0)$ is $k_0 = k_0 k_1 \cdots k_{n-1} * k_n$.
(x) For all $j \in \mathcal{E}_\lambda$, $\mathcal{A}_j \cong \mathcal{A}$, $\mathcal{P}_j \cong \mathcal{P}$.

Remarks. (i)–(iii) are quickly proved; for (iii), it may be seen that $p^{(n)} \geq x^{(n)}$ and for sufficiently large n, $x^{(n)} \geq q$. (iv) is Dehornoy's theorem in [2]. The linear orderings of $\mathcal{P}$ and $\mathcal{A}$ both have order type $\omega \cdot (1 + \eta)$. (v) immediately implies (vi) and (vii). In [5], (viii) is derived from the normal form theorem; McKenzie derived (viii) from (v). (ix) and (x) are proved in [5], (ix) plus (v) yields (x).

Results connected with critical points of members of $\mathcal{A}_j$ appear in [4], [6], and [7].

For $a, b \in \mathcal{P}$, let the iterates $I_n(a, b)$ of $\langle a, b\rangle$ $(n \geq 1)$ be defined by $I_1(a, b) = a$, $I_2(a, b) = ab$, $I_{n+2}(a, b) = I_{n+1}(a, b) I_n(a, b)$.

Call a term $b_0 b_1 \cdots b_{n-1} * b_n$, with each $b_i \in \mathcal{P}$, prenormal (with respect to a given ordering $\prec$) if $b_2 \preceq b_0$, $b_3 \preceq b_0 b_1$, $b_4 \preceq b_0 b_1 b_2, \ldots, b_n \preceq b_0 b_1 \cdots b_{n-2}$, and in the case $* = \circ$ and $n \geq 2$, $b_n \prec b_0 b_1 \cdots b_{n-1}$.

The main theorem is that for each $p, q \in \mathcal{P}$, q can be expressed in "p-division form," the natural fact suggested by the normal forms of [5]. For $p \in \mathcal{P}$ the set of p-division form representations of members of $\mathcal{P}$, and its lexicographic linear ordering, are defined as follows.

LEMMA 2. *For each $p \in \mathcal{P}$ there is a unique set p-DF of terms in the language of $\cdot$ and $\circ$, in the alphabet $\{q \in \mathcal{P} : q \leq p\}$, and a linear ordering $<_{\text{Lex}}$ of p-DF, such that:*

(i) *For each $q \leq_L p$, q (as a term of length one) is in p-DF, and for $q, r \leq_L p$, $q <_{\text{Lex}} r$ if and only if $q <_L r$.*

(ii) *$w \in p$-DF iff either $w \leq_L p$, or $w = pa_1a_1 \cdots a_{n-1} * a_n$, where each $a_i \in p$-DF, is prenormal with respect to $<_{\text{Lex}}$.*

(iii) *For $w \in p$-DF define the associated sequence of w to be $\langle w \rangle$ if $w \leq_L p$, to be $\langle p, a_0, a_1, \ldots, a_n \rangle$ if $w = pa_0a_1 \cdots a_n$, and, if $w = pa_0a_1 \cdots a_{n-1} \circ a_n$, to be (letting $u = pa_0a_1 \cdots a_{n-1}$)*

$$\langle p, a_0, a_1, \ldots, a_{n-1}, a_n, u, ua_n, ua_nu, ua_nu(ua_n), \ldots \rangle,$$

that is, the sequence beyond a_n is $\langle I_m(u, a_n) : m \geq 1 \rangle$. Then for $w, v \in p$-DF with associated sequences $\langle w_i : i < \alpha \rangle$, $\langle v_i : i < \beta \rangle$ $(\alpha, \beta \leq \omega)$, $w <_{\text{Lex}} v$ iff either $\langle w_i : i < \alpha \rangle$ is a proper initial segment of $\langle v_i : i < \beta \rangle$ or there is a least i with $w_i \neq v_i$, and $w_i <_{\text{Lex}} v_i$.

Proof. As in [5, Lemma 8], one builds up p-DF and $<_{\text{Lex}}$ by induction; a term $pa_0a_1 \cdots a_{n-1} \circ a_n$ is put in the set p-DF (and its lexicographic comparison with terms previously put in is established) only after all the iterates $I_m(pa_0 \cdots a_{n-1}, a_n)$, $m \geq 1$ have been put in the set.

Remarks. The members of p-DF are terms, and p-DF is closed under subterms (for $w \leq_L p$, w is the only subterm of w, and for $w = pa_0a_1 \cdots a_{n-1} * a_n$, the subterms of w are w and the subterms of $pa_0 \cdots a_{n-1}, a_n$). We will associate these terms without comment with the members of $\mathcal{P}$ they stand for, when no confusion should arise. If $w \in p$-DF and u is a proper subterm of w, then $u <_{\text{Lex}} w$. Terms of the form $(u \circ v)w$ or $(u \circ v) \circ w$ are never in p-DF. When using phrases such as "$uv \in p$-DF," "$u \circ v \in p$-DF," it is assumed that $u = pa_0 \cdots a_{n-1}$, $v = a_n$ are as in the definition of p-DF—isolated exceptions where uv or $u \circ v$ are $\leq_L p$ and are to be considered as singleton terms, will be noted.

If $u \circ v \in p$-DF, then $u \circ v$ is the $<_{\text{Lex}}$-supremum of $\{I_n(u, v) : n \geq 1\}$.

LEMMA 3. *The transitivization of the relation $\{\langle u, v \rangle : u, v \in p$-DF and either u is a proper subterm of v, or $v = a \circ b$ and u is an $I_k(a, b)\}$ is a well-founded partial ordering $\prec^p$ of p-DF.*

Proof. Otherwise there would be a sequence $\langle u_n : n < \omega \rangle$ with, for each n, either u_{n+1} a proper subterm of u_n, or u_{n+1} an iterate of $\langle a, b \rangle$ with $u_n = a \circ b$, such that no proper subterm of u_0 begins such a sequence. Then $u_0 = r \circ s$, u_1 is an iterate of $\langle r, s \rangle$, and by the nature of such iterates, some u_n must be a subterm of r or of s, a contradiction.

LEMMA 4.

(i) *If $w, a, b_0, b_1, \ldots, b_n \in \mathcal{P}$, $wb_0b_1 \cdots b_{n-1} * b_n$ is prenormal with respect to $<_L$, and $b_0 <_L a$, then $wb_0b_1 \cdots b_{n-1} * b_n <_L wa$.*

(ii) *For $p \in \mathcal{P}$, $u, v \in p$-DF, $u <_{\text{Lex}} v$ iff $u <_L v$.*

Proof. (i) By induction on i we show $wa >_L wb_0b_1 \cdots b_{i-1} \circ b_i$. For $i = 0$, it is Theorem 1(i). For $i = k+1$, $wa = (wb_0 \cdots b_{k-1} \circ b_k)u_0 \cdots u_{m-1} * u_m \geq_L wb_0 \cdots b_{k-1}(b_k u_0) = wb_0 \cdots b_{k-1}b_k(wb_0 \cdots b_{k-1}u_0) \geq_L wb_0 \cdots b_{k-1}b_k\,(b_{k+1}r)$ for some r (since $b_{k+1} \leq_L wb_0 \cdots b_{k-1}$) $>_L wb_0 \cdots b_k \circ b_{k+1}$.

(ii) It suffices to show $u <_{\text{Lex}} v \Rightarrow u <_L v$ (the other direction following from that, the linearity of $<_{\text{Lex}}$, and the irreflexivity of $<_L$). By induction on ordinals α, suppose it has been proved for all pairs $\langle u', v' \rangle$, $u', v' \in p$-DF, such that u' and v' have rank less than α with respect to $\prec^p$. If either of u, v is $\leq_L p$, or if the associated sequence of u is a proper initial segment of the associated sequence of v, the result is clear. So, passing to a truncation $p, a_0, a_1, \ldots, a_n$ of u's associated sequence if necessary, we have $u \geq_{\text{Lex}} pa_0a_1 \cdots a_n$, $v = pa_0a_1 \cdots a_{n-1}v_nv_{n+1} \cdots v_{m-1} * v_m$, some $m \geq n$, with $v_n <_{\text{Lex}} a_n$ (the reason why v cannot be $pa_0 \cdots a_{i-1} \circ a_i$ for some $i < n$ is that $a_n \leq_{\text{Lex}} v_n$ would then hold). Thus $u \geq_L pa_0a_1 \cdots a_n$ (clear), $v_n <_L a_n$ (by the induction hypothesis), and for each i, $v_{i+1} \leq_L pa_0a_1 \cdots a_{n-1}v_n \cdots v_{i-1}$ (by the induction hypothesis). Then apply part (i) of this lemma.

Thus, for $p, q \in \mathcal{P}$, to determine which of $q <_L p$, $q = p$, $p <_L q$ holds, lexicographically compare $|q|^x$ and $|p|^x$.

Write $<_L$ for $<_{\text{Lex}}$ below. "Prenormal," below, will be with respect to $<_L$. For $q, p \in \mathcal{P}$, let $|q|^p$ be the p-DF representation of q, if it exists.

Recall that the main theorem is that $|q|^p$ exists for all $q, p \in \mathcal{P}$. From Lemma 4, this may be stated as a type of division algorithm: if $q, p \in \mathcal{P}$ and $p <_L q$, then there is a $<_L$-greatest $a_0 \in \mathcal{P}$ with $pa_0 \leq_L q$, and if $pa_0 <_L q$, then there is a $<_L$-greatest $a_1 \in \mathcal{P}$ with $pa_0a_1 \leq_L q$, etc., and for some n, $pa_0a_1 \cdots a_n = q$ or $pa_0a_1 \cdots a_{n-1} \circ a_n = q$. And, if this process is repeated for each a_i, getting either $a_i \leq_L p$ or $a_i = pa_i^0a_i^1 \cdots a_i^{m-1} * a_i^m$, and then for each a_i^k, etc., then the resulting tree is finite. The normal form theorems in [5] correspond to similar algorithms—they were proved there just for $p \in \mathcal{A}$, and the present form has their generalizations to all $p \in \mathcal{P}$ as a corollary.

In certain cases on $u, v \in p$-DF (when "$u \sqsupset^p v$"), the existence of $|uv|^p$ and $|u \circ v|^p$ can be proved directly. We define $u \sqsupset^p v$ by induction: suppose $u' \sqsupset^p w$ has been defined for all proper subterms u' of u and all $w \in p$-DF.

(i) If $u <_L p$, then $u \sqsupset^p v$ iff $u >_L v$ and $u \circ v \leq_L p$.

(ii) $p \sqsupset^p v$ for all v.

(iii) $pa \sqsupset^p v$ iff $v \leq_L p$ or $v = pa_0a_1 \cdots a_{n-1} * a_n$ with $a \sqsupset^p a_0$; $p \circ a \sqsupset^p v$ for all v.

(iv) For $n \geq 1$, $pa_0a_1 \cdots a_n \sqsupset^p v$ iff either $v \leq_L pa_0a_1 \cdots a_{n-1}$ or

$$v = pa_0a_1 \cdots a_{n-1}\; v_nv_{n+1} \cdots v_{i-1} * v_i$$

with $a_n \sqsupset^p v_n$ and $a_n \circ v_n \leq_L pa_0a_1 \cdots a_{n-2}$.

(v) For $n \geq 1$, $pa_0a_1 \cdots a_{n-1} \circ a_n \sqsupset^p v$ iff $a_n \sqsupset^p v$ and $a_n \circ v \leq_L pa_0a_1 \cdots a_{n-2}$.

LEMMA 5. *If $u \sqsupset^p v$, $w \in p$-DF, and $v \geq_L w$, then $u \sqsupset^p w$.*

Proof. By induction on u in p-DF.

LEMMA 6. *If $u \sqsupset^p v$, then $|uv|^p$ and $|u \circ v|^p$ exist, and $|uv|^p \sqsupset^p u$.*

Proof. Assume the lemma has been proved for all $\langle u', w\rangle$, $w \in p$-DF and u' a proper subterm of u, and for all $\langle u, v'\rangle$, v' a proper subterm of v. Suppose $u \sqsupset^p v$.

(i) $u <_L p$. Then $uv <_L u \circ v \leq_L p$, so uv and $u \circ v$, as terms of length one, are in p-DF, and $uv \circ u = u \circ v$, so similarly $uv \sqsupset^p u$.

(ii) $u = p$. Then $|pv|^p = pv$, $|p \circ v|^p = p \circ v$, and $pv \sqsupset^p p$.

(iii) $u = pa$. Then if $v \leq_L p$ it is clear, so assume $v = pb_0 b_1 \cdots b_{n-1} * b_n$, where $a \sqsupset^p b_0$. The cases are:

(a) $v = pb$. Then $|uv|^p = p|ab|^p$, $|u \circ v|^p = p|a \circ b|^p$, when $|ab|^p$, $|a \circ b|^p$ exist by induction. And since by induction $|ab|^p \sqsupset a$, we have $|uv|^p \sqsupset u$.

(b) $v = p \circ b$. Then $|uv|^p = |pa(p \circ b)|^p = p|a \circ b|^p p \circ p|ab|^p$ by the induction hypothesis and Theorem 1(i). Similarly $|u \circ v|^p = |pa \circ (p \circ b)|^p = p \circ |a \circ b|^p$. To see $|uv|^p \sqsupset^p u$, we have $p|ab|^p \sqsupset_p pa$, as $|ab|^p \sqsupset^p a$ holds by the induction hypothesis, and $p(ab) \circ pa = p(ab \circ a) = p(a \circ b)$.

(c) $v = pb_0 b_1 \cdots b_{n-1} * b_n$ for $n \geq 1$. Then

$$|uv|^p = p|a \circ b_0|^p b_1 |pab_2|^p \cdots |pab_{n-1}|^p * |pab_n|^p$$

by the induction hypothesis and Theorem 1(i) and Lemma 4(ii). And in the case $* = \cdot$, $|u \circ v|^p = |uv \circ u|^p = p|a \circ b_0|^p b_1 |pab_2|^p \cdots |pab_n|^p \circ pa$. In the case $* = \circ$, $|u \circ v|^p = |uv \circ u|^p = p|a \circ b_0|^p b_1 |pab_2|^p \cdots |pab_{n-1}|^p \circ |pab_n \circ pa|^p$, namely, $|pab_n \circ pa|^p = |pa \circ b_n|^p$ exists by induction and is $<_L p(a \circ b_0) b_1 (pab_2) \cdots (pab_{n-2})$ by $b_n <_L pb_0 \cdots b_{n-2}$ and Theorem 1(i). To see $|uv|^p \sqsupset^p u$, it is immediate if $* = \cdot$, and if $* = \circ$, $pab_n \sqsupset^p pa$ by induction, and $pab_n \circ pa = pa \circ b_n <_L pa(pb_0 \cdots b_{n-2}) = p(a \circ b_0) b_1 (pab_2) \cdots (pab_{n-2})$, as desired.

(iv) $u = pa_0 a_1 \cdots a_n$, $n \geq 1$. Then the case where the induction hypothesis is used is where $v = pa_0 a_1 \cdots a_{n-1} b_n \cdots b_{m-1} * b_m$, where $a_n \sqsupset^p b_n$ and $a_n \circ b_n \leq_L pa_0 a_1 \cdots a_{n-2}$. The cases and computations are similar to (iii).

(v) $u = pa_0 a_1 \cdots a_{n-1} \circ a_n$, $n \geq 1$. Then $a_n \sqsupset^p v$, so $|a_n v|^p$, $|a_n \circ v|^p$ exist, and $a_n v <_L a_n \circ v \leq_L pa_0 \cdots a_{n-2}$. Thus $|uv|^p = pa_0 a_1 \cdots a_{n-1} |a_n v|^p$, $|u \circ v|^p = |pa_0 a_1 \cdots a_{n-1} \circ |a_n \circ v|^p|^p$ which is $pa_0 \cdots a_{i-1} \circ |a_i \circ v|^p$, where $i \leq n$ is greatest such that $i = 1$ or $a_i \circ v <_L pa_0 \cdots a_{i-1}$. And for $|uv|^p \sqsupset^p u$, we have $|a_n v|^p \sqsupset^p a_n$, and $a_n v \circ a_n = a_n \circ v \leq_L pa_0 \cdots a_{n-2}$, as desired.

LEMMA 7. *Suppose $p, q \in \mathcal{P}$, $w \in q$-DF. Then*

(i) $|pw|^{pq}$ exists, and $|pw|^{pq} \sqsupset^{pq} p$.

(ii) If $|pw|^{p \circ q}$ exists, then $|pw|^{p \circ q} \sqsupset^{p \circ q} p$.

Proof. We check part (ii), part (i) being similar. Assume the lemma is true for all proper components w' of w. If $w \leq q$, then $pw < p \circ w \leq p \circ q$ and, by $pw \circ p = p \circ w$, we have $pw \sqsupset^{p \circ q} p$. So assume the most general case on w, $w = qa_0 a_1 \cdots a_{n-1} \circ a_n$. Then $pw = (p \circ q) a_0 (pa_1) \cdots (pa_{n-1}) \circ (pa_n)$ is prenormal, so if $|pw|^{p \circ q}$ exists, then by Lemma 4(i) and (ii) $|pw|^{p \circ q} = (p \circ q)|a_0|^{p \circ q} |pa_1|^{p \circ q} \cdots |pa_{n-1}|^{p \circ q} \circ |pa_n|^{p \circ q}$. Then $|pa_n|^{p \circ q} \sqsupset^{p \circ q} p$ by the induction assumption, and $pa_n \circ p = p \circ a_n <_L p(qa_0 \cdots a_{n-2}) = (p \circ q) a_0 (pa_1) \cdots (pa_{n-2})$.

So $|pw|^{p\circ q} \sqsupset^{p\circ q} p$. The case $n = 0$ yields $p(q \circ a) = p(qa \circ q) = (p \circ q)a \circ (pq)$ and is similarly checked, using that $pq \sqsupset^{p\circ q} p$.

Note, for F a finite subset of $\mathcal{P}$, the following induction principle: if $S \subseteq \mathcal{P}$, $S \neq \emptyset$, then there is a $w \in S$ such that for all u, if $pu \leq w$ for some $p \in F$, then $u \notin S$. Otherwise some $w \in S$ would be $\geq$ arbitrarily long compositions of the form $p_0 \circ p_1 \circ \cdots \circ p_n$, each $p_i \in F$. By Theorem 1(ii), some $p \in F$ would occur at least m times in one of these compositions, where $p^m > p^{(m)} > w$, and applications of the $a \circ b = ab \circ a$ law would give $p^m \leq p_0 \circ \cdots \circ p_n \leq w$, a contradiction to Theorem 1(v) and (viii).

THEOREM. *For all $w, r \in \mathcal{P}$, $|w|^r$ exists.*

Proof. We show that $T = \{r \in \mathcal{P} : \text{for all } w \in \mathcal{P},\ |w|^r \text{ exists}\}$ contains x and is closed under $\cdot$ and $\circ$.

(i) $x \in T$. Suppose, letting $F = \{x\}$ in the induction principle, that $|w|^x$ does not exist but $|u|^x$ exists for all u such that $xu \leq w$. Pick $v \leq w$ such that $|v|^x$ does not exist, and, subject to that, the (x, x)-normal form of v ([5], Lemmas 25, 27, Theorem 28) has minimal length. The (x, x)-normal form of v is a term $xa_0a_1 \cdots a_{n-1} * a_n$, which is prenormal, where a_0 is in the normal form of [5] (see the corollary below), and for $i > 0$, each a_i is in (x, x)-normal form. Then for $i > 0$, each $|a_i|^x$ exists, and since $xa_0 \leq w$, $|a_0|^x$ exists. Thus $|v|^x$ exists, $|v|^x = x|a_0|^x \cdots |a_{n-1}|^x * |a_n|^x$.

(ii) $p, q \in T$ implies $pq \in T$. For $u \in p$-DF, define the $\langle p, q\rangle$-DF of u as follows. If $u \leq p$, the $\langle p, q\rangle$-DF of u is u. If $u = pa_0a_1 \cdots a_{n-1} * a_n$, the $\langle p, q\rangle$-DF of u is $p\bar{a}_0\bar{a}_1 \cdots \bar{a}_{n-1} * \bar{a}_n$, where $\bar{a}_0 = |a_0|^q$ and for $i > 0$, $\bar{a}_i$ is the $\langle p, q\rangle$-DF of a_i. Then by assumption every $r \in \mathcal{P}$ has a $\langle p, q\rangle$-DF representation. Pick v such that $|v|^{pq}$ does not exist, and subject to that, the $\langle p, q\rangle$-DF representation of v has minimal length. If $v \leq pq$, we are done. So assume v's $\langle p, q\rangle$-DF representation is $p(qa_0a_1 \cdots a_{n-1} * a_n)b_0b_1 \cdots b_{m-1} * b_m$, where the proof for $n \geq 0$ and the first $*$ being $\circ$ will cover all cases. Then $v = pq(pa_0)(pa_1) \cdots (pa_{n-1})(pa_nb_0)b_1 \cdots b_{m-1} * b_m$. Then $|pa_0|^{pq} \cdots |pa_{n-1}|^{pq}, |pa_n|^{pq}, |b_0|^{pq} \cdots |b_m|^{pq}$ all exist by the minimality of v's $\langle p, q\rangle$-DF representation. And since $b_0 \leq p$, $|pa_n|^{pq} \sqsupset^{pq} b_0$ by Lemma 7(i), and $|pa_nb_0|^{pq}$ exists by Lemma 6. The sequence

$$(pq), (pa_0) \cdots (pa_{n-1}), (pa_nb_0), b_1 \cdots b_{n-1}, b_n$$

need not be prenormal. But we claim

$$|p(qa_0 \cdots a_{n-1} \circ a_n)|^{pq} = pq|pa_0|^{pq} \cdots |pa_{n-1}|^{pq} \circ |pa_n|^{pq} \sqsupset^{pq} |b_0|^{pq}.$$

The equality is clear. For the $\sqsupset^{pq}$ relation, we have $|pa_n|^{pq} \sqsupset^{pq} |b_0|^{pq}$ and $pa_n \circ b_0 \leq pa_n \circ p = p \circ a_n \leq pq(pa_0) \cdots (pa_{n-2})$ since $a_n < pa_0 \cdots a_{n-2}$, giving the claim. So by Lemma 6, $|p(qa_0 \cdots a_{n-1} \circ a_n)b_0|^{pq} \sqsupset^{pq} |p(qa_0 \cdots a_{n-1} \circ a_n)|^{pq} \geq b_1$. By Lemma 5, $|p(qa_0 \cdots a_{n-1} \circ a_n)b_0|^{pq} \sqsupset^{pq} |b_1|^{pq}$. With this as the first step, iterate Lemma 6 and Lemma 5, m times, to get that $|p(qa_0 \cdots a_{n-1} \circ a_n)b_0b_1 \cdots b_{m-1} * b_m|^{pq}$ exists.

(iii) $p, q \in T$ implies $p \circ q \in T$. Letting $F = \{q\}$ in the induction principle, suppose $|w|^{p\circ q}$ does not exist but $|a|^{p\circ q}$ exists for all a such that $qa \leq w$.

Pick $v \leq w$ such that $|v|^{p \circ q}$ does not exist and, subject to that, the $\langle p, q\rangle$-DF representation of v has minimal length. If $v \leq p \circ q$, then again the cases on the $\langle p, q\rangle$-DF representation of v are covered by the proof where that representation is $p(qa_0 \cdots a_{n-1} \circ a_n)b_0 b_1 \cdots b_{m-1} * b_m$.

Then $v = (p \circ q)a_0(pa_1)\cdots(pa_{n-1})(pa_n b_0)b_1 \cdots b_{m-1} * b_m$. As in case (ii), $|pa_1|^{p \circ q}, \ldots, |pa_{n-1}|^{p \circ q}, |pa_n|^{p \circ q}, |b_0|^{p \circ q}, \ldots, |b_m|^{p \circ q}$ exist, and using Lemma 7(ii) and Lemma 6, $|pa_n b_0|^{p \circ q}$ exists. And since $qa_0 \leq v$, $|a_0|^{p \circ q}$ exists by the induction principle. Thus $|p(qa_0 \cdots a_{n-1} \circ a_n)|^{p \circ q}$ exists, and, as in case (ii), is $\sqsupset^{p \circ q} |b_0|^{p \circ q}$. Then iterate Lemmas 6 and 5 as in case (ii) to obtain the existence of $|v|^{p \circ q}$. This completes the proof of the theorem.

For $p \in \mathcal{P}$, say that a term w in the alphabet $\{q : q <_L p\} \cup \{p^{(i)} : i < \omega\}$ is in p-normal form (p-NF) if either $w <_L p$ is a term of length one, or $w = p^{(i)}a_0 a_1 \cdots a_{n-1} * a_n$, where each $a_k \in p$-NF, $p^{(i)}a_0 a_1 \cdots a_{n-1} * a_n$ is prenormal, and $a_0 <_L p^{(i)}$. Let $|w|_p$ be the p-NF representation of w if it exists. As in [5], Lemmas 9 and 12, such a representation is unique. It is proved in [5] that for all $p \in \mathcal{A}$ and $w \in \mathcal{P}$, $|w|_p$ exists. The DF theorem allows this to be extended to $p \in \mathcal{P}$.

COROLLARY. *If $p, w \in \mathcal{P}$, then $|w|_p$ exists.*

Proof. By induction on $w \in p$-DF. If $w <_L p$, we are done; so assume w is the p-DF term $pa_0 a_1 \cdots a_{n-1} * a_n$. Then each $|a_i|_p$ exists, and if $a_0 <_L p$, we are done. Also, if $a_0 = p$, then the p-NF expression for w is $p^{(1)}|a_1|_p \cdots |a_{n-1}|_p * |a_n|_p$. Without loss of generality assume a_0's p-NF representation is $p^{(m)}b_0 b_1 \cdots b_{k-1} \circ b_k$. Then it is easily checked that $|pa_0|_p = p^{(m+1)}|pb_0|_p \cdots |pb_{k-1}|_p \circ |pb_k|_p$. Thus $w = p^{(m+1)}(pb_0)\cdots(pb_{k-1})(pb_k a_1)a_2 \cdots a_{n-1} * a_n$. In [4, Theorem 16], a $\sqsupset_p$ theorem is proved for p-NF (for $p \in \mathcal{A}$, but a similar result holds for all $p \in \mathcal{P}$). We may use a version of it, and an analog of Lemma 7 above, as Lemmas 6 and 7 were used in Theorem 8, to obtain $|pa_0|_p \sqsupset_p a_1$, and then iterate to get the existence of $|w|_p$. The details are left to the reader.

REFERENCES

[1] P. DEHORNOY, *Free distributive groupoids*, **Journal of pure and applied algebra**, vol. 61 (1989), pp. 123–146.

[2] P. DEHORNOY, *Sur la structure des gerbes libres*, **C.R.A.S. Paris**, vol. 309, Série I (1989), pp. 143–148.

[3] P. DEHORNOY, *Braid groups and left distributive structures*, preprint.

[4] R. DOUGHERTY, *On critical points of elementary embeddings*, handwritten notes, 1988.

[5] R. LAVER, *On the left distributive law and the freeness of an algebra of elementary embeddings*, **Advances in mathematics**, vol. 91 (1992), pp. 209–231.

[6] R. LAVER, *On the algebra of elementary embeddings of a rank into itself*, **Advances in mathematics**, to appear.

[7] J. STEEL, *The well-foundedness of the Mitchell order*, ***The journal of symbolic logic***, vol. 58 (1993), pp. 931–940.

[8] D. LARUE, *On braid words and irreflexivity*, ***Algebra universalis***, to appear.

University of Colorado, Boulder

GENTZEN-TYPE SYSTEMS AND RESOLUTION RULE
PART II. PREDICATE LOGIC

G. MINTS

§0. Introduction. This paper is a sequel to Mints [17] where complete resolution-type calculi were constructed for several propositional modal logics including S5, S4, T, and K. Here we extend this to the predicate logic using the same method, which provides a general scheme for transforming a cutfree Gentzen-type system into a resolution-type system preserving the structure of derivations. This is a direct extension of the method introduced by Maslov [10] for the classical predicate logic. To make this paper self-contained, we recapitulate some material from Mints [17], i.e., Part I.

The main idea of Maslov's method can be summarized as follows. Resolution derivation of the goal clause g from a list X of input clauses can be obtained as the result of deleting X from the Gentzen-type cutfree derivation of the sequent $X \Rightarrow g$.

We show here how to treat predicate logics for which cutfree formulations with the subformula property are known. These include intuitionistic logic and quantified modal systems S4, T, K4, K, and only predicate logic S4 is presented in detail. Our resolution formulation of the predicate logic S5 illustrates here the treatment of systems possessing cutfree formulations in terms of semantic tableaux. We do not consider here the resolution formulation of the intuitionistic predicate logic since such formulation containing a device to avoid Skolemization was presented in Mints [13], [14].

Resolution method for a formal system C is determined by specifying
(i) a class of formulas called clauses,
(ii) a method of reduction of any formula F of the system C to a finite list X_F of clauses,
(iii) an inference rule (or rules) R called the resolution rule for deriving clauses, and
(iv) a derivation process by forward chaining so that all derivable objects are consequences of initial clauses and garbage removal from the search space is possible.

The resolution method is said to be sound and complete iff for any formula F the derivability of F in C is equivalent to derivability of the goal clause g from X_F using the rule R.

For systems based on classical logic the goal can be taken to be the empty clause $\emptyset$ (constant false). Indeed derivability of an atom g is equivalent to derivability of $\emptyset$ from the negation $\sim g$. There are several important features of the standard resolution method for classical logic which are highly desirable for any extension to the non-classical case deserving the name of resolution.

(i) Clauses should be much simpler than formulas in general with respect to complexity measure suitable for given systems C. In the classical case standard clauses are quantifier-free (with implicit universal quantification) disjunctions of literals. So there is neither nesting of Boolean connectives (in the propositional case) nor alternating quantifiers. For the intuitionistic propositional calculus clauses were defined in [13], [14] as implications with nesting at most 2. In the modal case one cannot deal only with clauses of modal depth 1 (except in S5), unless PSPACE = co-NP. Indèed, for most non-classical propositional logics below S5 the derivability problem is PSPACE-complete, but for formulas of depth 1 it is in co-NP. Nevertheless *non-initial* derivable clauses are disjunctions of modal literals, i.e., of the form l, $\Box l$, $\Diamond l$, where l is a variable or its negation.

(ii) Reduction of an arbitrary formula F to the form $X_F \rightarrow g$, where X_F is a finite list of clauses and g is a goal (propositional variable or $\emptyset$) should be much easier than the decision problem for the systems C in question. Some authors advise using distributivity for transforming classical formula to a set of clauses. Such a procedure is efficient for relatively small formulas and sometimes allows one to restrict the search space, especially for the classical predicate logic. On the other hand, this procedure is exponential in the worst case, that is, it has the same order of complexity as the existing decision procedures for classical propositional logic. Moreover, it destroys the structure of the original formula and is not applicable to non-classical systems, since they do not allow Skolemization.

We use instead a familiar depth-reducing transformation by introducing new propositional variables. It is linear in time, universally applicable, and preserves the structure of the original formula. The list of relations defining new variables can be considered as a new encoding of the original formula or as a presentation of the data structure of its subformulas.

(iii) It is natural to require that the resolution rule R for a given system be as close as possible to the standard resolution rule for the classical propositional calculus. For propositional systems based on classical logic we were able to preserve this rule completely. Differences between various modal systems were expressed by special rules for handling modalities, which can be used only together with the resolution rule, and so can be considered to be analogues of factorization or unification for the classical resolution.

For *predicate logic* we use ideas from Mints [13], [14] and Zamov [20], where resolution-type systems for predicate logics were formulated for non-Skolemized formulas. Here reduction of the formula-depth also plays an essential part.

(iv) Our requirement that the inference process should proceed by forward chaining corresponds to Maslov's [11] distinction between local methods (such as resolution or Maslov's inverse method) and global methods (such as semantic tableau methods with introduction of dummy variables and finding their values by searching through closure conditions for all branches of the semantic table). Both resolution and Maslov's inverse method work by deriving consequences from the negation of the original formula. This allows one to use subsumption:

if C is derived, then in most cases one can throw away C' when $C \to C'$ is valid, and so save space and time.

Reducing formulas to resolution (clause) form by introducing new predicates can lead to considerable growth of the search space due to resolution over literals containing these variables. One can restrict this disadvantage using the connection between resolution and Gentzen-type derivation discovered by Maslov's [10] and Maslov's [8] idea of decomposition (*razbivka*) of a formula. The latter simulates to a certain degree introduction of new variables. We propose a *strategy* which for the propositional case can be roughly formulated as follows: if a variable P was introduced to replace an occurrence of a given sign, then no clause containing occurrences of P with opposite sign can be derived by the resolution rule. Related ideas were employed by Voronkov. References to other work were given in Part I.

We begin with the resolution formulation due to N. Zamov of the first-order predicate logic which does not require Skolemization. The main new feature of Zamov's formulation is the presence of initial clauses $(\exists y)(L_1 \vee \cdots \vee L_n)$ and corresponding resolution rule

$$(R_\exists) \qquad \frac{(\exists y)(L \vee M);\ \sim L' \vee N}{(M \vee N)\sigma} \qquad \sigma = \mathrm{MGU}(L, L')$$

where σ does not contain substitution for y and the result $(M \vee N)\sigma$ does not contain y. It is this formulation which will be generalized to the modal case. The proofs will follow the pattern of Part I (i.e., Mints [17]) which used ideas by Maslov.

§1. Classical predicate calculus.

Material presented in this section is partly familiar from the literature. The aim of this exposition is to collect it, streamline the proofs, and prepare for further treatment of more complicated systems. New material here is a Maslov-type completeness proof and treatment of introduced predicates in Section 1.5.

We begin in Section 1.1 with the extended language of clauses [Zamov 20] and show that it is sufficiently general in the sense of the introduction, i.e., that any formula can be reduced to a system of clauses in linear time and space. A cutfree Gentzen-type system GK for classical logic together with a normal form theorem for derivations is described in the Section 1.2. Section 1.3 gives a detailed description of the resolution strategy corresponding to Gentzen-type derivability and a structure-preserving translation GR from GK into the resolution system. Section 1.4 is devoted to inverse translation. Section 1.5 contains an example of treatment of strategies under this approach.

1.1. Non-Skolemized clauses. Recall that in the usual formulation of the resolution method derivable objects are *clauses* which are disjunctions $L_1 \vee \cdots \vee L_k$ $(k \geq 0)$ of *literals* i.e., of atoms (propositional letters) $p, q, r, p_1, \ldots$ and their negations $\sim p, \sim q, \ldots$. Here $t_1, \ldots, t_k$ are *terms* constructed from variables by function symbols. Clauses are treated modulo order and number of occurrences of literals. $\emptyset$ means the empty clause (interpreted as the constant *false*). Complement $\sim L$ of a literal L is defined in a standard way so that $\sim(\sim p) = p$.

We will preserve the same class of derivable objects, i.e., of possible results of derivations, but extend the total set of clauses.

Initial clauses are disjunctions of literals as well as the formulas of the form $(\exists y)(L_1 \vee \cdots \vee L_k)$ where $k \geq 2$ and $L_1, \ldots, L_k$ are literals. We shall be interested in derivability relations $X \vdash C$, where X is a set of initial clauses and C is a clause. Recall that free individual variables in initial clauses are understood to be universally quantified.

Recall the following well-known fact.

THEOREM 1.1. *For any propositional formula F a set X_F of initial clauses of length < 3 can be constructed linearly in F by introduction of new variables such that F is valid iff X_F is inconsistent, i.e., the sequent $\forall X_F \Rightarrow \emptyset$ is derivable.*

Proof. The main idea is to introduce new predicate variables for subformulas of F by equivalences; for example replace subformula $P(x) \vee Q(x)$ by $X(x)$, and express equivalence $\forall x(X(x) \leftrightarrow P(x) \vee Q(x))$ by the set of clauses $\forall x(\sim X(x) \vee P(x) \vee Q(x))$, $\forall x(\sim P(x) \vee X(x))$, $\forall x(\sim Q(x) \vee X(x))$. Then take as X_F the union of introduced sets of clauses completed by the negation of the predicate letter introduced for the formula F itself.

More precisely, assign to every non-atomic subformula A of F a new predicate variable P_A with the number of arguments equal to the number of free individual variables in A. Define A^* to be A for atomic A, and to be the result of replacing the immediate non-atomic subformulas B of A by $P_B(\mathbf{y})$, where $\mathbf{y}$ is the list of free individual variables in B. Put

$$E_A \equiv (P_A(\mathbf{y}) \leftrightarrow A^*) \tag{1}$$

where $\mathbf{y}$ is the list of all individual variables free in A.

We express E_A as set of (universal closures) of clauses $C_A = C_A^+ \cup C_A^-$, where C_A^+ corresponds to the implication $A^* \rightarrow P_A(\mathbf{y})$, and C_A^- corresponds to inverse implication $P_A(\mathbf{y}) \rightarrow A^*$. Assuming to simplify notation that immediate subformulas of A are non-atomic and free individual variables of immediate subformulas are the same, we make the following definitions:

If $A \equiv B \,\&\, D$ then

$$C_A^+ \equiv \{\sim P_B(\mathbf{y}) \vee \sim P_D(\mathbf{y}) \vee P_A(\mathbf{y})\}$$
$$C_A^- \equiv \{\sim P_A(\mathbf{y}) \vee P_B(\mathbf{y}), \sim P_A(\mathbf{y}) \vee P_D(\mathbf{y})\}. \tag{$2_\&$}$$

If $A \equiv B \vee D$ then

$$C_A^+ \equiv \{\sim P_B(\mathbf{y}) \vee P_A(\mathbf{y}), \sim P_D(\mathbf{y}) \vee P_A(\mathbf{y})\}$$
$$C_A^- \equiv \{\sim P_A(\mathbf{y}) \vee P_B(\mathbf{y}) \vee P_D(\mathbf{y})\}. \tag{$2_\vee$}$$

If $A \equiv \sim B$ then

$$C_A^+ \equiv \{P_A(\mathbf{y}) \vee P_B(\mathbf{y})\}$$
$$C_A^- \equiv \{\sim P_A(\mathbf{y}) \vee \sim P_B(\mathbf{y})\}. \tag{$2_\sim$}$$

If $A \equiv \forall x B$ then

$$C_A^+ \equiv \{(\exists x)(\sim P_B(x,y) \vee P_A(y)\}$$
$$C_A^- \equiv \{\sim P_A(y) \vee P_B(x,y)\}. \quad (2_\forall)$$

If $A \equiv (\exists x)B$ then

$$C_A^+ \equiv \{P_A(y) \vee \sim P_B(x,y)\}$$
$$C_A^- \equiv \{(\exists x)(\sim P_A(y) \vee P_B(x,y))\}. \quad (2_\exists)$$

If one of the subformulas, say B, is atomic we write B instead of $P_B(\boldsymbol{y})$, etc. Note that unusual clauses containing existential quantifiers arise in positive clauses for $\forall$-quantifiers and in negative clauses for $\exists$-quantifiers, i.e., exactly in the situations where Skolemization is used in the standard approach to resolution.

Now put:

$$Y_F = \{C_A : A \text{ is non-atomic subformula of } F\} \quad (3)$$

$$X_F = Y_F \cup \{\sim P_F(\mathbf{c})\} \quad (4)$$

where $\mathbf{c}$ is a list of new constants and P_F is the predicate assigned to F. Inconsistency of X_F is equivalent to the derivability of

$$\forall Y_F \vdash P_F(\mathbf{c}) \quad (5)$$

1. Derivability of (5) implies derivability of F. Indeed, substitution of A for P_A in (5) gives

$$Y_F' \rightarrow F(\mathbf{c}) \quad (6)$$

where all formulas in Y' are easily derivable, since they result (by decomposition into clauses) from the formulas of the form $B \rightarrow B$. So (6) implies $F(\mathbf{c})$ and substituting back free variables $\boldsymbol{x}$ of F for $\mathbf{c}$ we have $F(\boldsymbol{x})$, i.e., F.

2. Derivability of F implies derivability of (5). Assuming for simplicity that F is closed, we derive

$$F \rightarrow (\forall Y_F \rightarrow P_F) \quad (7)$$

from

$$\forall Y_F \rightarrow (P_F \leftrightarrow F) \quad (8)$$

which is obtained by repeated use of the replacement of equivalents

$$\forall \boldsymbol{x}(A \leftrightarrow B) \rightarrow (G[A] \leftrightarrow G[B]) \quad (9)$$

where $\boldsymbol{x}$ contains all free variables of A, B, and G is any formula. This concludes the proof of the theorem.

Example 1. Let $F \equiv (\exists x)(\forall y)(P(x) \vee \sim P(y))$. Then denoting $P(x) \vee \sim P(y)$ by 1 and $\forall y(P(x) \vee \sim P(y))$ by 2 we have

$$E_1 \equiv P_1(x,y) \leftrightarrow (P(x) \vee \sim P(y)),$$
$$E_2 \equiv P_2(x) \leftrightarrow \forall y P_1(x,y),$$
$$E_F \equiv P_F \leftrightarrow (\exists x)P_2(x)$$

and so

$$X_F \equiv \{ \sim P_1(x,y) \lor P(x) \lor \sim P(y), \sim P(x) \lor P_1(x,y), P(y) \lor P_1(x,y),$$
$$\sim P_2(x) \lor P_1(x,y), (\exists y)(\sim P_1(x,y) \lor P_2(x)), (\exists x)(\sim P_F \lor P_2(x)),$$
$$\sim P_2(x) \lor P_F, \sim P_F\}.$$

1.2. Gentzen-type predicate calculus GK. Consider a Gentzen-type formulation of the classical predicate calculus suitable for pruning superfluous formulas (cf. below). Its derivable objects are *sequents* $X \Rightarrow Y$, where X, Y are finite (possibly empty) lists of formulas of the language considered. The order of formulas in X, Y will always be disregarded.

Gentzen-type system GK. Axioms: $A \Rightarrow A$.

Inference rules:

$$\frac{A_1, X_1 \Rightarrow Y_1; \ldots; A_n, X_n \Rightarrow Y_n}{A_1 \lor \cdots \lor A_n, X \Rightarrow Y} \quad (\lor \Rightarrow)$$

where $X_1 \cup \cdots \cup X_n = X$, $Y_1 \cup \cdots \cup Y_n = Y$.

$$\frac{X \Rightarrow Y, A_1, \ldots, A_m}{X \Rightarrow Y, A_1 \lor \cdots \lor A_n} (m \leq n)(\Rightarrow \lor) \qquad \frac{X \Rightarrow Y}{X', X \Rightarrow Y, Y'} \text{(thinning)}$$

$$(\Rightarrow \sim)\frac{A, X \Rightarrow Y}{X \Rightarrow Y, \sim A} \qquad \frac{X \Rightarrow Y, A}{\sim A, X \Rightarrow Y}(\sim \Rightarrow)$$

$$(\forall \Rightarrow) \frac{A[x := t], (\forall x A)^0, X \Rightarrow Y}{\forall x A, X \Rightarrow Y} \qquad \frac{X \Rightarrow Y, A[x := b]}{X \Rightarrow Y, \forall x A} (\Rightarrow \forall)$$

$$(\exists \Rightarrow) \frac{A[x := b], X \Rightarrow Y}{(\exists x)A, X \Rightarrow Y} \qquad \frac{X \Rightarrow Y, ((\exists x)A)^0, A[x := t]}{X \Rightarrow Y, (\exists x)A} (\Rightarrow \exists)$$

where superscript 0 means possible absence of the formula. So in fact we have two versions of the rule $(\forall \Rightarrow)$:

$$\frac{A[x := t], X \Rightarrow Y}{\forall x A, X \Rightarrow Y} \qquad \frac{A[x := t], \forall x A, X \Rightarrow Y}{\forall x A, X \Rightarrow Y}$$

and similarly for $(\Rightarrow \exists)$.

Let us fix terminology concerning Gentzen-type systems. In the sequent $X \Rightarrow Y$ the *left-hand side* X and *right-hand side* Y are sometimes called *antecedent* and *succedent*. In each inference rule the sequent written under the line is the conclusion, and the sequents over the line are premises. The formula shown explicitly in the conclusion, for example $A_1 \lor \cdots \lor A_n$ in $(\lor \Rightarrow)$ or X', Y' in (thinning), is the *main formula*, the formulas shown explicitly in the premises, for example $A_1, \ldots, A_n$ in the rule $(\lor \Rightarrow)$, are *side* formulas, and the remaining formulas, for example $X_1, Y_1, \ldots, X_n, Y_n, X, Y$ in $(\lor \Rightarrow)$ or X, Y in (thinning), are *parametric* formulas.

Recall the following facts from predicate logic.

THEOREM 1.2. *(a) Predicate formula F is valid iff the sequent $\Rightarrow F$ is derivable in GK.*

(b) The derivation of a sequent S uses only rules for connectives occurring in S, or more precisely, the succedent and antecedent rules corresponding to positive or negative occurrences of connectives.

(c) A list X of formulas is inconsistent iff the sequent $X \Rightarrow$ (with empty right-hand side) is derivable in GK.

(d) All thinning inferences can be moved downward (with possible deletion of some formulas and sequents) so that the thinning rule occurs only immediately preceding the last sequent of the derivation.

(e) Propositional inferences can be moved downward so that such inference L occurs only in a series of propositional inferences immediately preceding the last sequent of the derivation or immediately above the inference L' having the main formula of L as its side formula.

Proof. Cf. Kleene [4], [5].

A derivation satisfying the conditions in (d) (respectively in (e)) of Theorem 1.2 is called ***pruned*** (or p-***inverted***, respectively), and the operation of moving inferences downward mentioned there is called ***pruning*** (or p-***inversion***, respectively).

*1.3. **Resolution calculus corresponding to GK. Operation GR.*** Let us return to the clausal formulation (cf. Theorem 1.1). We consider a deduction relation $X \vdash C$ where X is a set of initial clauses and C is a clause. A positive (negative) occurrence of a predicate P in $X \vdash C$ is an occurrence in $P(t_1, \ldots, t_n)$ (in $\sim P(t_1, \ldots, t_n)$, respectively) as a member of C and in $\sim P(t_1, \ldots, t_n)$ (in $P(t_1, \ldots, t_n)$, respectively) as a member of one of the clauses in X.

Substitution is an expression of the form $[x_1 := t_1, \ldots, x_n := t_n]$ where t_i are terms, x_i are distinct variables which do not occur in t_i. If this substitution is denoted by σ, then the result $E\sigma$ of its execution is obtained by replacing all occurrences of x_i in E by t_i, $i = 1, \ldots, n$.

The ***unifier*** of the expressions E, F is a substitution σ unifying E and F, i.e., such that $E\sigma = F\sigma$. The ***most general unifier*** MGU(E, F) is the simplest unifier, that is $E\sigma' = F\sigma'$ implies $\sigma' = \sigma''\text{MGU}(E, F)$ for some substitution σ''. The substitution in the right-hand side of the last equation is the result of the successive execution of MGU(E, F) and σ''. MGU($E_1, \ldots, E_n$) is the most general unifier of all expressions $E_1, \ldots, E_n$.

The general formulation of the resolution rule is

$$(R) \qquad \frac{L \vee E;\ \sim L' \vee D}{(E \vee D)\sigma} \qquad \sigma = \text{MGU}(L, L')$$

where σ is the most general unifier of the literals L, L'. Alphabetic renaming of free variables is assumed throughout. More precisely, inference according to the rule (R) and all resolution-like rules below, like $(R_\exists)$, (RP), $(RP_\exists)$—but not in propositional-type rules $(R'_\exists)$, $(RP'_\exists)$—includes implicitly such a renaming to make all variables in $L \vee E$ distinct from all variables in $\sim L' \vee D$.

This form of the resolution is complete together with the factorization rule (cf. below).

To deal with (initial) clauses beginning with ($\exists$) we add the following rule introduced by Zamov [20]:

$$(R_\exists) \qquad \frac{(\exists y)(L \vee E);\ \sim L' \vee D}{(E \vee D)\sigma} \qquad \sigma = \mathrm{MGU}_\exists(L, L')$$

where subscript $\exists$ in MGU means that an additional proviso is imposed: σ is the most general substitution unifying L, L' which does not contain element $y := t$ and such that the resolvent $(E \vee D)\sigma$ does not contain y. It is understood as always that σ does not introduce collision of variables (here with the quantifier $(\exists y)$), i.e., y does not occur in elements $x := t$ with x free in $L \vee E$.

In other words σ does not change variable y and can introduce y only into L'. This restriction corresponds to the proviso in $\exists$-elimination rule for natural deduction

$$\frac{E \to (\exists y)L;\ L \to D}{E \to D}$$

where y should not occur in E, D. This remark is made precise in the following statement, where $\forall$ means universal closure.

LEMMA 1.3. *Under proviso of the rule* $(R_\exists)$ *the formula*

$$\forall(\exists y)(L \vee E) \,\&\, \forall(\sim L' \vee D) \to (E \vee D)\sigma$$

is derivable, so the rule is sound.

Proof. We have by $\forall$-elimination: $\forall(\sim L' \vee D) \to \forall y((\sim L' \vee D)\sigma)$. Since σ restricted to $L \vee E$ does not contain y we have $((\exists y)(L \vee E))\sigma \equiv (\exists y)((L \vee E)\sigma)$. By general properties of substitution

$$(L \vee E)\sigma \equiv L\sigma \vee E\sigma \quad \text{and} \quad (\sim L' \vee D)\sigma \equiv \sim L'\sigma \vee D\sigma.$$

Since $E\sigma$, $D\sigma$ do not contain y, equivalences

$$(\exists y)(L\sigma \vee E\sigma) \leftrightarrow (\exists y)L\sigma \vee E\sigma; \ \forall y(\sim L'\sigma \vee D\sigma) \leftrightarrow \forall y \sim L'\sigma \vee D\sigma$$

are derivable. Finally, since σ is a unifier of L and L', we have $\forall y \sim L'\sigma \leftrightarrow \sim(\exists y)L\sigma$ and so

$$\forall(\exists y)(L \vee E) \,\&\, \forall(\sim L' \vee D) \to ((\exists y)L\sigma \vee E\sigma) \,\&\, (\sim(\exists y)L\sigma \vee D\sigma) \to E\sigma \vee D\sigma$$

which is to be proved.

We introduce now a calculus for deriving $X \vdash C$ which has two kinds of axioms and three inference rules.

Resolution calculus RK. Axioms: Clauses belonging to the list X (input clauses) and $L \vee \sim L$ where the predicate symbol of the literal L occurs both positively and negatively in $X \vdash C$ (purity restriction).

Inference rules:

$$(RP) \qquad \frac{L_1 \vee \cdots \vee L_n \text{ (input)};\ \sim L_1' \vee D_1; \ldots; \sim L_n' \vee D_n}{(D_1 \vee \cdots \vee D_n)\sigma}$$

where $\sigma = \mathrm{MGU}(L_1, L'_1; \ldots; L_n, L'_n)$ i.e., σ is the most general substitution unifying each of the pairs (L_i, L'_i), $i = 1, \ldots, n$.

$$(RP_\exists) \qquad \frac{(\exists y)(L_1 \vee \cdots \vee L_n) \text{ (input)};\ \sim L'_1 \vee D_1; \ldots; \sim L'_n \vee D_n}{(D_1 \vee \cdots \vee D_n)\sigma}$$

$\sigma = \mathrm{MGU}_\exists(L_1, L'_1; \ldots; L_n, L'_n)$, $n \geq 2$, where $\mathrm{MGU}_\exists$ means the proviso similar to the one in $R_\exists$ above: σ does not change the variable y and can introduce it only into $L'_1, \ldots, L'_n$.

$$(F) \qquad \frac{L_1 \vee \cdots \vee L_n \vee D}{L \vee D\sigma}$$

where $n > 1, \sigma = \mathrm{MGU}(L_1, \ldots, L_n)$ and $L = L_1\sigma = \cdots = L_n\sigma$.

Comments. The rule RP can be thought of as a series of n inferences according to the standard resolution rule (R) presented at the beginning of this section. It is closely connected with the clash rule [Chang and Lee, 1]. Our axioms of the form $L \vee \sim L$ are introduced to ensure a complete clash form of RP. It is easy to see that deletion of all such tautologies from a derivation in RK results in a derivation of the same clause as before by a series of inferences which can be thought of as multiple applications of the standard resolution rule to an initial clause. Assuming to simplify notation that the axiom premises of the form $L \vee \sim L$ are the last ones, we can write such a series of the standard resolution rules in the form:

$$(R_{si}) \qquad \frac{L_1 \vee \cdots \vee L_n \text{ (input)};\ \sim L'_1 \vee D_1; \ldots; \sim L'_k \vee D_k}{(D_1 \vee \cdots \vee D_k \vee L_{k+1} \vee \cdots \vee L_n)\sigma}$$

$\sigma = \mathrm{MGU}(L_1, L'_1; \ldots; L_k, L'_k)$.

We can call this rule *semi-input* resolution. Recall that the input strategy of the standard resolution rule (R) is the requirement that at least one of the premises should be an input clause. This strategy is known to be incomplete [Chang and Lee, 1]. We shall see that the semi-input strategy is complete and corresponds to Gentzen-type derivability up to the structure of derivations for Skolemized formulas. For the case when the existential quantifier is present in clauses $(\exists y)D$ one should add a semi-input version of the rule $(RP_\exists)$.

Allowing axioms $L \vee \sim L$ without purity restriction would result in the admission of the substitution rule, since for every substitution σ we would have the following (RP)-inference:

$$\frac{L_1 \vee \cdots \vee L_n;\ L_1\sigma \vee \sim L_1\sigma; \ldots; L_n\sigma \vee \sim L_n\sigma}{(L_1 \vee \cdots \vee L_n)\sigma}$$

Purity restriction means that the substitution is allowed in all non-pure literals, i.e., ones which have a chance to be main literals in an axiom. An even more reasonable restriction is one used in Maslov's inverse method [Maslov, 8]: an axiom is a tautology of the form $L\sigma \vee \sim L'\sigma$ where L, $\sim L'$ are literals occurring positively in $X \vdash C$ and σ is their most general unifier with the natural $\exists$-proviso. Our completeness proof will in fact establish completeness of this restriction, but later we shall prove that all axioms of this kind are redundant (cf. Theorem 1.5.2).

One can get rid of the factorization rule by means of a familiar trick: replace resolution by its combination with factoring into the resolved literal. The completeness of this rule is proved by moving factorization downward in the derivation by resolution and factorization.

Proviso $n \geq 2$ in the rule $(RP_\exists)$ is made to simplify formulations, since it is satisfied for our initial clauses. We shall see that the completeness theorem is valid without this restriction and its proof requires only minor modification.

The calculus RK is used to derive clauses from some set X of input clauses. The leftmost premise of each of the rules (RP), $(RP_\exists)$, $(F_\exists)$ should be one of these input clauses and is called the *nucleus*, while remaining premises $\sim L'_i \vee D_i$ are *electrons*.

Notation $X \vdash C$ means that clause C is derivable in RK from the set X of clauses.

THEOREM 1.4. *A set X of clauses is inconsistent iff $\forall X \vdash \emptyset$ where $\forall$ means universal closure.*

This statement can be easily obtained from well-known results but we are interested here in presenting ideas of Maslov's [10] proof and the construction of Zamov [20] which we generalize in subsequent sections. Theorem 1.4 is obtained as a consequence of Theorem 1.2 (c), (d) and the properties of transformation GR of a Gentzen-type derivation d of a sequent $\forall X, Y \Rightarrow Y'$ (abbreviated $d: X, Y \Rightarrow Y'$) into a resolution derivation GR(d): $X \vdash \sim Y \vee Y'$, i.e., a derivation of $\sim Y \vee Y'$ from the initial clauses X. Here Y, Y' are lists of literals and $\sim Y \vee Y'$ is a clause consisting of members of Y' and complemented members of Y with obvious modifications when Y, Y' or both are empty. More precisely, GR(d) will derive a subclause of $\sim Y \vee Y'$, i.e., the result of deleting some (possibly no) literals from that clause.

We describe GR in detail to make possible references below.

In fact, GR(d) will be constructed in two steps. First we construct a derivation GR$'$(d) from the substitution instances of initial clauses by the following two rules:

$$(RP')\frac{(L_1 \vee \cdots \vee L_n) \text{ (input)}; \ \sim L'_1 \vee D_1; \ldots ; \sim L'_n \vee D_n}{D_1 \vee \cdots \vee D_n}$$

where $\quad L_i\sigma = L'_i$ for $i = 1, \ldots, n$

$$(RP'_\exists)\frac{(\exists y)(L_1 \vee \cdots \vee L_n) \text{ (input)}; \ \sim L'_1 \vee D_1; \ldots ; \sim L'_n \vee D_n}{D_1 \vee \cdots \vee D_n}$$

where $L_i\sigma = L'_i$ for $i = 1, \ldots, n$ and the $\exists$-proviso should be satisfied: neither substitution σ nor the resolvent $D_1 \vee \cdots \vee D_n$ can contain the variable y. This corresponds to the propositional part of the familiar completeness proof for the resolution rule via Herbrand's theorem, but now even this step contains the quantifier rule $(RP'_\exists)$. The final derivation GR(d) will be produced by a standard lifting construction, that is, by replacing arbitrary substitutions in the rules (RP'), $(RP'_\exists)$ by the most general unifiers and adding factorizations when necessary.

Definition of the transformation GR′. Let $d : X, Y \Rightarrow Y'$ be a derivation in GK, X be a list of formulas of the form $\forall x_1 \cdots \forall X_n C$ where $n \geq 0$, C is an initial clause, and Y, Y' are lists of atoms. Then every sequent in d is of the form

$$X_1, Z \Rightarrow Z' \tag{10}$$

where X_1 is a list of instances of formulas from X and Z, Z' are lists of atoms. Replace each of these sequents by

$$\sim Z \vee Z' \tag{11}$$

and delete repetition of adjacent clauses, i.e., passages $C\,/\,C$. To transform the obtained figure into a derivation by rules (RP'), $(RP'_{\exists})$ note that every inference in d belongs to one of the following types.

1. $(\vee \Rightarrow)$ with the main formula originating from a formula $\forall C$ in X, i.e., being an instance $C\theta$ of it where clause C does not contain the existential quantifier.
2. $(\vee \Rightarrow)$ with the main formula $C\theta[y := b]$ originating from a formula $\forall \mathbf{x}(\exists y)C$ in X, where θ is a substitution for variables in $\mathbf{x}$. Assume to simplify notation that in this case $b \equiv y$.
3. $(\sim\Rightarrow)$ with the main formula $C\sigma$ for $\forall C$ in X.
4. $(\sim\Rightarrow)$ with the main formula in Y.
5. $(\Rightarrow\sim)$ with the main formula in Y'.

Now for any $(\vee \Rightarrow)$-inference in d add the main formula of this inference as an additional premise (nucleus), i.e.,

$$\frac{L_1, X_1, Z_1 \Rightarrow Z_1'; \ldots ; L_n, X_n, Z_n \Rightarrow Z_n'}{L_1 \vee \cdots \vee L_n, Z \Rightarrow Z'} \tag{12}$$

is transformed into

$$\frac{(\exists y)^0 C;\ \sim L_1 \vee \sim Z_1 \vee Z_1'; \ldots ; \sim L_n \vee \sim Z_n \vee Z_n'}{\sim Z \vee Z'} \tag{13}$$

where $C\sigma \equiv L_1 \vee \cdots \vee L_n$ originates from $\forall C$ or $\forall(\exists y)C$; $(\exists y)^0$ stands for $\exists y$ in the second case and is empty in the first case.

For any $(\sim\Rightarrow)$-inference with conclusion (10) and main formula from X_1, add its main formula as an additional premise (nucleus), i.e., the inference

$$\frac{X, Z \rightarrow Z', A}{\sim A, X, Z \rightarrow Z'} \tag{14}$$

is transformed to

$$\frac{C\sigma;\ \sim Z \vee Z' \vee A:}{\sim Z \vee Z'} \tag{15}$$

where $\sim A \equiv C\sigma$.

This concludes the description of $\mathrm{GR}'(d)$.

Note. The conclusion of GR′ uses the restriction $n \geq 2$ for initial clauses $(\exists y)(L_1 \vee \cdots \vee L_n)$. It is easy to remove this restriction by treating the rule $(\exists \Rightarrow)$ with the main formula $(\exists y)L$

$$\begin{array}{rc} d': & L, X, Y \Rightarrow Y' \\ \hline d: & (\exists y)L, X, Y \Rightarrow Y' \end{array}$$

in a special way. After constructing $\mathrm{GR}'(d')$: $L, X \vdash \sim Y \vee Y'$ remove the initial clause L and add $\sim L$ to all clauses that depend on the removed L. This will give a derivation $X \vdash L \vee \sim Y \vee Y'$ and $\mathrm{GR}'(d)$ is concluded by $(RP'_{\exists})$ with nucleus $(\exists y)L$. So most of the following results are preserved with their proofs without the restriction $n \geq 2$ for initial $(\exists)$-clauses.

THEOREM 1.5. *(a) If X is a set of formulas of the form $\forall x_1 \cdots x_n C$, where $n \geq 0$ and C are initial clauses, Y, Y' are lists of literals and $d: X, Y \Rightarrow Y'$ is a pruned p-inverted GK-derivation containing no thinnings then*

$$GR'(d): X^{\forall} \vdash \sim Y \vee Y' \qquad \text{by the rules } (RP'), (RP'_{\exists}) \tag{16}$$

where $X^{\forall}$ is the result of dropping initial $\forall$-quantifiers from the formulas in X, or

$$X \text{ is empty and } Y = Y' \text{ is one and the same literal.} \tag{17}$$

(b) If X, C are lists of initial clauses, Y is a list of literals, and $\forall X, Y \Rightarrow C$ is GK-derivable, then $X \vdash (\sim Y \vee C)^{-}$ by the rules (RP'), $(RP'_{\exists})$ where minus means deletion of some (possibly no) literals from $\sim Y$, C.

(c) In particular GK-derivability of $\forall X \Rightarrow D$ for a clause D implies $X \vdash D^{-}$ and if X is inconsistent (i.e., $X \Rightarrow$ is GK-derivable), the $X \vdash \emptyset$.

Proof. (a) Apply induction on (the number of inferences in) the pruned derivation d of sequent $X, Y \Rightarrow Y'$.

Induction base. If Y' is X then Y is empty and $\sim Y \vee Y' = Y' = X$ is the initial clause. If Y' is Y the X is empty and we have (17).

Induction step. Only the rules $\sim\Rightarrow$, $\Rightarrow\sim$, $\vee\Rightarrow$, $\forall\Rightarrow$ and final thinning can be applied in the derivation.

Case 1. The inference $\Rightarrow\sim$: $A, X, Y \Rightarrow Y'/X, Y \Rightarrow Y', \sim A$ is transformed into repetition $\sim Y \vee Y' \vee \sim A/ \sim Y \vee Y' \vee \sim A$ (recall that we identify clauses differing only by permutation).

Case 2. The rule $\sim\Rightarrow$. If the main formula does not belong to X, then $\sim\Rightarrow$ is transformed into repetition since $\sim\sim A = A$. In the opposite case the $\sim\Rightarrow$ is transformed into the figure (15) which is the application of RP. The only non-trivial case is where (7) holds for the premise of the rule. Then the electron $\sim Y \vee Y$ should satisfy the purity restriction. Indeed, our derivation is pruned, so any atom (including Y) is traceable to an axiom $Y \Rightarrow Y$ and the purity restriction is satisfied.

Case 3. The rule $\vee\Rightarrow$. It has the form (12) and is transformed into (13) where $(\exists y)$ is absent and C is $L_1 \vee \cdots \vee L_n$. The purity condition for the premises of (13) satisfying (17) is valid for the same reason as in case 2.

Case 4. The rule $(\exists\Rightarrow)$. This is the main new point of our proof compared to Theorem 1.3 of Part I. The main formula of the rule $\exists\Rightarrow$ is the initial clause $(\exists y)C$, so C should be a disjunction. Since the given Gentzen-type derivation d

is p-inverted, its final part is of the following form:

$$\begin{array}{rc} & L_1, Z_1, Y_1 \to Y_1'; \ \ldots \ ; L_n, Z_n, Y_n \Rightarrow Y_n' \\ \hline d': & L_1 \vee \cdots \vee L_n, Z, Y \Rightarrow Y' \\ \hline d: & (\exists y)(L_1 \vee \cdots \vee L_n), Z, Y \Rightarrow Y' \end{array}$$

By the definition GR$'(d')$ ends in RP':

$$(RP') \qquad \frac{L_1 \vee \cdots \vee L_n; {\sim}L_1 \vee {\sim}Y_1 \vee Y_1'; \ldots ; {\sim}L_n \vee {\sim}Y_n \vee Y_n'}{{\sim}Y \vee Y'} \qquad (18)$$

In view of the proviso for the rule $(\exists \Rightarrow)$ the variable y does not occur in Y, Y', so prefixing $(\exists y)$ to the nucleus $L_1 \vee \cdots \vee L_n$ in (18) transforms our (RP) into a correct application of $(RP_\exists')$.

Note for the following that the $(RP_\exists')$-inferences in GR$'(d)$ exactly correspond to $(\exists \Rightarrow)$-inferences in d.

Case 5. The rule $(\forall \Rightarrow)$.

$$\begin{array}{rc} d': & C[x := t], \forall x C, X, Y \Rightarrow Y' \\ \hline d: & \forall x C, X, Y \Rightarrow Y' \end{array}$$

The derivation GR$'(d')$ differs from GR$'(d)$ only by replacement of some nuclei $C[x := t]$ in rules (RP'), $(RP_\exists')$ by C with corresponding addition of the element $[x := t]$ to the substitution σ. The only thing to check is the preservation of the $\exists$-proviso in each $(RP_\exists')$-inference

$$(RP_\exists') \qquad \frac{(\exists y)(L_1 \vee \cdots \vee L_n); \ {\sim}L_1' \vee D_1; \ldots ; {\sim}L_n' \vee D_n}{D_1 \vee \cdots \vee D_n} \qquad L_i\sigma = L_i'$$

If its nucleus is different from $C[x := t]$, this inference is not changed at all. Otherwise the nucleus is changed to C, i.e., the substitution $x := t$ is removed from $L_1 \vee \cdots \vee L_n$ and added to σ. As was just noted, the variable y of $(RP_\exists')$ in GR$'(d')$ is an eigenvariable of some $(\exists \Rightarrow)$-inference in d'. So it cannot occur free in the last sequent of d', in particular in the substitution $x := t$, and new substitution in $(RP_\exists')$ does not contain y as required.

This concludes the proof of (a).

(b) Let X, C be lists of clauses, $C = C_1, \ \ldots \ , C_n$. Let Y be a list of literals, and $\forall X, Y \Rightarrow C$ be GK-derivable. Let $C^\wedge$ be the result of replacing all disjunctions in C by commas. Then $\forall X, Y \Rightarrow C^\wedge$ is GK-derivable (use cuts with derivable sequents $C_i \Rightarrow C_i^\wedge$) and ${\sim}Y \vee C \equiv {\sim}Y \vee C^\wedge$. The pruned p-inverted GK-derivation of $X, Y \Rightarrow C^\wedge$, which exists by Theorem 1.2(d), contains thinning only at the very end: $\forall X^-, Y^- \Rightarrow C^{\wedge -} \ / \ \forall X, Y \Rightarrow C$. Applying part (a) of the present theorem to $\forall X^-, Y^- \Rightarrow C^\wedge$, we have $X \vdash {\sim}Y^- \vee C^-$ as required in (b). Additional information is that taking subclauses corresponds to thinning inferences in Gentzen-type derivation.

Part (c) of the theorem immediately follows from (b).

Now we define GR(d) as the result of the lifting, i.e., moving substitutions maximally down the derivation GR$'(d)$. For any derivation

$$d : X \vdash C \text{ by the rules } (RP'), (RP_\exists')$$

we define by induction on d the derivation $d : X \vdash C^\frown$ in the system RP such that $C = C^\frown\theta$ for some substitution θ.

If C is in X then $d' = d$. If C is $L \vee \sim L$, then $C^\frown$ is $P(x_1, \ldots, x_n) \vee \sim P(x_1, \ldots, x_n)$ where P is the predicate symbol of L and $x_1, \ldots, x_n$ are pairwise distinct variables.

Let d end in the rule $(RP'_\exists)$:

$$\frac{(\exists y)(L_1 \vee \cdots \vee L_n)(\text{input});\ \sim L'_1 \vee D_1 : \ldots; \sim L'_n \vee D_n}{D_1 \vee \cdots \vee D_n} \qquad \begin{array}{l} L_i\sigma \equiv L'_i, \\ i = 1, \ldots, n \end{array}$$

By the induction assumption we have derivations $d'_i : X \vdash (\sim L'_i \vee D_i)^\frown$ for $i = 1, \ldots, n$, such that $\sim L'_i \vee D_i \equiv (\sim L_i \equiv D_i)^\frown \theta_i$. Slightly abusing $^\frown$-notation we write $(\sim L'_i \vee D_i)^\frown \equiv \sim L'^\frown_i \vee D^\frown_i$. Note that $\sim L'^\frown_i$ can contain more than one literal, but it is unified in one literal $\sim L'_i$ by the substitution θ_i. So applying if necessary the factorization rule (F) we will assume that all $\sim L'^\frown_i$ $(i = 1, \ldots, n)$ are literals. In view of the proviso for $(RP'_\exists)$ the substitution σ does not contain the variable y, and the substitutions θ_i do not introduce it into $D_1 \vee \cdots \vee D_n$. Renaming if necessary free variables in $\sim L'^\frown_i \vee D^\frown_i$, we can assume that renaming conditions are satisfied for $(\exists y)(L_1 \vee \cdots \vee L_n)$, $\sim L'^\frown_1 \vee D^\frown_1, \ldots, \sim L'^\frown_n \vee D'^\frown_n$. Collecting θ_i into the common substitution θ we have

$$L_i\sigma\theta \equiv L_i\sigma \equiv L'_i \equiv L'^\frown_i\theta \equiv L'^\frown_i\ \sigma\theta.$$

So $\sigma\theta$ is a general unifier of all pairs $(L_i, L'_i), i = 1, \ldots, n$ satisfying these conditions for the following $(RP_\exists)$-inference:

$$(RP_\exists) \qquad \frac{(\exists y)(L_1 \vee \cdots \vee L_n);\ \sim L'^\frown_1 \vee D^\frown_1; \ldots; \sim L'^\frown_n \vee D^\frown_n}{(D^\frown_1 \vee \cdots \vee D^\frown_n)\sigma^\frown}$$

Since $\sigma^\frown$ is the most general unifier, we have $\sigma\theta \equiv \sigma^\frown\theta'$ for some substitution θ'.

Since $D^\frown_i\theta \equiv D^\frown_i\theta_i \equiv D_i$ $(i = 1, \ldots, n)$, and $\sigma\theta$ on $D^\frown_1, \ldots, D^\frown_n$ coincides with θ, we have $D_i = D^\frown_i\theta = D^\frown_i\sigma^\frown\theta'$ $(i = 1, \ldots, n)$, so the former resolvent $D_1 \vee \cdots \vee D_n$ is indeed a substitution instance of the new resolvent. That concludes the treatment of the rule $(RP'_\exists)$.

If d ends in the rule (RP'), the treatment is similar, but easier since there are no $\exists$-restrictions. This concludes description of GR(d).

THEOREM 1.6. *(a) If X is a set of initial clauses, Y, Y' are lists of literals, and $d : \forall X, Y \Rightarrow Y'$ is a pruned p-inverted GK-derivation containing no thinnings, then GR(d) : $X \vdash \sim Y \vee Y'$ in the system RP, or X is empty and $Y = Y'$ is one and the same literal.*

(b) If X, C are lists of initial clauses, Y is a list of literals, and $\forall X, Y \Rightarrow C$ is GK-derivable, then $X \vdash (\sim Y \vee C)^-$ in the system RP where minus means subsumption.

(c) In particular GK-derivability of $\forall X \Rightarrow D$ for a clause D implies $X \vdash D^-$, and if X is inconsistent (i.e., $X \Rightarrow$ is GK-derivable), then $X \vdash \emptyset$.

The proof is the same as for Theorem 1.5 using operation GR instead of GR$'$.

1.4. Translation RG of semi-input resolution into Gentzen-type derivations. Operation RG will be applied to a derivation by semi-input resolution $d : X \vdash D$ of a clause D from a set of clauses X and arbitrary partition $D \equiv D_1 \vee D_2$ of D into (possibly empty) clauses D_1, D_2. The result $\mathrm{RG}(d, D_1, D_2)$ (which we shall usually write as $\mathrm{RG}(d)$) is a Gentzen-type derivation of the sequent $\forall X, \sim D_1 \Rightarrow D_2$ where $\sim D_1$ consists of complements of literals in D_1 (if any).

Definition of $\mathrm{RG}(d)$ by induction on d.

If $D_1 \vee D_2 = D$ is a member of X, then (assuming for simplicity that D_1, D_2 are both unit clauses), $\mathrm{RG}(d)$ is of the form

$$\cfrac{\cfrac{\cfrac{D_1 \Rightarrow D_1}{D_1, \sim D_1 \Rightarrow} \qquad D_2 \Rightarrow D_2}{D_1 \vee D_2, \sim D_1 \Rightarrow D_2}}{X, D_1 \vee D_2, \sim D_1 \Rightarrow D_2} \text{(thinning)}$$

If $D_1 \vee D_2 = \sim L \vee L$ (in some order) then $\mathrm{RG}(d)$ is the obvious derivation from the axiom $L \Rightarrow L$.

If the last inference of d is RP, choose the partition in each of the premises in accordance with the partition of the conclusion, i.e., put literals from D_1 in the first part, and the literals from D_2 in the second part. Then the RP in question can be written as

$$\frac{L_1 \vee \cdots \vee L_n;\ \sim L'_1 \vee D_{11} \vee D_{12}; \ldots ; \sim L'_n \vee D_{n1} \vee D_{n2}}{(D_{11} \vee \cdots \vee D_{n1} \vee D_{12} \vee \cdots \vee D_{n2})\sigma \equiv (D_1 \vee D_2)\sigma}$$

Putting the resolved literals $\sim L'_1, \ldots, \sim L'_n$ in the first part of the partitioning and applying the inductive assumption we can construct GKp-derivations of the sequents

$$\forall X, L'_1, \sim D_{11} \Rightarrow D_{12}; \ldots ; \forall X, L'_n, \sim D_{n1} \Rightarrow D_{n2}.$$

After making substitution σ we can conclude $\mathrm{RG}(d)$ by the following $(\vee \Rightarrow)$ and $(\forall \Rightarrow)$-inferences taking into account that $L'_i\sigma \equiv L_i\sigma$:

$$\cfrac{\cfrac{\forall X, L_1\sigma, \sim D_{11}\sigma \Rightarrow D_{12}\sigma; \ldots ; \forall X, L_n\sigma, \sim D_{n1}\sigma \Rightarrow D_{n2}\sigma}{\forall X, (L_1 \vee \cdots \vee L_n)\sigma, \sim D_1\sigma \Rightarrow D_2\sigma}(\vee \Rightarrow)}{\forall X, \sim D_1\sigma \Rightarrow D_2\sigma}(\forall \Rightarrow) \qquad (19)$$

If the last inference of d is (F), one should make a factorizing substitution, erase superfluous copies of identical literals, and add (if necessary) the $\sim$ inferences to complete former axioms $L \Rightarrow L$ which became $\sim L, L \Rightarrow$ or $\Rightarrow \sim L, L$.

If the last inference of d is $(RP_\exists)$, the treatment is similar to the case of (RP). One has only to add $(\exists \Rightarrow)$-inference in the figure (19). The proviso of the rule $(\exists \Rightarrow)$ is satisfied in view of the proviso for the rule $(RP_\exists)$. This concludes the definition of $\mathrm{RG}(d)$.

THEOREM 1.7. *If $d : X \vdash D_1 \vee D_2$ then $RG(d, D_1, D_2) : \forall X, \sim D_1 \Rightarrow D_2$. In particular $d : X \vdash D$ implies $RG(d) : \forall X \Rightarrow D$.*

Proof by induction on d is in fact contained in the definition of RG(d).

Important note. Operations RG and GR (defined in the previous section) preserve much of the structure of the derivation. In particular there is a close correspondence between the $(\vee \Rightarrow)$, $(\exists \Rightarrow)$-inferences in a Gentzen-type derivation and (RP), $(RP_\exists)$-inferences in corresponding RP-derivation. This enabled Maslov [10] to transfer many results on strategies for one of these formulations to another one.

1.5. Completeness of strategies. We illustrate the use of the apparatus of the preceding section in proving completeness of two strategies. A much more general result was established by Maslov [9].

V. Neiman and V. Orevkov noted that hyperresolution is incomplete for input clauses with $\exists$, as the following example due to V. Orevkov shows:

$$(\exists x)P(x),\ \forall y(\sim P(y) \vee Q(y)), \forall u \sim Q(u) \vdash \emptyset$$

On the other hand, V. Orevkov noticed that it is possible to require that all $(RP_\exists)$-inferences follow (i.e., be situated below) all (RP)-inferences.

From now until the end of this section we will be interested in derivability relations $X \vdash g$ encoding (according to a refinement of Theorem 1.1) the derivability of a predicate formula F. It was noted very early that the introduction of new predicates for this encoding leads to an increase in the search space. We shall prove the completeness in the Skolemized case of a strategy combining hyperresolution with a device essentially restricting this defect.

Let us consider first a more economical encoding. We use the notation from the proof of Theorem 1.1, especially formulas (1), (2) and symbols C_A^+, C_A^-. Instead of including into the encoding set all clauses C_A^+, C_A^- as was done in (3) there, we include only C_A^+ if the replaced occurrence of A is positive, and C_A^- if the replaced occurrence is negative. Instead of the predicates P_A for non-atomic formulas A we use P_A^+, P_A^- respectively for positive and negative occurrences. Instead of the equivalence (1) we use implications

$$I_A^- \equiv (P_A^-(y) \to A); \qquad I_A^+ \equiv (A \to P_A^+(y)) \tag{21}$$

and put

$$Z_F = \bigcup\{I_A^\sigma; A \text{ is a non-atomic subformula having sign } \sigma \text{ in } F\}.$$

Note that it is not necessary to introduce new predicates for all non-atomic subformulas of F. For example one can treat literals as atoms, and encode multiple disjunction or conjunction of literals by a single predicate. The same is true for chains of negative quantifiers, for example for positive occurrences of $(\exists x)(\exists y) \sim(\forall z)$ or negative occurrences of $\forall x \forall y \sim(\exists z)\ldots$, etc. Similar optimization for positive quantifiers would be possible if we introduced a special resolution rule for chains of existence quantifiers.

We call the relation $Z_F \vdash P_F^+$ obtained in this way (possibly with optimization of the kind mentioned above) a *standard encoding* of the formula F.

Instead of the deductive equivalence stated in Theorem 1 we shall describe a more close connection between derivations of F and its standard encoding

$Z_F \vdash P_F^+$. Without loss of generality we consider closed formulas F since free variables can be replaced by new constants.

Definition. Let d be a GK-derivation of the formula F. Then d^c will be a derivation of the clause form of F

$$d^c : \forall Z_F \vdash P_F^+$$

constructed as follows. Replace each occurrence of non-atomic subformulas $A[\mathbf{x} := \mathbf{t}]$ as a member of a sequent in d by $P_A^\sigma(\mathbf{t})$. Here $\mathbf{x}$ is the list of free variables A which are bound in F, and σ is $+$ if the replaced occurrence is in the succedent (i.e., is positive in the whole sequent); if the replaced occurrence of $A[\mathbf{x} := \mathbf{t}]$ is in the antecedent, then put $\sigma \equiv -$. Then add $\forall Z_F$ to the antecedents of all sequents of the resulting figure except the uppermost ones, and make insertions to turn the figure into the derivation.

To simplify the description we assume that the formula A in any axiom $A \Rightarrow A$ of the derivation d is atomic. Then no insertion is made for thinnings. Suppose that L is a logic inference.

1. L is $\Rightarrow\sim$:

$$\frac{A[\mathbf{x} := \mathbf{t}], X \rightarrow Y}{X \rightarrow Y, \sim A[\mathbf{x} := \mathbf{t}]}$$

By the steps already described it is transformed in the figure:

$$\frac{P_A^-(\mathbf{t}), X' \Rightarrow Y'}{X' \Rightarrow Y', P_{\sim A}^+(\mathbf{t})} \tag{22}$$

The presence of the positive occurrence of $\sim A$ means that the list Z_F of clauses describing new variables introduced for encoding of the formula F contains the clause $P_{\sim A}^+(\mathbf{x}) \vee P_A^-(\mathbf{x})$, and so the antecedent X' above contains its universal closure. Now we make an insertion transforming the figure (22) into the following deduction:

$$\cfrac{\cfrac{P_{\sim A}^+(\mathbf{t}) \Rightarrow P_{\sim A}^+(\mathbf{t}); \quad P_A^-(\mathbf{t}), X' \Rightarrow Y'}{P_{\sim A}^+(\mathbf{t}) \vee P_A^-(\mathbf{t}), X' \Rightarrow Y', P_{\sim A}^+(\mathbf{t})}\,(\vee\Rightarrow)}{X' \rightarrow Y', P_{\sim A}^+(\mathbf{t})}\,(\forall\Rightarrow)$$

Note that the leftmost sequent is an axiom.

2. L is $\sim\Rightarrow$:

$$\frac{X \Rightarrow Y, A[\mathbf{t}]}{\sim A[\mathbf{t}], X \Rightarrow Y}$$

Using the corresponding clause in Z_F it is transformed into the following deduction:

$$\cfrac{\cfrac{\cfrac{P_{\sim A}^-(\mathbf{t}) \Rightarrow P_{\sim A}^-(\mathbf{t})}{\sim P_{\sim A}^-(\mathbf{t}), P_{\sim A}^-(\mathbf{t}) \Rightarrow;}\,(\sim\Rightarrow) \qquad \cfrac{X' \Rightarrow Y', P_A^+(\mathbf{t})}{\sim P_A^+(\mathbf{t}), X' \Rightarrow Y'}\,(\sim\Rightarrow)}{\sim P_A^-(\mathbf{t}) \vee \sim P_A^+(\mathbf{t}), P_{\sim A}^-(\mathbf{t}), X' \Rightarrow Y'}\,(\vee\Rightarrow)}{P_{\sim A}^-(\mathbf{t}), X' \Rightarrow Y'}\,(\forall\Rightarrow)$$

3. Now we list the results of transforming $\vee$-rules, assuming to simplify notation that the $(\Rightarrow \vee)$ rule has the form $X \Rightarrow Y,\ A\ /\ X \Rightarrow Y,\ A \vee B$. It is

transformed into the deduction

$$\cfrac{\cfrac{X' \Rightarrow Y', P_A^+(t)}{\sim P_A^+(t), X' \Rightarrow Y'} \qquad P_{A\vee B}^+(t) \Rightarrow P_{A\vee B}^+(t)}{\cfrac{\sim P_A^+(t) \vee P_{A\vee B}^+(t), X' \Rightarrow Y', P_{A\vee B}^+(t)}{X' \Rightarrow Y', P_{A\vee B}^+(t)}}$$

The rule ($\vee \Rightarrow$) is transformed into the deduction

$$\cfrac{\cfrac{P_{A\vee B}^-(t) \Rightarrow P_{A\vee B}^-(t)}{\sim P_{A\vee B}^-(t), P_{A\vee B}^-(t) \Rightarrow;} \qquad P_A^-(t), X_1' \Rightarrow Y_1'; \quad P_B^-(t), X_2' \Rightarrow Y_2'}{\cfrac{\sim P_{A\vee B}^-(t) \vee P_A^-(t) \vee P_B^-(t), P_{A\vee B}^-(t), X' \Rightarrow Y'}{P_{A\vee B}^-(t), X' \Rightarrow Y'}}$$

4. The rule $\Rightarrow \forall$:

$$\frac{X \Rightarrow Y, A[x := t,\, y := b]}{X \Rightarrow Y, \forall y A[x := t]}$$

is transformed into the deduction:

$$\cfrac{\cfrac{X' \Rightarrow Y', P_A^+(t,b)}{\sim P_A^+(t,b), X' \Rightarrow Y';} \qquad P_{\forall yA}^+(t) \rightarrow P_{\forall yA}^+(t)}{\cfrac{\sim P_A^+(t,b) \vee P_{\forall yA}^+(t), X' \Rightarrow Y', P_{\forall yA}^+(t)}{X' \Rightarrow Y', P_{\forall yA}^+(t)}}$$

using clause $\forall x(\exists y)(\sim P_A^+(x,y) \vee P_{\forall yA}^+(x))$ and the rules ($\forall \Rightarrow$), ($\exists \Rightarrow$). The same proviso for the variable b is required in both cases.

5. The rule ($\forall \Rightarrow$) is transformed similarly using clause $\forall x \forall y(P_A^-(x,y) \vee \sim P_{\forall yA}^-(x))$.

6. The rules for $\exists$ are treated similarly using clauses $\forall x \forall y(P_{(\exists y)A}^+(x) \vee \sim P_A^+(x,y))$ and $\forall x(\exists y)(\sim P_{(\exists y)A}^-(x) \vee P_A^-(x,y))$.

This concludes description of the derivation d^c.

Let us describe a strategy for resolution derivation of the canonical encoding which corresponds to Gentzen-type derivation of the encoded formula F. Recall that each clause in the antecedent Z_F of the standard encoding of F belongs to a set I_A^σ for some non-atomic subformula A of F. So such a clause contains a unique literal beginning with the predicate P_A^σ for $\sigma \equiv +$ or $-$. We call it the *leading literal* of the clause.

The resolution inference

$$\frac{(\exists y)^0(L_1 \vee \cdots \vee L_n);\ \sim L_1' \vee D_1; \ldots ; \sim L_n' \vee D_n}{(D_1 \vee \cdots \vee D_n)\sigma}$$

with the nucleus $(\exists y)^0(L_1 \vee \cdots \vee L_n)$ from Z_F will be called *G-inference* (or *G-resolution*) if the electron $\sim L_i' \vee D_i$ corresponding to the leading literal of the nucleus is a tautology $\sim L_i' \vee L_i'$. In other words, the leading literal is in fact not resolved in G-resolution, but preserved in the conclusion in the form $L_i'\sigma$, i.e., possibly with some substitution.

THEOREM 1.8. *Let d be a GK-derivation of a predicate formula F. Then:*
(a) $d^c : Z_F \Rightarrow P_F^+$ is a GK-derivation of the canonical encoding of F;
(b) $GR(d^c) : Z_F \vdash P_F^+$ is the derivation of the canonical encoding of F by G-resolution.

Proof. Part (a) was verified in the definition of d^c.

Part (b) immediately follows from the following facts.

(b1) In each $(\vee \Rightarrow)$-inference in d^c with the main formula being the result of dropping quantifiers from a clause in Z_F, the premise containing the leading literal is an axiom for this literal. This is verified by inspection of the definition of d^c.

(b2) If a premise of an $(\vee \Rightarrow)$-inference in a GK-derivation d is an axiom for some side formulas, then the corresponding electron of the resolution rule in GR(d) is a tautology. This is verified by inspection of the definition of GR(d). This concludes the proof.

Let us now prove that G-resolution is compatible with the hyperresolution for the canonical encodings of Skolemized formulas F in *positive normal form*, i.e., constructed from literals by $\vee, \&, \exists$. Without this latter restriction even the encoding of $\sim\sim a \vee \sim a$:

$$(\sim a \vee \sim \underline{n}), n \vee \underline{d}, a \vee \underline{d} \vdash d$$

with leading literals underlined, does not have a hyperresolution G-derivation. The restriction to positive normal form is inessential for Skolemized formulas since elimination of implication and moving negation inside is done in linear time and preserves the structure of a formula.

THEOREM 1.9. *Hyperresolution together with G-resolution is complete for canonical encodings of formulas in positive normal form.*

The proof uses the idea employed in Section 1.5 of Part I. Note that for a formula F in positive normal form all leading literals from the clauses in Z_F are positive, since they correspond to non-atomic subformulas of F, and the latter occur in F positively. We call a clause $P \vee N$, where P (the positive part of the clause) consists of positive literals and N (negative part) of negative literals, to be *essentially negative* in a sequent $S \equiv (P \vee N, X \Rightarrow)$, if X contains as separate clauses the negations of all literals in P. For example, in the sequent $a \vee b \vee c, \sim a \vee b \vee c, \sim b, \sim c, b \vee \sim d \vee \sim e \Rightarrow$ the second clause and the last clause are essentially negative, but the first clause is not. Our strategy (call it essentially negative) allows us to apply $\vee \Rightarrow$ only with an essentially negative main formula, i.e., to analyze in the process of the proof search only essentially negative clauses.

LEMMA 1.10. *Essentially negative strategy is complete for propositional calculus.*

Proof (reproduced from Part I). It is sufficient to prove that each provable sequent $X \Rightarrow$ where X is a list of clauses, either contains a complementary pair p, $\sim p$, or contains an essentially negative clause of length > 1. Indeed, in the latter case we can apply $(\vee \Rightarrow)$ bottom-up according to our strategy and diminish the length of the sequent. Assume for contradiction that X does not contain such a clause. Then each clause C in X of length > 1 contains a positive literal p_c such that $\sim p_c$ is not a member of X. Then the valuation making all p_c true validates all clauses of length > 1 and does not falsify any clause of length 1. Putting all the latter true validates X, which contradicts derivability of $X \Rightarrow$.

Proof of Theorem 1.9. Let X be the contradictory set of clauses and d be its derivation in GK according to essentially negative strategy. We show that GR(d) is an essentially negative derivation of $\emptyset$ from X according to G-strategy. Recall that any application of the rule (RP') in GR(d) results from an $(\vee \Rightarrow)$-inference in the derivation d. To simplify notation, assume that all $(\vee \Rightarrow)$-inferences in d are below all $(\sim\Rightarrow)$-inferences. This is easy to achieve by simply moving $(\sim\Rightarrow)$-inferences up to axioms. So all sequents in d except axioms have empty succedents. To fix notation suppose that $L_1, \ldots, L_k$ in (RP') are negative, and $L_{k+1}, \ldots, L_n$ are positive. Since d satisfies essentially negative strategy, all premises of (RP') containing positive side formulas $L_{k+1}, \ldots, L_n$, contain as well their complement, i.e., can be obtained from axioms in one step. So one can assume that the only positive atomic members in the conclusion of any $(\vee \Rightarrow)$-inference are initial clauses. Now we can write $\vee \Rightarrow$ in the form

$$\frac{\sim L_1, X_1', \sim D_1 \Rightarrow; \ldots, \sim L_k, X_k', \sim D_k \Rightarrow; L_{k+1}, \sim L_{k+1} \Rightarrow; \ldots; L_n, \sim L_n \Rightarrow}{\sim L_1 \vee \cdots \vee \sim L_k \vee L_{k+1} \vee \cdots \vee L_n, X', \sim D \Rightarrow}$$

so (RP') is of the form $\sim L_1 \vee \cdots \vee \sim L_k \vee L_{k+1} \vee \cdots \vee L_n; L_1 \vee D_1; \ldots; L_k \vee D_k; L_{k+1} \vee \sim L_{k+1}; \ldots; L_n \vee \sim L_n/D$ and dropping the last $n-k$ tautological premises we have the derivation by hyperresolution, in which all positive literals in the nucleus are preserved. Lifting in the passage from GR$'(d)$ to GR(d) does not change this property, and this concludes the proof.

§2. Modal logic S4. In this and the following section we extend to modal logic material from the Part I, i.e., [Mints, 17]. We begin with quantified S4, i.e., with the result of adding to propositional S4 the usual quantifier postulates, which correspond to the semantics of growing domains. It is difficult to expect that our methods will be applicable to systems with the Barcan formula (except S5), since no cutfree Gentzen-type formulation is known for them.

2.1. *Modal clauses.* We again employ depth-reducing by introduction of new predicate variables to transform any formula into clause form using the equivalence

$$\forall\Box(A \leftrightarrow B) \rightarrow (F[A] \leftrightarrow F[B]) \tag{1}$$

which holds in S4 and its extensions. In fact it is sufficient to write $\forall$ only for variables free in A, B but bound in F.

We define *predicate literals* as atoms and their negations and denote them by $l, l_1, \ldots$. *Modal literals* are by definition expressions of the form $l, \Box l, \Diamond l$. They are denoted by $L, M, N, L_1, M_1, N_1, \ldots$. Complements are defined by $\sim\Box l = \Diamond\sim l$, $\sim\Diamond l = \Box\sim l$ in a natural way.

Predicate clauses are disjunctions of predicate literals. *Modal clauses* (or simply clauses) are disjunctions of modal literals. *Initial* modal clauses are expressions of the form $\Box\forall C$ or $\Box\forall(\exists y)C$ where C is a modal clause or has a form $\Box D$ where D is a disjunction of predicate literals. To simplify notation we require that D contain at least two terms, but this is as inessential as in Section 1.1.

We proceed as in Section 2.1.

THEOREM 2.1. *Let S be an extension of the system S4. Then for any formula F one can construct (by introduction of new variables) the list X_F of initial clauses and a propositional variable g such that*

$$\vdash_S F \text{ iff } \vdash_S \&\forall X_F \rightarrow g$$

Proof. Exactly as in the proof of Theorem 1.1 introduce predicates P_A for non-atomic subformulas A and write clauses C_A obtained from clauses in (2) Section 1.1 by prefixing $\Box\forall x$. For example if $A = (B \,\&\, D)$ with non-atomic A, B, we put

$$\begin{aligned} C_A^+ &\equiv \{\Box\forall y({\sim}P_B(y) \vee {\sim}P_D(y) \vee P_A(y))\} \\ C_A^- &\equiv \{\Box\forall y({\sim}P_A(y) \vee P_B(y)), \Box\forall y({\sim}P_A(y) \vee P_D(y))\} \end{aligned} \qquad (2_\&)$$

The definition is extended to $\Box$-case. If $A \equiv \Box B$ then we put

$$\begin{aligned} C_A^+ &\equiv \{\Box\forall y({\sim}P_B(y) \vee \Box P_A(y))\} \\ C_A^- &\equiv \{\Box\forall y(\Diamond {\sim}P_A(y) \vee P_B(y))\} \end{aligned} \qquad (2_\Box)$$

After this it remains only to repeat the proof of Theorem 1.1.

Since our modal systems are based on classical logic, it is easy to reduce derivability of an arbitrary formula to inconsistency of a set of clauses, i.e., to derivability of the constant $\emptyset$ or empty clause.

COROLLARY 2.2. *Under the assumptions of Theorem 2.1 provability of a formula F can be reduced to inconsistency of a set of clauses.*

Proof. Take X_F, g as in Theorem 2.1 and put $X'_F = X_F \cup \{{\sim}g\}$.

Note. Further simplification is possible when additional reduction axioms for modality are available. For example in S5 it is possible to consider only initial clauses of the forms:

$$\Box\forall(l_1 \vee \cdots \vee l_m)\ (m \leq 3);\ \Box\forall(\exists y)(l_1 \vee l_2);\ \forall(L_1 \vee L_2) \qquad (2')$$

Indeed clauses corresponding to propositional connectives have the first of the above forms, the quantifiers add the second of these forms and clauses $(2_\Box)$ are equivalent in S5 respectively to

$$\forall y(\Box {\sim}P_B(y) \vee \Box P_A(y)) \quad \text{and} \quad \forall y(\Diamond {\sim}P_A(y) \vee \Box P_B(y)).$$

2.2. Gentzen-type modal calculus GS4. Sequents are expressions of the form $X \Rightarrow Y$ where X, Y are (possibly empty) lists of formulas (in the language

$\Box, \vee, \sim$). Axioms and inference rules for classical connectives and quantifiers are the same as in GK (cf. Section 1.2).

Modal rules have the following form:

$$\frac{X, (\Box A)^0, A \Rightarrow Y}{X, \Box A \Rightarrow Y}\ (\Box \Rightarrow) \qquad \frac{\Box X \Rightarrow A}{Y, \Box X \Rightarrow \Box A, Z}\ (\Rightarrow \Box)$$

Pruned derivation is again one containing thinnings only immediately preceding the last sequent of the derivation. The proof of part (a) of the following statement can be found in [Curry, 2]; the proof of (b) is standard.

THEOREM 2.3. *(a) Formula F is derivable in S4 iff the sequent $\Rightarrow F$ is derivable in GS4.*

(b) Any provable sequent has a p-inverted pruned derivation.

2.3. *Resolution calculus RS4.* Derivable objects of this calculus are modal clauses

$$L_1 \vee \cdots \vee L_p \tag{3}$$

where $L_1 \vee \cdots \vee L_p$ are modal literals, and we are interested in the derivability relations $X \vdash C$ where X is a set of initial clauses and C is a modal clause. Clauses from X are input clauses.

Axioms are initial clauses as well as $L \vee \sim L$ for modal literals L with the obvious purity restriction.

There are five inference rules. The rules (RP), $(RP_\exists)$ are as in Section 1.4, but the *nucleus* $(\exists y)^0 L_1 \vee \cdots \vee L_n$ of the rules (RP), $(RP_\exists)$ should be one of the input clauses or the result of deleting $\Box\forall$ from it. The rule

$$\frac{\Box D}{D}\ (\Box^-)$$

is to be applied only together with RP, $RP_\exists$.

Various modal systems will differ mainly by additional rules for modalities. These rules play the role somewhat similar to unification for the predicate logic. The rules for S4 are the following:

$$\frac{l \vee D}{\Diamond l \vee D}\ (\Diamond) \qquad \frac{l \vee \Diamond D}{\Box l \vee \Diamond D}\ (\Box)$$

Note that all rules are obviously valid for derivability from $\Box$-formulas in S4.

2.4. *Intertranslations between Gentzen-type and resolution systems.* The translation GR into GS4-derivations is defined for pruned Gentzen-type derivation $d : X \Rightarrow g$ where X is a list of initial clauses and g is a propositional variable. We extend the definition from Section 1.3. Note that any sequent in d has the form

$$X', Y \Rightarrow Y' \tag{4}$$

where X' consist of the clauses in X and the results of deleting from them some initial occurrences of $\Box, \forall$, i.e., of clauses $\Box^0 D \equiv \Box^0(L_1 \vee \cdots \vee L_n)$, $n \geq 2$. Y

is the set of remaining antecedent members. Y and Y' are lists of literals of the form $\Box l, l$. We obtain GR(d) by replacing sequents (4) by

$$\sim Y \vee Y' \tag{5}$$

and adding necessary input clauses to construct correct inferences by RP, $RP_{\exists}$.

Proceed by induction on d. Axioms are replaced as in Section 1.3. Consider the last inference L in d. If L is $\vee \Rightarrow$ apply the same transformation as in the classical case, adding $(\Box^-)$ when necessary.

Let L be $(\Box \Rightarrow)$. If the main formula belongs to X', simply ignore the rule. If it belongs to Y, i.e., is of the form $\Box l$, then L is transformed into the rule $(\Diamond)$ of RS4: $\sim l \vee \sim Y \vee Y' / \Diamond \sim l \vee \sim Y \vee Y'$. If L is $(\Rightarrow \Box)$, then it is transformed into the rule $(\Box)$: $\Diamond \sim Y \vee l / \Diamond \sim Y \vee \Box l$.

The definition of the transformation GR is concluded.

The definition of the transformation RG from a derivation in RS4, $d : X_F \vdash g$, into the derivation in GS4 is modeled after Section 1.4.

We define for $d : X \vdash D$ and a given representation of D as $D_1 \vee D_2$ (modulo permutation of disjunctive members) the derivation RG$(d, D_1, D_2) : X', \sim D_1 \Rightarrow D_2$ where $\sim D_1$ consists of complements of literals in D_1.

Definition of RG(d) is given by induction on d. The main differences from Section 1.4 are in the modal rules, and we treat only them. Let L be the last inference of d. We proceed as in Part I.

Let L be $\Box^- : \Box D / D$. Here we could use the fact that $\Box D \Rightarrow D$ is derivable by $(\Box \Rightarrow)$. This introduces cut, so we proceed slightly more cautiously. If $\Box D$ is initial, we obviously have $D, \sim D_1 \Rightarrow D_2$, and use $\Box \Rightarrow$. If $\Box D$ is not initial, then D is a literal, and dropping $\Box$ from all predecessors of $\Box D$ in a given Gentzen-type derivation of $X \Rightarrow \Box D$ we have a derivation of $X \Rightarrow D$ where some axioms $C \vee \Box D \Rightarrow C \vee \Box D$ are replaced by $C \vee \Box D \Rightarrow C \vee D$, but these are easily derivable.

Let L be $(\Diamond)$: $l \vee C / \Diamond l \vee C$. Then it is transformed into $\Rightarrow \Diamond$ if l is in D_2, or into $\Box \Rightarrow$ if l is in D_1 according to given partition $\Diamond l \vee C = D_1 \vee D_2$.

Let L be $(\Box)$. It is transformed into $\Rightarrow \Box$ or $\Diamond \Rightarrow$. Proviso for antecedent members is satisfied, since all initial clauses begin with $\Box$.

The description of RG(d) is finished.

THEOREM 2.4. (Soundness and Completeness Theorem.) *Let F be a modal formula, and X_F is as in Theorem 2.1.*

(a) If $d : X_F \Rightarrow g$ is the derivation in LS4, then $GR(d) : X'_F \vdash g$ (or $X_F \vdash \emptyset$) is a derivation in RS4, where X' is a sublist of X_F.

(b) If $d : X_F \vdash g$ (or $X_F \vdash \emptyset$) in RS4 then $RG(d) : X_F \Rightarrow g$ is the derivation in GS4.

(c) $\vdash_{\mathrm{S4}} F$ iff $X_F \vdash_{\mathrm{RS4}} g$.

Proof. (a), (b) were established during the definitions of GR, RG, and (c) follows from (a), (b), and Theorem 2.1. Q.E.D.

§3. Modal logic S5. Since cutfree Gentzen-type formulations are known for the quantified systems T, K4, and K, there seems to be no difficulty in extending to them the formulations and results of Section 4 along the lines of Part I (more precisely Section 4 of [Mints, 17]). For the quantified S5 the situation is different from the propositional case, where there exists a cutfree formulation complete for modalized formulas [Shvartz, 18] which was used in Part I, as well as a formulation with analytic cut complete for all formulas. The best existing approximation to a cutfree system is the formulation in terms of systems of sequents (semantical tableaux) due to Kripke [6] and Kanger [3]. The formulation of Mints [12] in terms of systems of sequents is essentially equivalent. We present here the modification of our approach suitable for this situation. The general schema is as before: the resolution derivation is obtained by moving the atomic part of the Gentzen-like derivation in the succedent, but now the original objects are more complicated, and this will be reflected in the more complex structure of clauses.

1. System TS5 of semantic tableaux for S5. We describe a system TS5 which is similar to system LS5 in Mints [12]. The main difference is that the sequents will now consist not only of a succedent, as in LS5.

Let a *tableau* be any expression of the form $\{S\}$ where S is a sequent. The expression $\{\ \}$ is treated as the constant *false*. Capital Greek letters Γ, Π, Φ etc. stand for sequents.

Arbitrary lists of tableaux are called systems (of tableaux) and denoted by S, T, U, V etc. We disregard the order of tableaux in a system.

The non-modal postulates (i.e., axioms and inference rules) of the system TS5 will be essentially the same as in GK. More precisely they will be obtained from the corresponding postulates of GK by adding arbitrary tableaux. Modal rules correspond to the Kripke semantics of S5-modality.

We shall ignore the order of members in a tableau and the order of tableaux in a system.

The translation of a tableau $\{A_1, \ldots, A_n \Rightarrow B_1, \ldots, B_m\}$ is a formula $\Box(\sim A_1 \vee \cdots \vee \sim A_n \vee B_1 \vee \cdots \vee B_m)$. *The translation of a sequent* is the disjunction of translations of its member tableaux.

Axioms: $\{A \Rightarrow A\}$

Inference rules:

$$\frac{\{X \Rightarrow Y, A_1, \ldots, A_m\}, S}{\{X \Rightarrow Y, A_1 \vee \cdots \vee A_n\}, S}\ (\Rightarrow\vee)\ m \le n \qquad \frac{\{X \Rightarrow Y\}, S}{\{X', X \Rightarrow Y, Y'\}, S, S'}\ (\text{thinning})$$

$$(\Rightarrow\sim)\frac{\{A, X \Rightarrow Y\}, S}{\{X \Rightarrow Y, \sim A\}, S} \qquad \frac{\{X \Rightarrow Y, A\}, S}{\{\sim A, X \Rightarrow Y\}, S}(\sim\Rightarrow)$$

$$\frac{\{A_1, X_1 \Rightarrow Y_1\}, S_1; \ldots; \{A_n, X_n \Rightarrow Y_n\}, S_n}{\{A_1 \vee \cdots \vee A_n, X \Rightarrow Y\}, S}\ (\vee\Rightarrow)$$

where $X_1 \cup \cdots \cup X_n = X$, $Y_1 \cup \cdots \cup Y_n = Y$ and $S_1 \cup \cdots \cup S_n = S$, i.e., each of the tableaux S_i is obtained from S by deleting whole tableaux and/or members of tableaux, and each formula in S is retained in at least one of the S_i.

$$(\forall\Rightarrow)\ \frac{\{A[x := t], (\forall x A)^0, X \Rightarrow Y\}, S}{\{\forall x A, X \Rightarrow Y\}, S} \qquad \frac{\{X \Rightarrow Y, A[x := b]\}, S}{\{X \Rightarrow Y, \forall x A\}, S}\ (\Rightarrow\forall)$$

$$(\exists\Rightarrow)\ \frac{\{A[x:=b],X\Rightarrow Y\},S}{\{(\exists x)A,X\Rightarrow Y\},S} \qquad \frac{\{X\Rightarrow Y,((\exists x)A)^0,A[x:=t]\},S}{\{X\Rightarrow Y,(\exists x)A\},S}\ (\Rightarrow\exists)$$

with usual proviso for $(\Rightarrow\forall),(\exists\Rightarrow)$: the eigenvariable b does not occur free in the conclusion.

$$\frac{\{M\Rightarrow\},\{X\Rightarrow y\},S}{\{M,X\Rightarrow Y\},S}\quad (M)\quad \frac{\{\Rightarrow M\},\{X\Rightarrow Y\},S}{\{X\Rightarrow Y,M\},S}$$

where M is a modalized formula.

$$(\Box\Rightarrow)\ \frac{\{\Box A\Rightarrow\}^0,\{A,X\Rightarrow Y\},S}{\{\Box A\Rightarrow\},\{X\Rightarrow Y\},S} \qquad \frac{\{\Rightarrow A\},S}{\{\Rightarrow\Box A\},S}\ (\Rightarrow\Box)$$

$$(\Diamond\Rightarrow)\ \frac{\{A\Rightarrow\},S}{\{\Diamond A\Rightarrow\},S} \qquad \frac{\{\Rightarrow\Diamond A\}^0,\{X\Rightarrow Y,A\},S}{\{\Rightarrow\Diamond A\},\{X\Rightarrow Y\},S}\ (\Rightarrow\Diamond)$$

Comments. During the proof search process (or in the derivation viewed bottom-up) new tableaux in the system arise only in the rule (M), but they are used in an essential way only in the rules $(\Rightarrow\Box)$, $(\Diamond\Rightarrow)$. The tableaux in the system correspond to different worlds in the Kripke model. The rule (M) says that a modalized formula has the same value in all worlds. The rules $(\Box\Rightarrow)$ and $(\Rightarrow\Diamond)$ express that if $\Box A$ is true ($\Diamond A$ is false) in some world, then A is true (false, respectively) in any world.

Example 1. Let us derive the Barcan formula.

$$\cfrac{\cfrac{\cfrac{\cfrac{\{P(a)\Rightarrow P(a)\}}{\{\Box p(a)\Rightarrow\},\{\Rightarrow P(a)\}}}{\{\forall x\Box P(x)\Rightarrow\},\{\Rightarrow P(a)\}}}{\{\forall x\Box P(x)\Rightarrow\},\{\forall x P(x)\}}}{\{\forall x\Box P(x)\Rightarrow\Box\forall x P(x)\}}$$

The main step is the third one, where (if we view it bottom-up) the individual a from the second world appeared in the first world, i.e., the symmetry of the accessibility relation between worlds was implicitly used.

Equivalence of TS5 to more familiar formulations is easily established by reference to [Kripke, 6] or [Mints, 12], and we shall not go into details of this.

3.2. Resolution calculus RS5. According to Note 1 in Section 2.1 each formula F can be reduced in S5 to the sequent $X_F\Rightarrow g$ where X_F is a list of clauses of the form 2.1(2′). Using the Barcan formula to interchange $\Box$ and $\forall$, and dropping initial $\forall$ we can put them into the form:

$$(l_1\vee\cdots\vee l_m),m\le 3;\qquad (\exists y)(l_1\vee l_2);\qquad L_1\vee L_2 \tag{1}$$

where L_1,L_2 are modal literals containing $\Box$ or $\Diamond$.

Let us call (1) *initial clauses*, and define *modal clauses* (for S5) to be disjunctions

$$\Box D_1\vee\cdots\vee\Box D_n\vee\Diamond D_{n+1} \tag{2}$$

where each of the D_i is the disjunction of predicate (non-modal) literals and $\Diamond(l_1\vee\cdots\vee l_k)$ is understood as $\Diamond l_1\vee\cdots\vee\Diamond l_k$ and $\Box\emptyset\equiv\Diamond\emptyset\equiv\emptyset$. We disregard as before the order and repetitions of terms in disjunctions. Note that these objects are more complicated than modal clauses for S4 as defined in Section 3.

Let us describe a resolution system for relations $X \vdash C$, i.e., for deriving modal clauses C from the set X of initial clauses. The clauses from X are input clauses.

Resolution system RS5.

Axioms: Input clauses, $\Box(l \vee \sim l)$, $L \vee \sim L$ for modalized L with purity restriction.

Inference rules are (RP), $(RP_\exists)$ in each disjunctive member of (2) and S5-modal rule. More precisely:

$$(RP) \qquad \frac{\Box(l_1 \vee \cdots \vee l_n); \Box(\sim l'_1 \vee D_1) \vee \underline{D}_1; \Box(\sim l'_n \vee D_n) \vee \underline{D}_n}{\Box(D_1 \vee \cdots \vee D_n)\sigma \vee \underline{D}}$$

where $\sigma \equiv \mathrm{MGU}(l_1, l'_1; \ldots; l_n, l'_n)$, $\underline{D} \equiv \underline{D}_1\sigma \cup \cdots \cup \underline{D}_n\sigma$, as well as a $(L_1 \vee L_2)$-version:

$$(RP) \qquad \frac{L_1 \vee L_2; \sim L'_1 \vee D_1; \sim L'_2 \vee D_2}{(D_1 \vee D_2)\sigma}$$

$$(RP_\exists) \qquad \frac{\Box(\exists y)(l_1 \vee l_2); \Box(\sim l'_1 \vee D_1) \vee \underline{D}_1; \Box(\sim l'_2 \vee D_2) \vee \underline{D}_2}{\Box(D_1 \vee D_2)\sigma \vee \underline{D}}$$

where $\sigma \equiv \mathrm{MGU}_\exists(l_1, l'_1; \ldots; l_n, l'_n)$, $\underline{D} = D_1\sigma \cup D_2\sigma$.

$$(\Box \to \Diamond) \ \frac{\Box(l \vee D) \vee \underline{D}}{\Diamond l \vee \Box D \vee \underline{D}}$$

Factorization rule (F) is as usual.

Example 2. $p \to \Box\Diamond p$.

Introducing variables x for $\Diamond p$, y for $\Box x$, z for $p \to y$ and using reduction to a clause form taking signs into account, we have the problem:

$$(1)\ \Box{\sim}p \vee \Box x, (2)\Diamond{\sim}x \vee \Box y, (3)\Box({\sim}y \vee z), (4)\Box(p \vee z) \vdash \Box z$$

The derivation (by semi-input resolution is as follows.

$$\cfrac{\Box({\sim}y \vee z) \quad \cfrac{\Box(p \vee z) \quad \cfrac{\Box{\sim}p \vee \Box x \quad \Diamond{\sim}x \vee \Box y}{\Box{\sim}p \vee \Box y}}{\Box z \vee \Box y}}{\Box z \vee \Box z \equiv \Box z}$$

Example 3. Barcan formula. Introducing the variable p for $\forall x Px$ we have the problem

$$\Box(\exists y)({\sim}P(y) \vee p), \Box P(x) \vdash p$$

which is solved in one step of $(RP_\exists)$.

The description of the algorithm GR translating TS5-derivations into RS5-derivations is by now standard. Delete from the given derivation all modalized formulas which are not literals, move modalized literals into separate tables and replace each system

$$\{X_1 \to Z_1\}, \ldots, \{X_n \to Z_n\}$$

by the clause

$$\Box(\sim X_1 \vee Z_1) \vee \cdots \vee \Box(\sim X_n \vee Z_n)$$

where $\Box$ is not prefixed if a table consists of the only modalized formula.

Then the rule $(\vee \Rightarrow)$ is transformed into (a modal version of) (RP') or $(RP'_\exists)$ (cf. Section 1.3) depending on the initial clause which is an ancestor of the main formula of the $(\vee \Rightarrow)$ considered. The rules $(\Box \Rightarrow)$, $(\Rightarrow \Diamond)$ having a modal literal as main formula are transformed into $(\Box \rightarrow \Diamond)$. After this, lifting is applied as in Section 1.3 to assure the standard form of the rules.

Combined with soundness of the rules for RS5 this establishes the following.

THEOREM 3.1. *The system RS5 is complete for S5-derivability of relations* $\forall X \vdash C$ *where* X *is a list of initial clauses and* C *is a modal clause (modulo subsumption).*

REFERENCES

[1] C.-L. CHANG and R. LEE. ***Symbolic logic and mechanical theorem proving.*** Academic Press, New York, 1973.

[2] H. CURRY. ***Foundations of mathematical logic.*** McGraw-Hill, New York, 1963.

[3] S. KANGER. ***Handbook i logic.*** (Mimeographed.) Stockholm, 1959.

[4] S. C. KLEENE. ***Introduction to metamathematics.*** Van Nostrand, Princeton, 1952; reprinted by North-Holland, Amsterdam.

[5] S. C. KLEENE. *Permutability of inferences in Gentzen's calculi LK and LJ.* ***Two papers on the predicate calculus***, Memoirs of the American Mathematical Society, vol. 10, 1952.

[6] S. KRIPKE. *A completeness theorem in modal logic.* ***The journal of symbolic logic***, vol. 24 (1959), pp. 1–14.

[7] V. LIFSCHITZ. *What is the inverse method?* ***Journal of automated reasoning***, vol. 5 (1989), pp. 1–23.

[8] S. YU. MASLOV. *The inverse method of establishing deducibility* (Russian). ***Trudy Matematicheskogo Instituta imeni V. A. Steklova***, vol. 98 (1968), pp. 26–87. (Translated by the American Mathematical Society.)

[9] S. YU. MASLOV. *Proof search strategies based on the ordering in a favorable set.* ***Seminars in mathematics***, vol. 16, Plenum Press, New York, 1971.

[10] S. YU. MASLOV. *Connection between the strategies of the inverse method and the resolution method.* ***Seminars in mathematics***, vol. 16, Plenum Press, New York, 1971.

[11] S. YU. MASLOV. ***Theory of deductive systems and its applications.*** MIT Press, Cambridge, Mass, 1987.

[12] G. MINTS. *Lewis systems and the system T* (Russian). In: R. Feys,

Modal logic, Nauka, Moscow, 1974. English translation in G. Mints, ***Proof-theoretic transformations***, Bibliopolis, Napoli, 1990.

[13] G. MINTS. *Resolution calculi for the non-classical logics* (Russian). ***9th Soviet Symposium in Cybernetics***, VINITI, Moscow, 1981.

[14] G. MINTS. *Resolution calculi for the non-classical logics* (Russian). ***Semiotics and informatics***, vol. 25 (1985), pp. 120–135.

[15] G. MINTS. *Resolution calculi for modal logics* (Russian). ***Proceedings of the Estonian Academy of Science***, N3 (1986), pp. 279–290. (English translation in ***Translations of the AMS***, series 2, vol. 143, pp. 1–10, 1989.)

[16] G. MINTS. *Cutfree formalisations and resolution methods for propositional modal logic.* ***VIII International Congress for Logic, Methodology and Philosophy of Science***, Moscow, 1987, pp. 46–48.

[17] G. MINTS. *Gentzen-type systems and resolution rules. Part I. Propositional logic.* ***COLOG-88***, Lecture notes in computer science, vol. 417, Springer-Verlag, Berlin, Heidelberg, New York, 1990, pp. 198–231.

[18] G. SHVARTZ. *Gentzen style systems for K45 and K45D.* Lecture notes in computer science, vol. 363, Springer-Verlag, Berlin, Heidelberg, New York, 1989, pp. 245–256.

[19] N. ZAMOV. *Resolution without Skolemization.* ***Doklady Akademii Nauk SSSR***, vol. 293, N5 (1987), pp. 1046–1049.

[20] N. ZAMOV. *Maslov's inverse method and decidable classes.* ***Annals of pure and applied logic***, vol. 42 (1989), pp. 165–194.

Institute of Cybernetics
Estonian Academy of Sciences
Tallinn 200108
Estonia

AN INTUITIONISTIC THEORY OF LAWLIKE, CHOICE AND LAWLESS SEQUENCES

JOAN RAND MOSCHOVAKIS[1]

Dedicated to Stephen Cole Kleene

Abstract

In [12] we defined an extensional notion of relative lawlessness and gave a classical model for a theory of lawlike, arbitrary choice, and lawless sequences. Here we introduce a corresponding intuitionistic theory and give a realizability interpretation for it. Like the earlier classical model, this realizability model depends on the (classically consistent) set theoretic assumption that a particular Δ^2_1 well ordered subclass of Baire space is countable.

§1. Introduction.

1.1. Background. Infinitely proceeding sequences of natural numbers are the fundamental objects of L. E. J. Brouwer's intuitionistic theory of the continuum. Choice sequences are generated by more or less freely choosing one integer after another; at each stage, the chooser *may* also specify restrictions on future choices (compatible with previous restrictions, if any, and with the indefinite continuation of the process).

Brouwer called "lawlike" or "a sharp arrow" any sequence *all* of whose values are completely determined (restricted) according to some fixed law at some finite stage in the generation of the sequence. G. Kreisel [9] called "lawless" any sequence for which (i) "the *simplest kind of restriction on restrictions is made*, namely some finite initial segment of values is prescribed, and beyond this, no restriction is to be made." Kreisel and A. S. Troelstra developed a theory of lawlike and *intensionally* lawless sequences, based on (i), for which they were able to prove that every formula without free lawless variables is equivalent to one without any lawless variables and hence "it is possible to regard lawless sequences as a 'figure of speech'."[2]

Alternatively a sequence could be called lawless if (ii) it successfully evades description by any fixed law. The assumption that lawless sequences are real

[1]I wish to thank Yiannis Moschovakis, Anne Troelstra and Dirk van Dalen for constructive comments on earlier drafts of this paper, Occidental College for sabbatical support, and UCLA and MSRI for their generous hospitality in 1989–90 when much of this work was done.

[2][15, p. 639]. Kreisel [9, p. 225] asserts however that the equivalence result is *not* to be interpreted in this way, but rather as "a complete analysis of all known properties of lawless sequences in the given context."

objects of the intuitionistic continuum, whose properties are determined by their relationship to the lawlike sequences as suggested by (ii), leads to an entirely *extensional* theory of lawlike, general choice, and lawless sequences reminiscent of the theory of generic real numbers.[3] A classical model for such a theory appears in [12], under the classically consistent assumption that a particular Δ^2_1 well ordered subclass of Baire space is countable. The class of "definably lawless sequences" studied there satisfies Kreisel's Axiom of Open Data (suggested by (i) above) and a strong continuity principle (but not bar induction) and is a comeager subset of the continuum. In another paper (now in preparation) we show that it has classical measure zero and is simply definable in terms of a notion of forcing.

This paper introduces intuitionistic theories of definably lawless sequences incorporating S. C. Kleene's fundamental axiomatization **FIM** [8] of Brouwer's theory of the continuum and extends Kleene's function realizability interpretation of **FIM** to the new systems under the set-theoretic assumption appealed to in [12]. Whenever possible the reasoning used is constructive; however the realizability of some of the new axioms will be established only classically.

1.2. Motivation. Before discussing lawless sequences in context (ii) we need to know something about the lawlike ones. According to [6] Kleene did not introduce a special type of lawlike sequences because the class of general recursive functions was adequate for his purposes and was definable in his theory. Here we need a broader interpretation of "lawlike" which we shall try to motivate constructively.

What assumptions can reasonably be made about *all* lawlike sequences? We propose the following:

1. If $P(x, y)$ is a definite property of ordered pairs of natural numbers such that for each x there is exactly one y which makes $P(x, y)$ true, then there is a lawlike function ϕ such that for all x, y:

$$\phi(x) = y \text{ if and only if } P(x, y).$$

2. The class of all lawlike sequences is countably infinite in the classical sense, but has no lawlike enumeration.

For (1) we accept as "definite" only properties P all of whose sequence parameters are lawlike, and whose constructive and classical meanings essentially coincide modulo Markov's Principle. Subject to this restriction P may involve quantification over all choice sequences and over all lawlike sequences, as well as over the natural numbers.

As in [12] we next define a notion of "lawless" relative to any given notion of "lawlike" satisfying (1) and (2). One possibility would be to adopt (ii) as the definition, so a sequence α is lawless if for no lawlike sequence ϕ and for no natural number x is it the case that $\lambda t\ \alpha(x + t)$ is ϕ; however then α might be lawless even though e.g. $\lambda t\ \alpha(2t)$ was lawlike. This objection suggests something like "α is lawless if and only if for each lawlike injection ϕ, $\alpha \circ \phi$ satisfies (ii)." What we

[3] The context of this theory is somewhat wider than Kreisel and Troelstra's since it includes arbitrary choice sequences as well as lawlike and lawless ones; however, some of the axioms concerning properties specific to lawless sequences will be restricted to the narrower context.

seem to need for the proofs is a stronger notion of "lawless" whose definition and key properties appear in Section 3.

1.3. Sources. This paper is intended to be a direct sequel to Kleene and Vesley's [8] and may be read independently of all other sources. However anyone interested in the subject should surely read Kreisel's [9] and consult Kreisel and Troelstra's [10]. One may also wish to consult [12], although there the viewpoint was classical, the formalization cumbersome, and the presentation uneven.

Especially since the publication twenty-five years ago of Kleene's and R. E. Vesley's metamathematical investigation [8], much effort has been devoted to axiomatizing parts of intuitionistic mathematics beyond number theory. Troelstra's and D. van Dalen's two recent volumes [15], taken together with Vesley's address [16] to the 1979 Kleene Symposium, provide an excellent guide to the history and current state of this work. In particular, Chapters 4 and 12 of [15] give the background of Kreisel and Troelstra's work on lawlike and lawless sequences; Chapter 12 also describes other special classes of choice sequences which have recently been studied by Troelstra, van Dalen, G. F. van der Hoeven, and others.

§2. The formal theories.

2.1. The basic theory BD. This will be an extension of Kleene's basic formal system **B** for the common portion of intuitionistic and classical analysis [8, Sections 4–6]. The main syntactic difference is that **BD** has *two* sorts of variables for functions, i.e., choice sequences of natural numbers; in the intended interpretation, one sort ranges over lawlike (or definable) sequences and the other over arbitrary choice sequences.[4] We use the letters a, b, c, d, e, g, h (with or without subscripts) to denote variables over lawlike sequences, and i, j, k, l, m, ..., i_1, ... as number variables. As in [12, 8] $\alpha, \beta, \gamma, \ldots, \alpha_1, \ldots$ denote variables over arbitrary choice sequences.

The language includes the numerical equality constant = , Church's λ, a finite list $\mathrm{f}_0, \mathrm{f}_1, \ldots, \mathrm{f}_p$ of constants for primitive recursive functions, and the logical constants &, $\vee, \neg, \supset, \forall, \exists$. Each f_i expresses a function $f_i(x_1, \ldots, x_{k_i}, \alpha_1, \ldots, \alpha_{l_i})$ which, considered as a function of $x_1, \ldots, x_{k_i}$, is primitive recursive uniformly in $\alpha_1, \ldots, \alpha_{l_i}$. In particular, f_0 is 0, f_1 is $'$, f_2 is $+$, and f_3 is $\cdot$; see [8] and [7] for a suitable list.

Terms and **functors** are defined as in [8] except that now a, b, c, d, e, g, h, $\mathrm{a}_1, \ldots$ (as well as $\alpha, \beta, \ldots, \alpha_1, \ldots$) are functors while i, j, ..., i_1, ... are terms. Thus $\mathrm{f}_i(\mathrm{t}_1, \ldots, \mathrm{t}_{k_i}, \mathrm{u}_1, \ldots, \mathrm{u}_{l_i})$ is a term if t_j are terms and u_j are functors. If $k_i = 1$ and $l_i = 0$ then f_i is a functor. If u is a functor and t is a term, (u)(t) is a term. If x is a number variable and s is a term, λx(s) is a functor. A term t (functor u) is a **D-term** (**D-functor**) if it contains no occurrences of arbitrary function variables.

Prime formulas are of the form s = t where s, t are terms. **Formulas** are built up from these using the propositional connectives, and the quanti-

[4] While retaining Kleene's view of the primary importance of arbitrary choice sequences, we follow Kreisel [9] in adopting the notion of lawlike sequence as an additional primitive concept.

fiers $\forall$x, $\exists$x, $\forall$a, $\exists$a, $\forall\alpha$, $\exists\alpha$ over all three sorts of variables. A **D-formula** is one having free no arbitrary function variables. If u, v are functors, "u = v" abbreviates $\forall$x u(x) = v(x) where x is not free in u or v. For any formulas A and B, "A $\sim$ B" abbreviates (A $\supset$ B) & (B $\supset$ A). "$\exists$!yA(y)" abbreviates $\exists$yA(y) & $\forall$y$\forall$z(A(y) & A(z) $\supset$ y = z) and similarly for $\exists$!aA(a) and $\exists$!α A(α).

The substitution lemma (Lemma 3.1 of [8, p. 12]) has to be restated to allow substitution of D-functors (but not arbitrary functors) for definable function variables. In particular, if D-functors are substituted for *all* function variables occurring free in a term (functor) [formula], the result is a D-term (D-functor) [D-formula].

Lemma 3.3 of [8] has the following restatement: Let s be a term (u be a functor) [P be a prime formula] containing free no variables but $x_1, \ldots, x_k$, $a_1, \ldots, a_l$, $\alpha_1, \ldots, \alpha_m$. Then under the intended interpretation s (u(y) where y is another number variable) [P] expresses, as the ambiguous value, a function of $x_1, \ldots, x_k$, $a_1, \ldots, a_l$, $\alpha_1, \ldots, \alpha_m$ (function of $x_1, \ldots, x_k, y, a_1, \ldots, a_l, \alpha_1, \ldots, \alpha_m$) [predicate of $x_1, \ldots, x_k$, $a_1, \ldots, a_l$, $\alpha_1, \ldots, \alpha_m$] primitive recursive uniformly in $a_1, \ldots, a_l$, $\alpha_1, \ldots, \alpha_m$.

The new logical rules and axiom schemata needed are

9D. C $\supset$ A(a) / C $\supset$ $\forall$aA(a).

10D. $\forall$aA(a) $\supset$ A(g).

11D. A(g) $\supset$ $\exists$aA(a).

12D. A(a) $\supset$ C / $\exists$aA(a) $\supset$ C.

For 9D and 12D, a is not free in C. For 10D and 11D, g is any D-functor free for a in A(a).

Using these we can easily derive, for all formulas C, A(α) such that a is not free in C $\supset$ A(α), α is not free in C, and a is free for α in A(α) :

$$C \supset A(\alpha) \;/\; C \supset \forall a A(a) \quad \text{and} \quad A(\alpha) \supset C \;/\; \exists a A(a) \supset C.$$

Notice also that $\forall$a$\exists\alpha\forall$x a(x) = α(x) is a formal theorem.

As in [12] we follow Kleene's conventions for coding finite sequences of numbers and functions, although our notation differs somewhat from his.[5] Here $\langle x_0, \ldots, x_{k-1}\rangle$ abbreviates $\prod_{i<k} p_i^{x_i+1}$ and $(x_0, \ldots, x_{k-1})$ is $\prod_{i<k} p_i^{x_i}$ where p_i is the $(i+1)^{st}$ prime; $(m)_i$ is the exponent of p_i in the prime factorization of m; $\langle \alpha_0, \ldots, \alpha_{l-1}\rangle$ is $\lambda t\ \langle \alpha_0(t), \ldots, \alpha_{l-1}(t)\rangle$ (similarly with ( ) instead of $\langle\ \rangle$); and $(\alpha)_i$ is $\lambda t (\alpha(t))_i$. We follow Kleene in writing $\overline{\alpha}(x)$ for the standard code $\langle \alpha(0), \ldots, \alpha(x-1)\rangle$ for the sequence of the first x values of α.[6]

If w codes a finite sequence, its length is the number $lh(w)$ of non-zero exponents in the prime factorization of w and for each $i < lh(w)$ the $(i+1)$st term of the sequence is $(w)_i - 1$. The code for the concatenation of finite sequences with codes u and v is $u * v$, and $u * \alpha$ is the infinite sequence defined by

$$(u * \alpha)(t) = \begin{cases} (u)_t - 1 & \text{if } t < lh(u), \\ \alpha(t - lh(u)) & \text{otherwise.} \end{cases}$$

[5]Kleene uses $\langle\ \rangle$, [ ] where we use ( ), $\langle\ \rangle$ respectively. Here [ ] will be given a different meaning.
[6]The notation $\tilde{\alpha}(x)$ for $(\alpha(0), \ldots, \alpha(x-1))$ is seldom used.

Seq(w) is an almost negative formula expressing the primitive recursive predicate $Seq(w)$, "w is the code of a finite sequence of numbers," and $\alpha \in$ w abbreviates $\overline{\alpha}(\mathrm{lh}(\mathrm{w})) = \mathrm{w}$. The primitive recursive coding functions are among the initial functions $f_0, \dots, f_p$ and their properties are assumed formally. For future applications we assume the characteristic functions of the primitive recursive predicates $T(e, \alpha, a, x, y)$, $T_1(e, w, z, x, y)$ and $U(y)$ are among the initial functions.[7]

We adapt the number-theoretic postulates and recursion equations for the initial functions f_i of [8, pp. 14, 19ff.] to the current situation by writing x, y, z in place of a, b, c. Similarly with Kleene's postulates concerning functions:

$^{\times}0.1.\ \{\lambda \mathrm{x}\ \mathrm{r}(\mathrm{x})\}(\mathrm{t}) = \mathrm{r}(\mathrm{t})$.

$^{\times}1.1.\ \mathrm{x} = \mathrm{y} \supset \alpha(\mathrm{x}) = \alpha(\mathrm{y})$.

$^{\times}2.1.\ \forall \mathrm{x} \exists \alpha \mathrm{A}(\mathrm{x}, \alpha) \supset \exists \alpha \forall \mathrm{x} \mathrm{A}(\mathrm{x}, \lambda \mathrm{y} \alpha((\mathrm{x}, \mathrm{y})))$.

For $^{\times}0.1$, r(x), t are terms such that t is free for x in r(x). For $^{\times}1.1$ and $^{\times}2.1$, x and y are distinct number variables and x is free for α in $\mathrm{A}(\mathrm{x}, \alpha)$.

A formula is **almost negative** if it contains no $\vee$ and no $\exists$ except in parts of the form $\exists$xP, $\exists$aP, $\exists\alpha$P with P prime, and $\exists \mathrm{a} \forall \mathrm{x}\ \mathrm{a}(\mathrm{x}) = \mathrm{t}$ where t is a term not containing a free. Note that $\exists!\mathrm{y}\mathrm{B}(\mathrm{y})$, $\exists!\mathrm{a}\mathrm{B}(\mathrm{a})$ and $\exists!\alpha\mathrm{B}(\alpha)$ are almost negative if B is prime, and then $\forall \mathrm{x} \exists! \mathrm{y} \mathrm{B}(\mathrm{x}, \mathrm{y})$ is almost negative as well.

For each almost negative D-formula A(x, y) in which a and x are free for y we take as an axiom

$^{\times}2.2!\mathrm{D}.^{-}\ \forall \mathrm{x} \exists! \mathrm{y} \mathrm{A}(\mathrm{x}, \mathrm{y}) \supset \exists \mathrm{a} \forall \mathrm{x} \mathrm{A}(\mathrm{x}, \mathrm{a}(\mathrm{x}))$.

For any almost negative D-formula A(x, a) in which x is free for a it follows that[8]

$^{*}2.1!\mathrm{D}.^{-}\ \forall \mathrm{x} \exists! \mathrm{a} \mathrm{A}(\mathrm{x}, \mathrm{a}) \supset \exists \mathrm{a} \forall \mathrm{x} \mathrm{A}(\mathrm{x}, \lambda \mathrm{y}\ \mathrm{a}((\mathrm{x}, \mathrm{y})))$.

The Replacement Theorem (Lemma 4.2 of [8]) now holds with "$\mathrm{x}_1, \dots, \mathrm{x}_n, \mathrm{a}_1, \dots, \mathrm{a}_m, \alpha_1, \dots, \alpha_l$" in place of "$\mathrm{x}_1, \dots, \mathrm{x}_n, \alpha_1, \dots, \alpha_m$" in the version for formulas. As in Lemma 4.3 of [8] each term, functor and formula has a normal form (without superfluous λs).

Lemmas 5.3 and 5.5 of [8] now have additional lawlike parts. Thus for Lemma 5.3, if y, z are distinct number variables, a is any lawlike function variable, and p(y), q, r(y, z), and r(z) are D-terms not containing a free, with a and y free for z in r(y, z) and r(z), then

(a) $\vdash \exists \mathrm{a} \forall \mathrm{y}\ \mathrm{a}(\mathrm{y}) = \mathrm{p}(\mathrm{y})$.

(b) $\vdash \exists \mathrm{a}[\mathrm{a}(0) = \mathrm{q}\ \&\ \forall \mathrm{y}\ \mathrm{a}(\mathrm{y}') = \mathrm{r}(\mathrm{y}, \mathrm{a}(\mathrm{y}))]$.

(c) $\vdash \exists \mathrm{a} \forall \mathrm{y}\ \mathrm{a}(\mathrm{y}) = \mathrm{r}(\overline{\mathrm{a}}(\mathrm{y}))$ and $\vdash \exists \mathrm{a} \forall \mathrm{y}\ \mathrm{a}(\mathrm{y}) = \mathrm{r}(\tilde{\mathrm{a}}(\mathrm{y}))$.

Lemma 5.5 extends to allow definitions of lawlike functions by cases, combined with primitive or course-of-values recursion, provided the case descriptions are almost negative D-formulas. As an example, for almost negative D-formulas $\mathrm{Q}_1, \mathrm{Q}_2$ not containing a free and D-terms $\mathrm{r}_1, \mathrm{r}_2$:

---

[7] See [5] and [7] for details. $T(e, \alpha, a, x, y)$ expresses "y is the Gödel number of a proof of $\{e\}(\alpha, a, x) = u$ for some u," and then $U(y)$ is the u.

[8] Formal theorems *n are distinguished notationally from axioms $^{\times}$m.

$$\forall w[\mathrm{Seq}(w) \supset (Q_1(w) \vee Q_2(w)) \ \&\ \neg(Q_1(w) \ \&\ Q_2(w))] \vdash$$
$$\exists a \forall y\ a(y) = \begin{cases} r_1(\bar{a}(y)) & \text{if } Q_1(\bar{a}(y)), \\ r_2(\bar{a}(y)) & \text{if } Q_2(\bar{a}(y)). \end{cases}$$

The last axiom schema of **BD** is the "Bar Theorem" in Kleene's form

$^{\times}$26.3. $\forall\alpha\exists!xR(\overline{\alpha}(x)) \ \&\ \forall w[\mathrm{Seq}(w) \ \&\ R(w) \supset A(w)]]$
$\&\ \forall w[\mathrm{Seq}(w) \ \&\ \forall sA(w * \langle s\rangle) \supset A(w)] \supset A(1).$

Here A(w) and R(w) may be any formulas satisfying the obvious restrictions on the variables α, x, w, s. Observe that R is assumed to "bar" *all* choice sequences, not just the lawlike ones.

2.2. The theory BDLS$^-$. We now extend **BD** by adding axioms for lawless sequences. Here "DLS(α)" abbreviates a specific almost negative formula of the language of **BD** having free only the arbitrary function variable α; this formula will be given explicitly in the next section. (We purposely leave open the possibility of later interpreting "DLS(α)" as primitive, or as an abbreviation for another formula of this or an expanded language.) As in [12], $[\alpha, \beta]$ is the sequence defined by

$$[\alpha,\beta](2k) = \alpha(k), \quad [\alpha,\beta](2k+1) = \beta(k).$$

Similarly $[\alpha_1, \dots, \alpha_n]$ is the sequence obtained by meshing $\alpha_1, \dots, \alpha_n$. For future reference we introduce also the projection functions

$$^k[\beta]_i = \lambda t\ \beta(kt+i)$$

for $0 \le i < k$. These notions have the obvious formal equivalents.

The class of **restricted** formulas is defined by induction as follows. Prime formulas are **restricted**. If A, B are **restricted**, x is a number variable, and a is a definable function variable then A & B, A$\vee$B, $\neg$A, A $\supset$ B, $\forall$xA(x), $\exists$xA(x), $\forall$aA(a) and $\exists$aA(a) are all **restricted**. If $A(\beta, \gamma_1, \dots, \gamma_n)$ is **restricted** and contains free no arbitrary function variables but $\beta, \gamma_1, \dots, \gamma_n$ then $\forall\beta[\mathrm{DLS}([\beta, \gamma_1, \dots, \gamma_n]) \supset A(\beta, \gamma_1, \dots, \gamma_n)]$ and $\exists\beta[\mathrm{DLS}([\beta, \gamma_1, \dots, \gamma_n]) \ \&\ A(\beta, \gamma_1, \dots, \gamma_n)]$ are **restricted**.

The axioms for lawless sequences are then

$^{\times}$DLS1.$^-$ $\forall w[\mathrm{Seq}(w) \supset \neg\forall\alpha\neg(\mathrm{DLS}(\alpha) \ \&\ \overline{\alpha}(\mathrm{lh}(w)) = w)]$.

$^{\times}$DLS2.$^-$ $\forall\alpha[\mathrm{DLS}(\alpha) \supset \forall w[\mathrm{Seq}(w) \supset \neg\forall\beta\neg(\mathrm{DLS}([\alpha,\beta]) \ \&\ \overline{\beta}(\mathrm{lh}(w))= w)]]$.

$^{\times}$DLS3.$^-$ $\forall\alpha[\mathrm{DLS}(\alpha) \ \&\ A(\alpha) \supset \exists x\forall\beta[\overline{\beta}(x) = \overline{\alpha}(x) \ \&\ \mathrm{DLS}(\beta) \supset A(\beta)]]$.

$^{\times}$DLS4.$^-$ $\forall\alpha[\mathrm{DLS}(\alpha) \supset \exists xA(\alpha, x)] \supset \exists e\forall\alpha[\mathrm{DLS}(\alpha) \supset \exists!y\ e(\overline{\alpha}(y)) > 0 \ \&$
$\forall y(e(\overline{\alpha}(y)) > 0 \supset A(\alpha, e(\overline{\alpha}(y)) - 1))]$.

For $^{\times}$DLS3$^-$ A(α) is restricted and almost negative and contains free no arbitrary function variables but α. For $^{\times}$DLS4$^-$ A(α, x) satisfies the same conditions and in addition e, α, y are free for x in A(α, x).

2.3. The intuitionistic theory IDLS$^-$. Kleene's basic theory **B** and his intuitionistic theory **FIM** differed only by a single continuity axiom, "Brouwer's Principle for Functions" [8, $^{\times}$27.1]. Similarly, but with an important difference: **IDLS$^-$** comes from **BDLS$^-$** by adjoining the axiom schema we call "Kleene's Principle for Functions":

$^\times$KL1. $\forall\alpha[A(\alpha) \supset \exists\beta B(\alpha,\beta)] \supset \exists\tau\forall\alpha[A(\alpha) \supset \forall x\exists! y\ \tau(\langle x\rangle * \overline{\alpha}(y)) > 0\ \&$
$\forall\beta[\forall x\exists y\ \tau(\langle x\rangle * \overline{\alpha}(y)) = \beta(x)+1 \supset B(\alpha,\beta)]]$,

for all almost negative formulas $A(\alpha)$ and all formulas $B(\alpha,\beta)$ where α,β must be distinct arbitrary choice sequence variables.[9] An immediate consequence is "Kleene's Principle for Numbers" for A almost negative and τ, y, α free for x in $B(\alpha, x)$:

$^\bullet$KL2. $\forall\alpha[A(\alpha) \supset \exists x B(\alpha,x)] \supset \exists\tau\forall\alpha[A(\alpha) \supset \exists y\tau(\overline{\alpha}(y)) > 0\ \&$
$\forall y[\tau(\overline{\alpha}(y)) > 0 \supset B(\alpha, \tau(\overline{\alpha}(y)) - 1)]]$.

Kleene observed in [8, p. 74] that the special case $^\bullet$27.4 of $^\times$KL1 in which $A(\alpha)$ is $\forall x\ \sigma(\overline{\alpha}(x)) = 0$, with the additional assumption that σ is a spread-law [8, p. 56], follows from Brouwer's Principle for Functions; he also showed [8, p. 80, $^\bullet$27.16] that $^\bullet$KL2 (hence $^\times$KL1) fails if $A(\alpha)$ is not required to be almost negative.

Two important consequences of $^\times$KL1 are

$^\bullet$KL3. $\forall a\exists\beta B(a,\beta) \supset \exists\tau\forall a[\forall x\exists! y\tau(\langle x\rangle * \overline{a}(y)) > 0\ \&$
$\forall\beta[\forall x\exists y\tau(\langle x\rangle * \overline{a}(y)) = \beta(x)+1 \supset B(a,\beta)]]$

and the corresponding consequence $^\bullet$KL4 of $^\bullet$KL2, both proved by taking the almost negative formula $\exists a(a = \alpha)$ as the $A(\alpha)$. Since $DLS(\alpha)$ will be almost negative also, we conclude that in **IDLS$^-$** every function completely defined on either the species of all lawlike functions or the species of all lawless functions is continuous on that domain, though it may have no continuous extension to **B**.[10]

Since $^\times$KL1 is Kleene function-realizable it can consistently replace $^\times$27.1 in **FIM**. It is obvious, but worth emphasizing, that Kleene's formal systems **B** and **FIM** are subsystems of **BD** (*a fortiori* **BDLS$^-$**) and **IDLS$^-$** respectively.

2.4. Strengthening the density axioms. The theories **BDLS** and **IDLS** are extensions of **BDLS$^-$** and **IDLS$^-$** respectively, obtained by replacing $^\times$DLS1$^-$ and $^\times$DLS2$^-$ by

$^\times$DLS1. $\forall w[Seq(w) \supset \exists\alpha(DLS(\alpha)\ \&\ \overline{\alpha}(lh(w)) = w)]$,

$^\times$DLS2. $\forall\alpha[DLS(\alpha) \supset \forall w[Seq(w) \supset \exists\beta(DLS([\alpha,\beta])\ \&\ \overline{\beta}(lh(w)) = w)]]$.

Because of the "almost negativity" condition on $^\times$DLS3$^-$ and $^\times$DLS4$^-$, even **IDLS** is not an entirely satisfactory intuitionistic theory of lawlike, choice and lawless sequences, yet it is the strongest system whose consistency will be established in this paper. The condition can in fact be relaxed somewhat without strengthening the axioms.

2.5. Sidestepping almost negativity. A formula is **mildly assertive** if it is almost negative or obtainable from almost negative formulas using only disjunction and existential quantification over number and definable function variables; **feebly assertive** if only disjunction and existential number quantification are allowed. In **IDLS$^-$** we can prove the extension $^\bullet$DLS5 of $^\times$DLS3$^-$ to restricted mildly assertive

[9]This extension of Brouwer's Principle is called the "Generalized Continuity Principle" by Troelstra [14] who has used it to characterize Kleene's realizability. Brouwer's Principle follows trivially from it when $A(\alpha)$ is $0 = 0$.

[10]A similar situation arises in the theory of constructive real numbers, where local continuity holds but uniform continuity on [0, 1] (Brouwer's "Fan Theorem") may fail.

$A(\alpha)$ and the extension *DLS6 of $^{\times}$DLS4$^-$ to restricted feebly assertive $A(\alpha, x)$, always assuming no arbitrary choice sequence variables but α are free in A.

A formula is **assertive** if it is almost negative or obtainable from almost negative formulas using only disjunction and existential quantification. In **IDLS** we can prove the extension *DLS7 of $^{\times}$DLS3$^-$ to restricted assertive formulas $A(\alpha)$ containing free no arbitrary function variables but α.

$^{\times}$KL1 cannot consistently be similarly extended. For a counterexample let $A(\alpha)$ be $\forall x\alpha(x) = 0 \vee \neg\forall x\alpha(x) = 0$ and $B(\alpha, \beta)$ be $(\beta(0) = 0 \supset \forall x\alpha(x) = 0)$ & $(\beta(0) \neq 0 \supset \neg\forall x\alpha(x) = 0)$.

§3. Lawlessness relative to D.

3.1. The informal notion. Let N be $\{0, 1, 2, \ldots\}$ and B be Baire space N^N. Assume D is a given subclass of B which is closed under relative recursion; we think of D as the class of **lawlike** sequences.

If $\beta \in B$ maps sequence numbers to sequence numbers, β is called a **predictor**. If $\gamma, \delta \in B$ and

$$\delta(n) = \begin{cases} 0 & \text{if } \gamma(m) \neq n \text{ for all } m, \\ \mu m(\gamma(m) = n) + 1 & \text{otherwise} \end{cases}$$

then δ is called the **converse** of γ. A sequence γ is **strongly lawlike** if both γ and its converse are lawlike.

A sequence $\alpha \in B$ will be called **lawless (relative to D)** if for each lawlike predictor π and each strongly lawlike injection γ, there is an x so that

$$\alpha \circ \gamma \in (\overline{\alpha \circ \gamma})(x) * \pi((\overline{\alpha \circ \gamma})(x)).$$

Here $\alpha \circ \gamma$ can be thought of as a **subpermutation** of α, so α is lawless if and only if every lawlike predictor is eventually correct (and hence very often wrong) on every strongly lawlike subpermutation of α.[11]

Note added in proof: A simpler, but equivalent, definition of "lawless (relative to D)" appears in [13].

A finite list of sequences $\alpha_0, \ldots, \alpha_{k-1}$ is **independent** if $[\alpha_0, \ldots, \alpha_{k-1}]$ is lawless. This convention, which was also used by Michael Fourman in [4], is incompatible with Kreisel and Troelstra's strongly intensional treatment of lawless sequences; however, it greatly simplifies the extensional theory.

All the lemmas of Section 1 of [12] hold for the present notion. We summarize them here, providing constructive proofs (modulo Markov's Principle, for the finite injury priority argument for Lemma 4) of the density lemmas.[12]

LEMMA 1. (Technical Lemma.) *Every strongly lawlike subpermutation of a lawless sequence is lawless. In particular:*

(a) If $[\alpha_0, \ldots, \alpha_{k-1}]$ is lawless, so is $[\alpha_{\sigma(0)}, \ldots, \alpha_{\sigma(k-1)}]$ where σ is any permutation of $\{0, \ldots, k-1\}$.

[11] Every strongly lawlike γ is lawlike with lawlike range, and conversely; thus this definition of "lawless relative to D" is equivalent to the one in [12].

[12] The proofs in [12] of the Technical and Uniformity Lemmas were already effective.

(b) If $\alpha_0, \ldots, \alpha_{k-1}$ is an independent list, then each α_i is lawless, and if $0 \leq i < j < k$ then α_i and α_j are independent. If α is lawless so is ${}^k[\alpha]_i$ for each $k \in N$ and each $0 \leq i < k$.

(c) If α is lawless, so is $\lambda y\ \alpha((x,y))$ for each $x \in N$.

(d) If w is any sequence number and α is lawless then $w * \alpha$ is lawless.[13]

LEMMA 2. (First Density Lemma.) *Assume D is countable and let L be the class of all sequences lawless relative to D. Then L is dense in B.*

PROOF. By Lemma 1(d) we need only produce one lawless sequence β. Let $T = \{\tau_0, \tau_1, \ldots\}$ be an enumeration of D^3 which is recursive in the given enumeration of D. Call a triple $\tau = ((\tau)_0, (\tau)_1, (\tau)_2)$ **good** if $(\tau)_0$ is a predictor and $(\tau)_1$ is a strongly lawlike injection with converse $(\tau)_2$.

Call a triple τ **nice at** w **for** n when both w and $(\tau)_0(w)$ are sequence numbers and if

$$p = lh(w * (\tau)_0(w)) \quad \text{and} \quad m = max(n, max\{(\tau)_1(i) : 0 \leq i < p\}) + 1$$

then for each $0 \leq i < j < p$:

$$(\tau)_1(i) \neq (\tau)_1(j)$$

and for each $0 \leq j < m$:

$$\text{if} \quad 0 < (\tau)_2(j) \quad \text{then} \quad (\tau)_1((\tau)_2(j) - 1) = j.$$

Observe that niceness (unlike goodness) is effectively decidable, and τ is good if and only if τ is nice at *every* sequence number w for *every* n.

By induction on k we define x_k, w_k, n_k (with $n_0 < n_1 < \cdots$) and $\overline{\beta}(n_k)$ as follows. For convenience set $n_{-1} = 0$. In general, let x_k be the least $x \geq 0$ such that for all $0 \leq j < n_{k-1}$, $(\tau_k)_2(j) \leq x$. (In particular, $x_0 = 0$.) Let w_k be the sequence number of length x_k such that for each $i < x_k$:

$$(w_k)_i = \begin{cases} \beta((\tau_k)_1(i)) + 1 & \text{if } (\tau_k)_1(i) < n_{k-1}, \\ 1 & \text{otherwise.} \end{cases}$$

If τ_k is not nice at w_k for n_{k-1}, let $n_k = n_{k-1} + 1$ and $\beta(n_k - 1) = 0$. Otherwise, let $p_k = x_k + lh((\tau_k)_0(w_k))$ and $n_k = max(n_{k-1}, max\{(\tau_k)_1(i) : 0 \leq i < p_k\}) + 1$, and for each $n_{k-1} \leq j < n_k$ define

$$\beta(j) = \begin{cases} (w_k * (\tau_k)_0(w_k))_i - 1 & \text{if } (\tau_k)_2(j) = i + 1 \text{ so } (\tau_k)_1(i) = j, \\ 0 & \text{if } (\tau_k)_2(j) = 0. \end{cases}$$

The reader may verify that if τ_k is nice at w_k for n_{k-1} then $(\overline{\beta \circ (\tau_k)_1})(p_k) = w_k * (\tau_k)_0(w_k)$, so β is lawless.

LEMMA 3. (Uniformity Lemma.) *If α is lawless, π is a lawlike predictor, and γ is a strongly lawlike injection, then for each $x_0 \in N$ there is some $x \geq x_0$ such that*

$$\alpha \circ \gamma \in (\overline{\alpha \circ \gamma})(x) * \pi((\overline{\alpha \circ \gamma})(x)).$$

[13]Troelstra's distinction between lawless and protolawless sequences is lost here.

LEMMA 4. (Second Density Lemma.) *If D is countable and α is lawless then the class of all β such that $[\alpha, \beta]$ is lawless is dense in B.*

PROOF. Assume α is lawless relative to D, and let T be as in the proof of Lemma 2. Call a triple τ **α-nice at w for n** if τ is nice at w for n, and if $i < lh(w * (\tau)_0(w))$ and $(\tau)_1(i) = 2q$ then $(w * (\tau)_0(w))_i = \alpha(q) + 1$. We define β in stages.

Stage 0. Let $n_0 = 0$ so $\overline{\beta}(n_0) = \langle\ \rangle = 1$. For notational convenience set $n_{-1} = 0$.

In general, at the conclusion of stage m we have $n_0 < n_1 < \cdots < n_m$ and values $\overline{\beta}(n_m)$. For $k \leq m$ we say $\overline{\beta}(n_k)$ **is permanent at** m if for each $j \leq k$ either

(i) for some sequence number $w \leq m$, τ_j is not nice at w for m, or

(ii) for some $s \leq m$, if $w = \overline{[\alpha, \overline{\beta}(n_k) * \lambda t\ 0] \circ (\tau_j)_1}(s)$ and $p = lh(w * (\tau_j)_0(w))$ then τ_j is α-nice at w for m and for each $i < p$: if $(\tau_j)_1(i) = 2q + 1$ then $q < n_k$ and $(w * (\tau_j)_0(w))_i = \beta(q) + 1$. Observe that in this case $w * (\tau_j)_0(w) = \overline{[\alpha, \overline{\beta}(n_k) * \gamma] \circ (\tau_j)_1}(p)$ for every $\gamma \in B$.

Stage $m+1$: Consider the least $k \leq m+1$ such that $\overline{\beta}(n_k)$ is not permanent at m. *Case 1.* If for some $s \leq m$ τ_k is α-nice at $w = \overline{[\alpha, \overline{\beta}(n_{k-1}) * \lambda t\ 0] \circ (\tau_k)_1}(s)$ for n_{k-1}, let w_k be the least such w and (re)define $n_k = max(n_{k-1}, max\{q : (\tau_k)_1(i) = 2q + 1 \text{ for some } 0 \leq i < p_k\}) + 1$ where $p_k = lh(w_k * (\tau_k)_0(w_k))$. For $n_{k-1} \leq j < n_k$ (re)define

$$\beta(j) = \begin{cases} (w_k * (\tau_k)_0(w_k))_i - 1 & \text{if } (\tau_k)_2(2j+1) = i+1, \\ 0 & \text{if } (\tau_k)_2(2j+1) = 0. \end{cases}$$

If $n_k < m + 1$, (re)define $n_{k+i} = n_k + i$ and $\beta(n_{k+i}) = 0$ for $i = 1, \ldots, m + 1 - k$. Observe that $\overline{\beta}(n_k)$ is permanent at $m + 1$ in this case. *Case 2.* Otherwise, set $n_{m+1} = n_m + 1$ and $\beta(n_m) = 0$.

Relative to α and T the construction is effective and for each $k \leq m$ one can decide effectively whether $\overline{\beta}(n_k)$ is permanent at m. By Markov's Principle with Lemma 3 and the lawlessness of α, for each k there is a stage at which $\overline{\beta}(n_k)$ becomes permanent, and if τ_k is good then $w_k * (\tau_k)_0(w_k) = \overline{[\alpha, \beta] \circ (\tau_k)_1}(p_k)$; so $[\alpha, \beta]$ is lawless. If u is any sequence number then $[\alpha, u * \beta]$ is lawless by Lemma 1, and the proof is complete.

3.2. The formal predicate. In the language of **IDLS** (or **IDLS$^-$**) we may express "α is lawless relative to D" by the almost negative formula

$$\mathrm{DLS}(\alpha) \equiv \forall b \forall c \forall d[\mathrm{Pred}(b) \ \&\ \mathrm{Inv}(c, d) \supset \exists x\ \alpha \circ c \in (\overline{\alpha \circ c})(x) * b((\overline{\alpha \circ c})(x))]$$

where

$$\mathrm{Pred}(b) \equiv \forall w[\mathrm{Seq}(w) \supset \mathrm{Seq}(b(w))]$$

and

$$\mathrm{Inv}(c, d) \equiv \forall x \forall y[c(x) = y \sim d(y) = x + 1].^{14}$$

The assumption "D is countable" may be expressed formally by $\exists\delta \mathrm{ED}(\delta)$ where

$$\mathrm{ED}(\delta) \equiv \forall n \exists a(a = (\delta)_n) \ \&\ \forall a \exists n(a = (\delta)_n).$$

---

[14]Note that $\exists$d Inv(c,d) economically expresses "c is a strongly lawlike injection."

We do *not* assume this formally. Eventually we will consider the weaker assertive assumption $\exists\delta ED^-(\delta)$ ("D is weakly countable") where

$$ED^-(\delta) \equiv \forall n \exists a(a = (\delta)_n) \ \&\ \forall a \neg\forall n \neg(a = (\delta)_n).$$

3.3. Consistency questions. By [12], under the assumption of a certain (classically consistent) set-theoretic axiom there is a classical model, with countably many lawlike sequences, for a theory **DLS** of which the current **BDLS** is (modulo notation) a proper subsystem.[15] Thus **BDLS** + $\exists\delta ED(\delta)$ is classically consistent.

To verify the constructive content as well as the consistency of **IDLS**$^-$ and **IDLS** it is natural to look for realizability interpretations analogous to the one developed by Kleene in [8] for **FIM**. The next section provides a classical function-realizability interpretation for each of the new systems, relative to a defined class D of "lawlike" sequences, under the assumption that D is countable.

§4. The realizability interpretations.

4.1. Definition of D. Let $E_0(x, y), E_1(x, y), \ldots$ be an enumeration of all almost negative D-formulas having free no number variables except the distinct variables x and y; in particular let $E_0(x, y) \equiv a(x) = y$. For each i let

$$F_i \equiv \forall x \exists! y E_i(x, y).$$

The primitive recursive function symbols $\lambda, 0, ', +, \cdot, \ldots, f_p, =$ will have their standard interpretations.

If $a_0, \ldots, a_{k-1}$ is the (possibly empty) list of the distinct variables occurring free in F_i in order of first free occurrence, and if $A \subset B$ and $\phi \in B$ and $\psi_0, \ldots, \psi_{k-1} \in A$, we say that E_i **defines** ϕ **over** A **from** $\psi_0, \ldots, \psi_{k-1}$ if and only if, when number variables range over N, definable function variables over A, and arbitrary function variables over B and $a_0, \ldots, a_{k-1}$ are interpreted by $\psi_0, \ldots, \psi_{k-1}$:

(i) F_i is true classically, and

(ii) for $x, y \in N$: $\phi(x) = y$ if and only if $E_i(x, y)$ is true.

We say E_i **defines** φ **uniformly over A** if and only if for *all* $\psi_0, \ldots, \psi_{k-1} \in A$, E_i defines $\varphi[\psi_0, \ldots, \psi_{k-1}] = \varphi$ over A from $\psi_0, \ldots, \psi_{k-1}$. Observe that E_0 defines φ uniformly over A, where $\varphi[\phi] = \phi$.

Now let $\mathsf{Def}(A)$ be the class of all $\phi \in B$ which are definable over A by some E_i, from some $\psi_0, \ldots, \psi_{k-1} \in A$. Let

$$D_0 = \emptyset, \ D_{\eta+1} = \mathsf{Def}(D_\eta),$$

and for limit ordinals λ :

$$D_\lambda = \bigcup_{\zeta<\lambda} D_\zeta.$$

We want this induction to close off at a countable ordinal. The key is to observe that $\bigcup_{\zeta \in OR} D_\zeta$ has a natural definable well-ordering.

[15] **DLS** omits the "!" in ^{x}2.2!D$^-$ and the requirements of almost negativity from all axioms, strengthens the present ^{x}DLS4$^-$ and asserts the countable axiom of choice for the class of lawlike functions. The classical model naturally fails to satisfy ^{x}KL1.

In general, if $\prec$ well-orders A, and $\phi, \psi \in \mathsf{Def}(A)$, set $\phi \prec^* \psi$ if and only if either

(i) $\phi, \psi \in A$ and $\phi \prec \psi$, or

(ii) $\phi \in A$ and $\psi \notin A$, or

(iii) $\phi \notin A, \psi \notin A$, and $\Delta_A(\phi) < \Delta_A(\psi)$, where $\Delta_A(\phi)$ is the smallest tuple $(i, \psi_0, \dots, \psi_{k-1})$ in the lexicographic well-ordering $<$ of $N \cup \bigcup_{k>0}(N \times A^k)$ determined by $<$ on N and $\prec$ on A such that E_i defines ϕ over A from $\psi_0, \dots, \psi_{k-1}$.

Now let

$$\prec_0 = \emptyset, \qquad \prec_{\eta+1} = (\prec_\eta)^*,$$

and for limit ordinals λ :

$$\prec_\lambda = \bigcup_{\zeta<\lambda} \prec_\zeta .$$

Clearly $\prec_\eta$ well-orders D_η for each ordinal η. Since each $D_\eta \subset D_{\eta+1} \subset B$, by cardinality considerations there is a least ordinal ξ such that $D_\xi = D_{\xi+1}$; for this ξ let

$$D = D_\xi \quad \text{and} \quad \prec = \prec_\xi .$$

Then $\prec$ is a definable well-ordering of D. In fact, both D and $\prec$ are Δ^2_1 definable over B. If E_i defines $\varphi[\psi_0, \dots, \psi_{k-1}]$ uniformly over D we naturally call $\lambda\Psi\ \varphi$ a **definable operator on** D.

4.2. The countability assumption. We now assume that D is countable, in accord with Brouwer's assertion [1] (see also [3]) that every well-ordered species is countable and with the discussion in Section 1.2 above. Levy [11] proved the classical consistency with **ZFC** (relative to **ZF**) of the assumption that every definably well-ordered subclass of Baire space is countable; hence our assumption is classically consistent as well as constructively plausible.

No enumerating function can itself be lawlike, since D is closed under recursive operations and if δ enumerates D then for no $n \in N$ is $\lambda t((\delta(t))_t + 1) = (\delta)_n$. All we are assuming is that *some* enumerating function exists (i.e., that $\exists\delta\ \mathrm{ED}(\delta)$ is classically true) so the conclusions of the density lemmas hold.

4.3. Realizability/$_D$ and realizability//$_D$. Following Kleene, if $\tau, \alpha \in B$ we say $\{\tau\}[\alpha]$ is **properly defined** if and only if $(t)(E!y)\tau(\langle t\rangle * \overline{\alpha}(y)) > 0$, and then

$$\{\tau\}[\alpha](t) \simeq \tau(\langle t\rangle * \overline{\alpha}(\mu y \tau(\langle t\rangle * \overline{\alpha}(y)) > 0)) - 1.$$

If $x \in N$ then $\{\tau\}[x] \simeq \{\tau\}[\lambda t\ x]$; and $\{\tau\} \simeq \{\tau\}[0]$. If $x_1, \dots, x_k \in N$ and $\alpha_1, \dots, \alpha_m \in B$ then

$$\{\tau\}[x_1, \dots, x_k, \alpha_1, \dots, \alpha_m] \simeq \{\tau\}[(x_1, \dots, x_k, \alpha_1, \dots, \alpha_m)].$$

As in [8], if $\varphi[\Theta, \alpha]$ is partial recursive then there is a primitive recursive functional $\Lambda\alpha\ \varphi[\Theta, \alpha]$ such that

$$\{\Lambda\alpha\ \varphi[\Theta, \alpha]\}[\alpha] \simeq \varphi[\Theta, \alpha]$$

and if $\varphi[\Theta, \alpha]$ is completely defined then $\{\Lambda\alpha\ \varphi[\Theta, \alpha]\}[\alpha]$ is properly defined. Also $\Lambda x\ \varphi[\Theta, x] = \Lambda\alpha\ \varphi[\Theta, \alpha(0)]$ and $\Lambda\ \varphi[\Theta] = \Lambda x\ \varphi[\Theta]$, so $\{\Lambda x\ \varphi[\Theta, x]\}[x] \simeq \varphi[\Theta, x]$ and $\{\Lambda\ \varphi[\Theta]\} \simeq \varphi[\Theta]$.

Kleene's Λ thus incorporates the meaning of his S^m_n theorem. There is an obvious relativized notion $\Lambda^{\Psi}\alpha$ for functionals recursive in Ψ, where Ψ is any list of functions from B.

An **appropriate** interpretation of a list $\alpha_0, \dots, \alpha_k, \mathrm{a}_0, \dots, \mathrm{a}_m, \mathrm{x}_0, \dots, \mathrm{x}_n$ of variables of the types indicated is any choice of functions $\alpha_0, \dots, \alpha_k \in B$, $\phi_0, \dots, \phi_m \in D$, and numbers $x_0, \dots, x_n$. We now define when $\pi \in B$ **realizes-Ψ** a formula E all of whose distinct free variables are interpreted appropriately by the functions and numbers Ψ. The reader acquainted with Kleene's function-realizability interpretations need look only at the new Clauses 8 and 9.

1. π **realizes-Ψ** a prime formula P, if P is true-Ψ.

2. π **realizes-Ψ** A & B, if $(\pi)_0$ **realizes-Ψ** A and $(\pi)_1$ **realizes-Ψ** B.

3. π **realizes-Ψ** A $\vee$ B, if $(\pi(0))_0 = 0$ and $(\pi)_1$ **realizes-Ψ** A, or $(\pi(0))_0 \neq 0$ and $(\pi)_1$ **realizes-Ψ** B.

4. π **realizes-Ψ** A $\supset$ B, if, if σ **realizes-Ψ** A, then $\{\pi\}[\sigma]$ (is properly defined and) **realizes-Ψ** B.

5. π **realizes-Ψ** $\neg$A, if π **realizes-Ψ** A $\supset 1 = 0$.

6. π **realizes-Ψ** $\forall$xA(x), if, for each $x \in N$, $\{\pi\}[x]$ **realizes-Ψ, x** A(x).

7. π **realizes-Ψ** $\exists$xA(x), if $(\pi)_1$ **realizes-$\Psi, (\pi(0)_0)$** A(x).

8. π **realizes-Ψ** $\forall$aA(a), if, for each $\phi \in D$, $\{\pi\}[\phi]$ is completely defined and **realizes-Ψ, ϕ** A(a).

9. π **realizes-Ψ** $\exists$aA(a), if $\{(\pi)_0\} \in D$ and $(\pi)_1$ **realizes-$\Psi, \{(\pi)_0\}$** A(a).

10. π **realizes-Ψ** $\forall\alpha$A(α), if, for each $\alpha \in B$, $\{\pi\}[\alpha]$ is completely defined and **realizes-Ψ, α** A(α).

11. π **realizes-Ψ** $\exists\alpha$A(α), if ($\{(\pi)_0\}$ is properly defined and) $(\pi)_1$ **realizes-$\Psi, \{(\pi)_0\}$** A(α).

We say a closed formula E is **realizable/$_D$** [**realizable//$_D$**], if a function π general recursive in finitely many functions of [and finitely many definable operators on] D realizes E. An open formula is **realizable/$_D$** [**realizable//$_D$**] if and only if its closure is.

Note that a formula E all of whose free variables occur among Ψ is realizable/$_D$ [realizable//$_D$] if and only if there is a function φ partial recursive in finitely many functions of [and definable operators on] D such that, for each appropriate Ψ, $\varphi[\Psi]$ is completely defined and realizes-Ψ E. Such a φ is called a **realization/$_D$** [**realization//$_D$**] **function** for E.

LEMMA 5. *Let Ψ be any list of variables and let Ψ_1 be those of Ψ which occur free in* E. *For each ε and each appropriate interpretation of the variables: ε realizes-Ψ_1* E *if and only if ε realizes-Ψ* E.

LEMMA 6. *For each assertive formula* E *containing free only the variables Ψ and each appropriate Ψ:*

(i) If E *is realized-Ψ by some function ε then* E *is true-Ψ.*

To each almost negative formula E *containing free only the variables Ψ there is a partial recursive function $\varepsilon_{\mathrm{E}}[\Psi] = \lambda t \varepsilon_{\mathrm{E}}(\Psi, t)$ such that for each appropriate interpretation Ψ of the free variables:*

(ii) If E *is true-Ψ then $\varepsilon_{\mathrm{E}}[\Psi]$ is completely defined and realizes-Ψ* E.

The proof is like that of Lemma 8.4 of [8], with three new cases for (ii). If E is $\forall$aA(a) then $\varepsilon_E[\Psi]$ is $\Lambda\phi\varepsilon_{A(a)}[\Psi,\phi]$. If E is $\exists$aA(a) where A(a) is prime and Ψ is β, c, x then $\varepsilon_E[\Psi]$ is $(\Lambda\ \lambda t(\mu s[Seq(s)\ \&\ T_1^{2,1}(\overline{\beta}(lh(s)),\overline{\varsigma}(lh(s)),s,n,x)])_t - 1, \lambda t\, 0)$, where n is a Gödel number of the primitive recursive predicate $P(\phi,\beta,\varsigma,x)$ expressed by A(a). And if E is $\exists$a$\forall$x(a(x) = t) where t = t[Ψ, x] is a term containing only Ψ, x free, representing the primitive recursive function $t[\Psi,x]$, then $\varepsilon_E[\Psi]$ is $(\Lambda\ \lambda x\ t[\Psi,x], \Lambda x\lambda s\, 0)$.

Since the predicate DLS(α) is almost negative, in particular α is lawless relative to D if and only if DLS(α) is realized-α by some function ε, if and only if $\varepsilon_{DLS(\alpha)}[\alpha]$ realizes-α DLS(α). This fact will be crucial to the proof of the main theorem.

LEMMA 7. *Lemma 8.5 of [8] on numeralwise representability (expressibility) of general recursive functions (predicates) is true for all the formal systems here, as are Lemmas 8.7 and 8.8 on formal decidability of the representing predicates. Hence D is closed under recursion.*

LEMMA 8. *Lemmas 9.1 and 9.2 of [8] on substitution of terms and functors carry over, with the new part for Lemma 9.1:*

(c) If g[Ψ_1, a] *is a D-functor, free for* a *in* A(a) *and containing free only the definable function and number variables* Ψ_1, a *where* $\Psi_1 \subset \Psi$, *then* g *represents a primitive recursive function* $g[\Psi_1,\phi]$, *and* ε *realizes-*Ψ, $g[\Psi_1,\phi]$ A(a) *if and only if* ε *realizes-*Ψ,ϕ A(g).

4.4. The consistency of ***IDLS*****⁻**.

THEOREM 1. *If* $\Gamma \vdash$ E *in* **IDLS⁻** *and the formulas* Γ *are all realizable*$//_D$ *then* E *is realizable*$//_D$.

PROOF. For each axiom E of **IDLS** which is "new" (in the sense that it is not an axiom of **FIM** extended to the language of **IDLS⁻**) we give a realization$//_D$ function $\varphi = \lambda\Psi\ \varphi[\Psi]$ where $\varphi[\Psi] = \lambda t\varphi(\Psi,t)$; and assuming that such a $\varphi'[\Psi']$ exists for each premise of a new rule of inference, we give a $\varphi[\Psi]$ for the conclusion.[16]

10D. $\forall$aA(a) $\supset$ A(g) where g is a D-functor free for a in A(a). Let $\varphi[\Psi]$ be $\Lambda\sigma\ \{\sigma\}[g[\Psi_1,\phi]]$ where $g[\Psi_1,\phi]$ is the interpretation of g[Ψ_1, a] and ϕ interprets a in Ψ.

11D. A(g) $\supset$ $\exists$aA(a) with the same conditions on g. Define $\varphi[\Psi]$ to be $\Lambda\sigma$ $(\Lambda\ g[\Psi_1,\phi],\sigma)$.

Rule 9D. C $\supset$ A(a) / C $\supset$ $\forall$aA(a) where a is not free in C. If $\varphi'[\Psi,\phi]$ realizes-Ψ,ϕ C $\supset$ A(a) for each $\phi \in D$, let $\varphi[\Psi]$ be $\Lambda\sigma\Lambda\phi\{\varphi'[\Psi,\phi]\}[\sigma]$.

Rule 12D. A(a) $\supset$ C / $\exists$aA(a) $\supset$ C. If $\varphi'[\Psi,\phi]$ realizes-Ψ,ϕ A(a) $\supset$ C for each $\phi \in D$, let $\varphi[\Psi]$ be $\Lambda\sigma\{\varphi'[\Psi,\{(\sigma)_0\}]\}[(\sigma)_1]$.

$^{\times}$2.2!D.- $\forall$x$\exists$!yA(x, y) $\supset$ $\exists$a$\forall$xA(x, a(x)) where A(x, y) is almost negative and all the free variables Ψ are lawlike function or number variables. If σ realizes-Ψ the hypothesis then $\forall$x$\exists$!yA(x, y) is true-Ψ (using Lemma 6), so A(x, y) defines

[16] In the proof of Kleene's corresponding theorem for **FIM** (Theorem 9.3 of [8]), recursive realization functions were provided for all the "old" axioms and rules.

$\psi = \lambda x(\{\sigma\}[x](0))_{0,0}$ over D from Ψ, so $\psi \in D$. The axiom is realized$//_D$ by $\varphi[\Psi] = \Lambda\sigma(\Lambda\ \lambda x(\{\sigma\}[x](0))_{0,0}, \Lambda x(\{\sigma\}[x])_{0,1})$.

^{x}DLS1$^-$ and ^{x}DLS2$^-$ are almost negative, and true by the density lemmas with the countability assumption, so Lemma 6 provides recursive realization functions for them.

^{x}DLS3$^-$. $\forall\alpha[\mathrm{DLS}(\alpha)\ \&\ \mathrm{A}(\alpha) \supset \exists x\forall\beta[\overline{\beta}(x) = \overline{\alpha}(x)\ \&\ \mathrm{DLS}(\beta) \supset \mathrm{A}(\beta)]]$, where $\mathrm{A}(\alpha)$ is restricted and almost negative and contains free no arbitrary sequence variables but α. For convenience denote the almost negative subformula $\forall\beta[\overline{\beta}(x) = \overline{\alpha}(x)\ \&\ \mathrm{DLS}(\beta) \supset \mathrm{A}(\beta)]$ by $\mathrm{F}(\overline{\alpha}(x))$, so the axiom $\mathrm{E}[\Psi]$ is $\forall\alpha[\mathrm{DLS}(\alpha)\ \&\ \mathrm{A}(\alpha) \supset \exists x\mathrm{F}(\overline{\alpha}(x))]$. By Lemma 6, there is a partial recursive function $\varepsilon_{\mathrm{F}}[\Psi,\alpha,x]$ which realizes-Ψ,α,x $\mathrm{F}(\overline{\alpha}(x))$ if and only if $\mathrm{F}(\overline{\alpha}(x))$ is true. By induction on the logical form of $\mathrm{A}(\alpha)$, we provide a function $\xi(\Psi,\alpha,\sigma)$ partial recursive in functions from and definable operators on D so that if σ realizes-Ψ,α $\mathrm{DLS}(\alpha)\ \&\ \mathrm{A}(\alpha)$, then $\xi(\Psi,\alpha,\sigma)$ is defined and $\mathrm{F}(\overline{\alpha}(x))$ is true-$\Psi,\alpha,\xi(\Psi,\alpha,\sigma)$ and hence $\rho[\Psi,\alpha,\sigma] = (\lambda t\xi(\Psi,\alpha,\sigma), \varepsilon_{\mathrm{F}}[\Psi,\alpha,\xi(\Psi,\alpha,\sigma)])$ realizes-Ψ,α $\exists x\mathrm{F}(\overline{\alpha}(x))$. Then $\varphi[\Psi] = \Lambda\alpha\Lambda\sigma\ \rho[\Psi,\alpha,\sigma]$ realizes-Ψ the axiom.

1. $\mathrm{A}(\alpha)$ is $\mathrm{s} = \mathrm{t}$, where s expresses $s(\Psi,\alpha)$ and t expresses $t(\Psi,\alpha)$ and Ψ consists only of numbers $z_0,\dots,z_{l-1}$ and elements $\psi_0,\dots,\psi_{k-1}$ of D. Since s and t are primitive recursive the (representing function of the) predicate $\mathrm{s} = \mathrm{t}$ has a Gödel number e from Ψ,α. Let

$$\xi(\Psi,\alpha,\sigma) \simeq \mu x T_1^{k+1,l}(\overline{\psi_0}(x),\dots,\overline{\psi_{k-1}}(x),\overline{\alpha}(x),e,z_0,\dots,z_{l-1}).$$

If σ realizes-Ψ,α $\mathrm{DLS}(\alpha)\ \&\ \mathrm{s} = \mathrm{t}$ then $\alpha \in L$ and $s(\Psi,\alpha) = t(\Psi,\alpha)$ is true, so $\xi(\Psi,\alpha,\sigma)$ is defined and has an appropriate value.

2. $\mathrm{A}(\alpha)$ is $\mathrm{B}(\alpha)\ \&\ \mathrm{C}(\alpha)$. By the induction hypothesis there are realization$//_D$ functions $\chi_{\mathrm{B}}, \chi_{\mathrm{C}}$ for the instances of ^{x}DLS3$^-$ with $\mathrm{B}(\alpha)$, $\mathrm{C}(\alpha)$ respectively as the $\mathrm{A}(\alpha)$. If σ realizes-Ψ,α $\mathrm{DLS}(\alpha)\ \&\ \mathrm{A}(\alpha)$, then $\nu_{\mathrm{B}} = ((\sigma)_0, (\sigma)_{1,0})$ realizes-Ψ,α $\mathrm{DLS}(\alpha)\ \&\ \mathrm{B}(\alpha)$ and $\nu_{\mathrm{C}} = ((\sigma)_0, (\sigma)_{1,1})$ realizes-Ψ,α $\mathrm{DLS}(\alpha)\ \&\ \mathrm{C}(\alpha)$, so take $\xi(\Psi,\alpha,\sigma)$ to be the larger of $(\{\{\chi_{\mathrm{B}}[\Psi]\}[\alpha]\}[\nu_{\mathrm{B}}](0)_0)$ and $(\{\{\chi_{\mathrm{C}}[\Psi]\}[\alpha]\}[\nu_{\mathrm{C}}](0)_0)$.

4. $\mathrm{A}(\alpha)$ is $\neg\mathrm{B}(\alpha)$. By the induction hypothesis there is a realization$//_D$ function χ_{B} for the instance of ^{x}DLS3$^-$ for $\mathrm{B}(\alpha)$. Recall that 1 is the smallest sequence number, coding the empty sequence $\langle\ \rangle$. Consider the almost negative predicates

$$\mathrm{D(w,v)} \equiv (\mathrm{Seq(w)} \supset \mathrm{Seq(v)})\ \&\ (\mathrm{v} = 1 \supset \forall\beta[\beta \in \mathrm{w}\ \&\ \mathrm{DLS}(\beta) \supset \neg\mathrm{B}(\beta)])$$
$$\&\ (\mathrm{v} > 1 \supset \forall\beta[\beta \in \mathrm{w} * \mathrm{v}\ \&\ \mathrm{DLS}(\beta) \supset \mathrm{B}(\beta)]),$$

$$\mathrm{E(w,v)} \equiv \mathrm{D(w,v)}\ \&\ \forall\mathrm{u}(\mathrm{u} < \mathrm{v} \supset \neg\mathrm{D(w,u)}).$$

Classically, $\forall\mathrm{w}\exists!\mathrm{vE(w,v)}$ is true-Ψ, by the following argument. If w is a sequence number and $\mathrm{DLS}(\gamma)\ \&\ \mathrm{B}(\gamma)$ is true-Ψ,γ for some γ with $\gamma \in w$ then by Lemma 6 some ε recursive in Ψ,γ realizes-Ψ,γ $\mathrm{DLS}(\gamma)\ \&\ \mathrm{B}(\gamma)$. Hence $\{\{\chi_{\mathrm{B}}[\Psi]\}[\gamma]\}[\varepsilon]$ realizes-Ψ,γ $\exists x\forall\beta[\overline{\beta}(x) = \overline{\gamma}(x)\ \&\ \mathrm{DLS}(\beta) \supset \mathrm{B}(\beta)]$, so this feebly assertive formula is true-Ψ,γ and there is a nontrivial sequence number v for which $\forall\beta[\beta \in \mathrm{w} * \mathrm{v}\ \&\ \mathrm{DLS}(\beta) \supset \mathrm{B}(\beta)]$ is true-Ψ. So $\mathrm{E(w,v)}$ defines a function ϕ uniformly from $\Psi \in D$, so $\phi \in D$. Since ϕ is a lawlike predictor, for each α making $\mathrm{DLS}(\alpha)$ true there is some x for which $\alpha \in \overline{\alpha}(x) * \phi(\overline{\alpha}(x))$, so if $\neg\mathrm{B}(\alpha)$ is true-Ψ then $\phi(\overline{\alpha}(x)) = 1$. Let

$$\xi(\Psi,\alpha,\sigma) \simeq \mu x\ \alpha \in \overline{\alpha}(x) * \phi(\overline{\alpha}(x)).$$

5. $A(\alpha)$ is $B(\alpha) \supset C(\alpha)$. Similarly, consider the almost negative predicates

$$G(w,v) \equiv (Seq(w) \supset Seq(v)) \;\&\; (v = 1 \supset \forall\beta[\beta \in w \;\&\; DLS(\beta) \supset (B(\beta) \supset C(\beta))]) \\ \&\; (v > 1 \supset \forall\beta[\beta \in w * v \;\&\; DLS(\beta) \supset B(\beta) \;\&\; \neg C(\beta)]),$$

$$H(w,v) \equiv G(w,v) \;\&\; \forall u(u < v \supset \neg G(w,u)).$$

By Cases 2 and 4 (already established) with the induction hypothesis on B and C, classically $H(w,v)$ defines a lawlike predictor ϕ uniformly from $\Psi \in D$. Define ξ from ϕ as in Case 4.

6. $A(\alpha)$ is $\forall x B(\alpha, x)$, where $B(\alpha, x)$ is almost negative. Consider the almost negative predicates

$$J(w, v, x) \equiv (v = 1 \supset \forall\beta[\beta \in w \;\&\; DLS(\beta) \supset \forall x B(\beta, x)]) \\ \&\; (v > 1 \supset \forall\beta[\beta \in w * v \;\&\; DLS(\beta) \supset \neg B(\beta, x)]),$$

$$K(w,v) \equiv (Seq(w) \supset Seq(v)) \\ \&\; \neg\forall x \neg J(w, v, x) \;\&\; \forall y(y < v \;\&\; Seq(y) \supset \forall x \neg J(w, y, x)).$$

Classically, by Case 4 and the induction hypothesis on B, $K(w,v)$ defines a lawlike predictor ϕ uniformly from $\Psi \in D$. Define ξ from ϕ as in the preceding two cases.

7. $A(\alpha)$ is $\exists x B(\alpha, x)$ where $B(\alpha, x)$ is prime. Let f be a Gödel number of the (representing function of the) primitive recursive predicate expressed by $B(\alpha, x)$, and suppose Ψ consists of k lawlike function variables and l number variables. Let

$$\xi(\Psi, \alpha, \sigma) \simeq \mu x\; T_1^{k+1, l+1}(\overline{\psi_0}(x), \ldots, \overline{\psi_{k-1}}(x), \overline{\alpha}(x), f, z_0, \ldots, z_{l-1}, ((\sigma)_1(0))_0).$$

If σ realizes-Ψ, α $DLS(\alpha) \;\&\; \exists x B(\alpha, x)$ then $B(\alpha, x)$ is true-$\Psi, \alpha, ((\sigma)_1(0))_0$ so $\xi(\Psi, \alpha, \sigma)$ is defined and $F(\overline{\alpha}(x))$ is true-$\Psi, \alpha, \xi(\Psi, \alpha, \sigma)$.

8. $A(\alpha)$ is $\forall a B(\alpha, a)$. Similar to Case 6.

9. $A(\alpha)$ is $\exists a B(\alpha, a)$ where $B(\alpha, a)$ is prime or of the form $\forall x(a(x) = t(\alpha, x))$ for a term t not containing a free. The induction hypothesis gives a realization$//_D$ function χ_B for the instance of ${}^{x}DLS3^{-}$ with $B(\alpha, a)$ as the $A(\alpha)$. If σ realizes-Ψ, α $DLS(\alpha) \;\&\; \exists a B(\alpha, a)$ then $\{(\sigma)_{1,0}\} \in D$ and $\nu_B = ((\sigma)_0, (\sigma)_{1,1})$ realizes-$\Psi, \{(\sigma)_{1,0}\}$ $DLS(\alpha) \;\&\; B(\alpha, a)$ so $\xi(\Psi, \alpha, \sigma) \simeq (\{\{\chi_B[\Psi, \{(\sigma)_{1,0}\}]\}[\alpha]\}[\nu_B](0))_0$.

10. $A(\alpha)$ is $\forall\gamma[DLS([\alpha, \gamma]) \supset B(\alpha, \gamma)]$. By Case 4 with the induction hypothesis there is a realization$//_D$ function $\varphi[\Psi]$ for $\forall\gamma[DLS(\gamma) \;\&\; \neg B({}^2[\gamma]_0, {}^2[\gamma]_1) \supset \exists x \forall\delta[\overline{\delta}(x) = \overline{\gamma}(x) \;\&\; DLS(\delta) \supset \neg B({}^2[\delta]_0, {}^2[\delta]_1)]]$. Consider the almost negative predicates

$$L(w,v) \equiv (Seq(w) \supset Seq(v)) \\ \&\; (v = 1 \supset \forall\beta[\beta \in w \;\&\; DLS(\beta) \supset \forall\gamma[DLS([\beta, \gamma]) \supset B(\beta, \gamma)]]) \\ \&\; (v > 1 \supset \forall\beta[\beta \in w * v \;\&\; DLS(\beta) \supset \neg\forall\gamma[DLS([\beta, \gamma]) \supset B(\beta, \gamma)]]),$$

$$M(w,v) \equiv L(w,v) \;\&\; \forall y\, (y < v \supset \neg L(w,y)).$$

Then $\forall w \exists! v M(w,v)$ is true-Ψ so $M(w,v)$ uniformly defines a $\phi \in D$ from which ξ is determined as before.

Since $A(\alpha)$ is restricted and almost negative, no other cases can occur.

${}^{x}DLS4.^{-}$ $\forall\alpha[DLS(\alpha) \supset \exists x A(\alpha, x)] \supset \exists e B(e)$ where $B(e) \equiv \forall\alpha[DLS(\alpha) \supset \exists! y\, e(\overline{\alpha}(y)) > 0 \;\&\; \forall y(e(\overline{\alpha}(y)) > 0 \supset A(\alpha, e(\overline{\alpha}(y)) - 1))]$ and $A(\alpha, x)$ is restricted and almost negative with no free function variables but α, so $B(e)$ is almost negative also. Consider the almost negative predicates

$$\begin{aligned}P(w, x) \equiv (\mathrm{Seq}(w) \supset \mathrm{Seq}((x)_0))\\ \& ((x)_0 > 1 \supset \forall\alpha[\mathrm{DLS}(\alpha) \& \alpha \in w * (x)_0 \supset A(\alpha, (x)_1)])\\ \& ((x)_0 = 1 \supset \forall\alpha[\mathrm{DLS}(\alpha) \& \alpha \in w \supset \forall y \neg A(\alpha, y)]),\end{aligned}$$

$$Q(w, x) \equiv P(w,x) \& \forall u < x \neg P(w,u).$$

Classically $\forall w \exists! x Q(w, x)$ is true-Ψ by Lemma 6 with the realizability of ${}^{\times}$DLS3^{-}, just established, so Q(w, x) defines uniformly from Ψ some $\psi = \psi[\Psi]$ in D. Define χ from ψ by

$$\chi(w) = \begin{cases} (\psi(w))_1 + 1 & \text{if } Seq(w) \text{ and } (\psi(w))_0 > 1, \\ 0 & \text{otherwise,} \end{cases}$$

Then $\chi = \chi[\Psi] \in D$, uniformly in Ψ, and $\Lambda\sigma(\Lambda\, \chi, \varepsilon_{B(e)}[\chi])$ is a realization$//_D$ function for the axiom.

${}^{\times}$KL1. A realization$//_D$ function is $\Lambda\sigma(\Lambda\, \tau, \Lambda\alpha\Lambda\rho\, (\varepsilon_{\forall x \exists! y \tau(\langle x\rangle * \overline{\alpha}(y))>0}, \pi))$, where $\tau = \Lambda\alpha\{(\{\{\sigma\}[\alpha]\}[\varepsilon_{A(\alpha)}])_0\}$ and $\pi = (\{\{\sigma\}[\alpha]\}[\varepsilon_{A(\alpha)}])_1$.
This completes the proof of the theorem.

*4.5. The consistency of **IDLS**.* Suppose δ enumerates the class D defined in Section 4.1 so δ classically satisfies ED(δ). The consistency of **IDLS** is an easy corollary of Theorem 1.

THEOREM 2. *If* $\Gamma \vdash$ E *in* ***IDLS*** *and the formulas* Γ *are all realizable*$//_{D\cup\{\delta\}}$ *then* E *is realizable*$//_{D\cup\{\delta\}}$.

PROOF. Realization$//_{D\cup\{\delta\}}$ functions must be provided for the axioms, and for the conclusions of the rules of inference (given realization$//_{D\cup\{\delta\}}$ functions for the hypotheses). Since we have not altered the definition of "ε realizes-Ψ E," lawlike function variables still range over D so by Theorem 1 we need only check the rules of inference and the new axioms. The rules present no problems, and the new axioms can be handled with the help of the density lemmas.

${}^{\times}$DLS1. $\varphi = \Lambda w \Lambda\sigma(\Lambda\, \psi_1[w], (\varepsilon_{DLS(\alpha)}[\psi_1[w]], \lambda t\, 0))$ realizes the axiom, where ψ_1 is recursive in δ by the proof of the first density lemma.

${}^{\times}$DLS2. $\varphi = \Lambda\alpha\Lambda\sigma\Lambda w\Lambda\rho(\Lambda\, \psi_2[\alpha], (\varepsilon_{DLS}([\alpha,\beta])[\alpha, \psi_2[\alpha]], \lambda t\, 0))$ realizes the axiom, where ψ_2 is recursive in δ by the proof of the second density lemma.

4.6. Remarks. No function can realize $\exists\delta$ED(δ), since there is no continuous way of obtaining from an arbitrary $\phi \in D$ an n for which $\phi = (\delta)_n$. Thus $\neg\exists\delta$ED(δ) is realizable (and hence realizable$//_{D\cup\{\delta\}}$) though false in our interpretation, while $\neg\exists$d ED(d) is provable in **BDLS**. On the other hand $\exists\delta$ED$^{-}(\delta)$ *is* realizable$//_{D\cup\{\delta\}}$ (but not realizable$//_D$).

As usual Markov's Principle for decidable formulas is (classically) realizable$//_D$ and realizable$//_{D\cup\{\delta\}}$, as is $\neg\forall a \neg(a = u) \supset \exists a(a = u)$ for u any functor not containing a free. The Bar Theorem for lawless sequences and Troelstra's Extension Principle fail in **IDLS**; for counterexamples see [12].

Some of the anomalies of intuitionistic analysis are smoothed out by the lawless subspecies. For example, if B(α) is $\forall x\alpha(x) = 0 \vee \neg\forall x\alpha(x) = 0$ then **IDLS** $\vdash \neg\forall\alpha B(\alpha)$ and **IDLS** $\vdash \neg\forall a B(a)$ but **BDLS** $\vdash \forall\alpha[\mathrm{DLS}(\alpha) \supset B(\alpha)]$. However $\forall\beta[\mathrm{DLS}(\beta) \supset A(\beta) \vee \neg A(\beta)]$ is not in general realizable$//_D$ even for A(β) almost negative; for a trivial counterexample let A(β) be $\forall x(a(x) = 0)$.

The general form $\forall\beta\exists\alpha(A(\beta) \sim \exists x\alpha(x) = 0)$ of Kripke's Schema conflicts with ^{x}KL1, even for $A(\beta) \equiv \forall x\beta(x) = 0$. However, if $A(\beta)$ is almost negative and contains free no arbitrary function variables but β then $\forall\beta[DLS(\beta) \supset \exists\alpha(A(\beta) \sim \exists x\alpha(x) = 0)]$ is realizable$//_D$ and hence (using Lemma 6) true under the interpretation. In fact, for such $A(\beta)$: $\exists a\forall\beta[DLS(\beta) \supset (\exists x\ a(\overline{\beta}(x)) = 0 \sim A(\beta))]$ is realizable$//_D$ and hence true.

References

[1] L. E. J. Brouwer [1918], *Begrundung der Mengenlehre unabhangig vom logischen Satz vom ausgeschlossenen Dritten. Erster Teil: Allgemeine Mengenlehre*, ***Ver. Kon. Akad. v. Wet. I***, **12**, no. 5; reprinted in [1975, 286-290].

[2] ——— [1975], ***Collected Works***, **1** (A. Heyting, editor), North-Holland, Amsterdam.

[3] ——— [1981], ***Brouwer's Cambridge Lectures on Intuitionism*** (D. van Dalen, editor), Cambridge Univ. Press, Cambridge.

[4] M. P. Fourman [1981], talk at the L. E. J. Brouwer Centenary Symposium.

[5] S. C. Kleene [1952], ***Introduction to Metamathematics***, North-Holland, Amsterdam.

[6] ——— [1967], *Constructive functions in "The Foundations of Intuitionistic Mathematics"*, in ***Logic, Methodology and Philos. of Science III***, North-Holland, Amsterdam, 137–144.

[7] ——— [1969], *Formalized recursive functionals and formalized realizability*, ***Mem. Amer. Math. Soc.***, **89**.

[8] S. C. Kleene and R. E. Vesley [1965], ***The Foundations of Intuitionistic Mathematics, Especially in Relation to Recursive Functions***, North-Holland, Amsterdam.

[9] G. Kreisel [1968], *Lawless sequences of natural numbers*, ***Compos. Math.***, **20**, 222–248.

[10] G. Kreisel and A. S. Troelstra [1970], *Formal systems for some branches of intuitionistic analysis*, ***Ann. Math. Logic***, **1**, 229–387.

[11] A. Levy [1970], *Definability in axiomatic set theory*, in ***Mathematical Logic and Foundations of Set Theory*** (Proceedings, Jerusalem 1968), North-Holland, Amsterdam, 129–145.

[12] J. R. MOSCHOVAKIS [1987], *Relative lawlessness in intuitionistic analysis*, ***Jour. Symb. Logic***, **52**, 68–88.

[13] J. R. MOSCHOVAKIS [1994], *More about relatively lawless sequences*, ***Jour. Symb. Logic***, to appear.

[14] A. S. TROELSTRA [1977], ***Choice Sequences, a Chapter of Intuitionistic Mathematics***, Clarendon Press, Oxford.

[15] A. S. TROELSTRA and D. VAN DALEN [1988], ***Constructivism in Mathematics: An Introduction***, **1** and **2**, North-Holland, Amsterdam.

[16] R. E. VESLEY [1980], *Intuitionistic analysis: The search for axiomatization and understanding*, in ***The Kleene Symposium***, North-Holland, Amsterdam, 317–331.

Department of Mathematics
Occidental College

SENSE AND DENOTATION AS ALGORITHM AND VALUE

YIANNIS N. MOSCHOVAKIS[1,2]

§1. Introduction.

In his classic 1892 paper *On sense and denotation* [12], Frege first contends that in addition to their **denotation** (*reference*, *Bedeutung*), proper names also have a **sense** (*Sinn*) "wherein the mode of presentation [of the denotation] is contained." Here proper names include common nouns like "the earth" or "Odysseus" and descriptive phrases like "the point of intersection of lines L_1 and L_2" which are expected by their grammatical form to name some object. By the second fundamental doctrine introduced by Frege in the same paper, they also include declarative sentences: "[a simple, assertoric sentence is] to be regarded as a proper name, and its denotation, if it has one, is either the True or the False." Thus every sentence *denotes* (or *refers to*) *its truth value* and *expresses its sense*, which is "a mode of presentation" of its truth value: this is all there is to the meaning of a sentence as far as logic is concerned. Finally, Frege claims that although sense and denotation are related (the first determines the second), they obey separate *principles of compositionality*, so that the truth value of a complex sentence ϕ is determined solely by the denotations of its constituent parts (terms and sentences), whatever the senses of these constituents parts may be. This is the basic principle which has made possible the development of sense-independent (two-valued, classical) *denotational semantics* for predicate logic, a variety of richer formal languages and (at least) fragments of natural language.

Consider, for example, the arithmetical sentences

$$\theta \equiv 1 + 0 = 1, \tag{1}$$

$$\eta \equiv \text{there exist infinitely many primes}, \tag{2}$$

$$\rho \equiv \text{there exist infinitely many twin primes}. \tag{3}$$

The first two are true, the third is a famous open problem. Surely we understand all three and we understand them differently, even as we recognize the first two to be true and concede that we do not know the truth value of the third. Frege would say that they mean different things because they express different senses, but he did not define sense: what exactly *is* the sense of θ and how does it

[1]During the preparation of this paper the author was partially supported by an NSF grant.

[2]I am grateful to Pantelis Nicolacopoulos for steering me to the van Heijenoort papers, to Alonzo Church for a copy of his still unpublished [1], to Tony Martin and the two referees for their useful remarks and especially to Kit Fine, whose extensive comments on a preliminary draft of the paper were very valuable and influenced substantially the final exposition.

differ from that of η? Our aim here is to propose a precise (mathematical, set theoretic) definition of Fregean sense and to begin the development of a rigorous *sense semantics* which explains differences in meaning like that between θ and η. The mathematical results of the paper are about formal languages, but they are meant to apply also to those fragments of natural language which can be formalized, much as the results of denotational semantics for formal languages are often applied to fragments of natural language. In addition to the language of predicate logic whose sense semantics are fairly simple, the theory also covers languages with description operators, arbitrary connectives and modal operators, generalized quantifiers, indirect reference and the ability to define their own truth predicate.

To explain the basic idea, consider a typical sentence of predicate logic like the arithmetical sentences above, but simpler:

$$\chi \equiv (\forall x)(\exists y)R(x,y) \vee (\exists z)Q(z). \tag{4}$$

Here R and Q are interpreted by relations R, Q on some domain A and we may suppose at first, for simplicity, that A is finite. If we know nothing about R and Q but their extensions, there is basically only one way to determine the truth value of χ, by the following procedure.

Step (1). Do steps (2) and (4). If one of them returns the value **t**, give the value t; if both of them return the value **f**, give the value **f**.

Step (2). For each $a \in A$, do step (3). If for every $a \in A$ the value **t** is returned, return the value **t**; if for some $a \in A$ the value **f** is returned, return the value **f**.

Step (3). Given $a \in A$, examine for each $b \in A$ the value $R(a,b)$. If one of these values is **t**, return the value **t**; if all these values are **f**, return the value **f**.

Step (4). For each $c \in A$, examine the value $Q(c)$. If one of these values is **t**, return the value **t**; if all these values are **f**, return the value **f**.

In contemporary computing terms, we have defined an *algorithm which computes the truth value of* χ. What if A is infinite? An attempt to "execute" (implement) these instructions will lead now to an infinitary computation, but still, the instructions make sense and we can view them as defining an ideal, infinitary algorithm which will compute ("determine" may be better now) the truth value of A. This algorithm is the *referential intension* or just **intension** of χ,

$$int(\chi) = \text{ the algorithm which computes the truth value of } \chi, \tag{5}$$

and we propose to model the sense of χ by its referential intension.

Plainly, we mean to reduce the notion of "sense" to that of "algorithm" and we must specify which precise notion of algorithm we have in mind. We will put this off until Section 3, where we will also review briefly the definitions and results we need from the *theory of recursive algorithms*[3] developed in [22, 23, 24]. From this simple example, however, we can already see two basic properties algorithms must have if we are to use them in this way.

1.1. *Algorithms are semantic* (mathematical, set theoretic) *objects*. This is almost self evident, but we list it because there is some general tendency in the literature to confuse algorithms with *programs*, like the instructions above. A program is a piece of text, it means nothing uninterpreted; its interpretation (or one of them) is precisely the algorithm it defines, and that algorithm is no longer a syntactic object. On the account we are giving, the relation between a program and the algorithm it defines is exactly that between a sentence and its sense.

1.2. *Algorithms are relatively effective*, but they need not be effective in an absolute sense. Consider step (4) above which calls for the examination of the entire extension of the relation Q and cannot be "implemented" by a real machine if A is infinite. It is a concrete, precise instruction and *it is effective relative to the existential quantifier*, which we surely take as "given" when we seek the sense of sentences of predicate logic. Put another way, the claim that we understand χ entails that we understand the quantifiers which occur in it and the specific instructions of the intension simply "codify" this understanding in computational terms. Effective computability relative to assumed, given, infinitary objects has been studied extensively, especially in the work of Kleene in *higher type recursion*.

There is a third, fundamental property of algorithms we need which will be easier to explain after we discuss the first of three applications we intend to make of the proposed modeling of sense.

1.3. Self referential sentences. One of the main reasons for introducing sense is so that we may assign meaning to non-denoting names. Frege makes it clear that sentences may also fail to denote, but he only gives examples where the lack of a truth value is due to a non-denoting constituent proper name. There is, however, another stock of such examples in the self-referential sentences which are usually studied in connection with developing a *theory of truth*. Consider the following bare bones version[4] of the *liar* which avoids the use of a truth predicate,

$$\text{it is not the case that (6),} \tag{6}$$

and the corresponding version of the *truthteller*,

$$\text{it is the case that (7).} \tag{7}$$

Here we understand "it is the case" and "it is not the case" as direct affirmation and negation, i.e., with less regard for readability we could rewrite these definitions

[3] More accurately, we should call it the *theory of pure, single-valued, recursive algorithms*, since [22, 23, 24] do not deal with interaction and non-determinacy. It is possible to extend the results of this theory to algorithms whose implementations start engines, send and receive messages and "choose" in arbitrary ways which actions to execute, but for our present purpose pure algorithms suffice.

[4] These examples are alluded to in Kripke [19].

simply as

$$\neg(8), \tag{8}$$

$$(9). \tag{9}$$

These sentences have no problematic constituents and appear to be meaningful, we certainly understand at least the *liar* well enough to argue that it cannot have a truth value. In a Fregean approach we would expect both of them to have sense, and in fact different senses, since we understand them differently. A natural way to read this version of the *liar* (6) as an algorithm for computing its truth value leads to the single instruction

Step (1). Do step (1); if the value **t** is returned, give the value **f**; if the value **f** is returned, give the value **t**.

Similarly understood, the *truthteller* leads to the instruction

Step (1). Do step (1) and give the value returned.

The circular nature of these instructions corresponds to the self reference of the sentences, but there is nothing unusual about circular clauses like these in programs. The algorithms defined by them are *recursive algorithms* and (in this case, as one might expect), they do not compute any value at all, they *diverge*: on this account, neither the *liar* nor the *truthteller* have a truth value. On the other hand they have distinct referential intensions, because the instructions above define different algorithms.

To define the referential intension of self referential sentences we need recursive algorithms. We state this condition somewhat cryptically here; it will be explained in Section 3.

1.4. *Recursive definitions define algorithms directly.*

Kripke [19] has developed a rigorous semantics (with *truth gaps*) for a formal language which can define its own truth predicate, so that it can express indirect self reference and the more familiar versions of the *liar* and the *truthteller*. We will introduce in Section 2 the language LPCR of *Lower Predicate Calculus with Reflection* (or self reference), a richer and easier to use version of the language of Kripke which we will then adopt as the basic vehicle for explaining referential intensions. The denotational semantics of LPCR are Kripke's *grounded semantics*, by which all reasonable expressions of the *liar* and the *truthteller* fail to denote and cannot be distinguished. On the other hand, every sentence of LPCR (and a fortiori the less expressive language of Kripke) will be assigned a recursive algorithm for its referential intension, a Fregean sense which distinguishes these sentences from one another.[5] Part of the interest in this application is that it "explains"

[5]Calling this notion of sense for LPCR "Fregean" stretches the point a bit, since Frege counted non-denoting signs an imperfection of natural language and wished them banished from the language of mathematics. Nevertheless, we will see that its properties match quite closely the basic properties of sense assumed by Frege.

truth gaps in languages with self reference in terms of non-convergence of computations, a phenomenon which is quite common and well understood in the theory of algorithms.

1.5. Faithful translations. There is a special word in Greek, "μπατζανάκηδες", for the important relationship between two men whose wives are sisters. Most speakers of both Greek and English would agree that

"Niarchos and Onassis married sisters" (10)

is a faithful translation of

"O Νιάρχος και ο Ωνάσσης ήταν μπατζανάκηδες", (11)

even if they had never heard of Niarchos or Onassis and knew nothing about their wives; and no reasonable person would claim that

"Niarchos and Onassis were rivals" (12)

is a faithful translation of (11), although (as it happens) it is true, as is (11), and it is much closer in grammatical structure to (11) than (10)—"were" literally translates "ήταν". What makes (10) a faithful translation of (11) is that they both express the same algorithm for determining their truth: check to see if at any time Niarchos and Onassis were married to two sisters. On this account, we can define *a faithful translation* (at the sentence level) as *an arbitrary syntactic transformation of sentences from one language to another which preserves referential intension*, and that would apply even to languages with radically different syntax and distinct primitives, provided only that they both have a grammatical category of "assertoric sentence." The idea is direct from Frege who says that "the same sense has different expressions in different languages, or even in the same language" and that "the difference between a translation and the original should properly [preserve sense]."

Some claim that faithful translation is impossible, though I have never understood exactly what this means, perhaps because I was unable to translate it faithfully into Greek. In any case, the precise definition proposed here provides a technical tool for investigating the question rigorously. It may be that only fragments of Greek can be translated faithfully into English, or that we can only find translations which preserve *some* properties of the referential intension, and then we could study just what these fragments and these properties are. A typical application to formal languages is that Kripke's language can be translated faithfully into LPCR but not conversely, essentially because direct self reference is expressible in LPCR while Kripke's language allows only indirect self reference. The languages are denotationally equivalent (on most interesting structures), i.e., the same relations are definable by their formulas.

1.6. Primitives with complex sense. In defining the appropriate structures for sense semantics, we will allow the primitives of a language to express arbitrary senses which may be complex, as we assumed above to be the case with "μπατζανάκηδες" in Greek. This is technically convenient, both in developing a practical theory of translation and also for dealing with definitions: if we get tired of referring to Onassis and Niarchos as "married to sisters", we may want to add the word "sisbands" to English with the stipulation that it expresses the same thing as "married to sisters", and then it will be a primitive of the new, extended English with a complex sense. We do this all the time both in natural and in formal languages. On the other hand, there are accounts of language which deal with this phenomenon in other ways and insist that the "ultimate" (atomic) primitives of a language cannot mean anything complex. It has also been argued that there is no coherent way to assign a non-trivial sense to some historical proper names like "Odysseus"or "Aristotle", cf. Kripke [18] and Donnellan [7]. If some of the primitives of the language stand for such atomic relations or directly denoting proper names, then in the modeling proposed here they will be assigned a trivial sense which is completely determined by their denotation. Our aim in this paper is to propose a logical mechanism which explains how complex senses can be computed from given (complex or simple) senses and there is nothing in the arguments and results of this paper which favors or contradicts the direct reference theory for names, or any theory of logical atomism for language.

1.7. Sense identity and intensional logic. Granting any precise, semantic definition of the sense of sentences of some language on some structure $\mathbf{A}$, put

$$\phi \sim_{\mathbf{A}} \psi \iff \mathit{sense}_{\mathbf{A}}(\phi) = \mathit{sense}_{\mathbf{A}}(\psi). \tag{13}$$

Do we always have

$$(\phi \mathbin{\&} \psi) \sim_{\mathbf{A}} (\psi \mathbin{\&} \phi) \quad \text{or} \quad \neg\neg\phi \sim_{\mathbf{A}} \phi?$$

Is the relation $\sim_{\mathbf{A}}$ of sense identity on a fixed structure decidable? Can we give a useful axiomatization for the class of sense identities valid in all structures and is this class decidable? Once we identify sense with referential intension, these questions become precise mathematical problems and they can be solved. The answers (even for a restricted language) will help us sharpen our intuition about sense, particularly as Frege is apparently quite obscure on the question of sense identity, cf. [29]. Incidentally, as one might expect, $(\phi \mathbin{\&} \psi)$ and $(\psi \mathbin{\&} \phi)$ are intensionally equivalent with the standard interpretation of & but $\neg\neg\phi$ has a more complex intension than ϕ.

Previous Work

Where sense is defined in the literature, it is often identified with some version of *proposition* in the sense of Carnap, i.e., the sense of ϕ is the function which assigns to each possible world W the truth value of ϕ in W. For example, Montague [20] adopts this definition. This does not explain the difference in meaning between arithmetic sentences like θ and η above which presumably have the same truth value in all possible worlds.

Church [2, 3, 4, 5] makes (essentially) two proposals for axiomatizing the relation of sense identity without giving a direct, semantic definition of sense. It is hard to compare such approaches with the present one, because they do not allow for specific, semantic import into the notion of sense, sense by them depends only on syntactic form. For example, according to Frege "John loves Mary" and "Mary is loved by John" should have the same sense. In a technical formulation, this should mean that if two basic relation symbols R and Q are interpreted by (senseless) converse relations R and Q in a structure so that

$$R(x,y) \iff Q(y,x),$$

then for any two terms a, b, the sentences R(a,b) and Q(b,a) have the same sense, and they certainly have the same referential intension by our definition. This, however cannot be deduced from their syntactic form.

Although they work in different contexts and they have different aims, Church in the more recent [1], Cresswell [6] and van Heijenoort [30] use variants of the same idea for getting the sense of a sentence, by replacing in it all the primitive syntactic constructs by their denotations. For example, the Cresswell sense and the Church *propositional surrogate* of R(a, b) are (essentially) the triple $\langle R, a, b\rangle$. Both Church and Cresswell worry about introducing semantic import into meanings, but (if I understand them correctly), both would assign to R(a,b) and Q(b,a) the distinct triples $\langle R, a, b\rangle$ and $\langle Q, b, a\rangle$. I cannot find in such mechanisms a general device which can catch accidental, complex semantic relationships like that between a relation and its converse.

On the other hand, the idea that the sense of a sentence determines its denotation is at the heart of the Frege doctrine and practically everyone who has written on the matter uses some form of computational analogy to describe it. Perhaps closest to the present approach is Dummett's interpretation of Frege, most clearly quoted by Evans in a footnote of [9]:

> "This leads [Dummett] to think generally that the sense of an expression is (not a way of thinking about its [denotation], but) a method or procedure for determining its denotation. So someone who grasps the sense of a sentence will be possessed of some method for determining the sentence's truth value.... The procedures in question cannot necessarily be effective procedures, or procedures we ourselves are capable of following."

Evans goes on to call this view "ideal verificationism" and says that "there is scant evidence for attributing it to Frege." Dummett himself says in [8] that

> "[Frege's basic argument about identity statements] would be met by supposing the sense of a singular term to be related to its reference as a programme to its execution, that is, if the sense provided an effective procedure of physical and mental operations whereby the reference would be determined."

If we replace "programme" and "execution" by "algorithm" and "value", then this comes very close to identifying sense with referential intension. What we add here, of course is a specific, mathematical notion of recursive algorithm which makes it possible to develop a rigorous theory of sense.

Outline of what follows

We will state the basic definitions in the body of the paper for the simple to describe language LPCR and we will include enough expository material about the theory developed in [22, 23, 24] so that the gist of what we are saying can be understood without knowledge of these papers. On the other hand, the natural domain for this theory is the *Formal Language of Recursion* FLR defined in [23] and in Section 4 we will prove the main result of the paper for this general case. This is *the decidability of intensional identity for terms of* FLR, *on any fixed structure*, provided only that the signature (the number of primitives) is finite. Intensional identity is relatively trivial on sentences of predicate logic, but we will establish a non-trivial lower bound for its (decidable) degree of complexity for sentences of LPCR: as one might expect, once you are allowed self reference you can say the same thing in so many different ways, that it is not always obvious that it is still the same thing. Finally, the proof of decidability for intensional identity has independent interest and it gives some insight into the concept of a faithful translation. It also suggests an axiomatization for the *logic of intensions*, but we will not consider this here.

§2. Languages with direct self reference.

A *relational* (first order) *signature* or *similarity type* τ is any set of formal *relation* (or predicate) *symbols*, each equipped with a fixed, integer (≥ 0) *arity*. We assume the usual inductive definition of *formulas* for the language LPC = LPC(τ) of *Lower Predicate Calculus* (with identity) on the signature τ, with formal individual variables v_1, v_2, ... and (for definiteness) logical symbols $=$, $\neg$, $\&$, $\vee$, $\supset$, $\exists$, $\forall$. To obtain the language LPCR = LPCR(τ) of *Lower Predicate Calculus with Reflection* (or self reference) on the same signature, we first add to the basic vocabulary an infinite sequence P_1^k, P_2^k, ... of formal k-ary *partial relation variables*, for each $k \geq 0$. The number k is the *arity* of P_i^k. We call these "partial" relation variables because we will interpret them by partially defined relations in the intended semantics, but syntactically they are treated like ordinary relation variables, so we add to the formation rules for formulas in LPC the following:

2.1. If P is a partial relation variable of arity k and $x_1, \ldots, x_k$ are variables, then the string $P(x_1, \ldots, x_k)$ is a formula with free variables $x_1, \ldots, x_k$ and P. In particular, for every *partial, propositional variable* P with arity 0, $P(\,)$ is a formula.

Self reference is introduced in LPCR by the following key, last rule of formula formation.

2.2. If $\phi_0, \ldots, \phi_n$ are formulas, $P_1, \ldots, P_n$ are partial relation variables and for each i, $\vec{u}^i$ is a sequence of individual variables of length the arity of P_i, then the string

$$\phi \equiv \phi_0 \text{ where } \{P_1(\vec{u}^1) \simeq \phi_1, \ldots, P_n(\vec{u}^n) \simeq \phi_n\} \qquad (14)$$

is a formula. The bound occurrences of variables in ϕ are the bound occurrences in *the head* ϕ_0 and *the parts* $\phi_1, \ldots, \phi_n$ of ϕ, all the occurrences of $P_1, \ldots, P_n$ and the occurences of the individual variables of each $\vec{u}^i$ in $P_i(\vec{u}^i) \simeq \phi_i$.

Notice that the sequence $\vec{u}^i$ must be empty here when P_i is a partial, propositional variable. The simplest examples of self referential sentences involve such "propositional reflection," e.g., the formal LPCR versions of the direct *liar* and *truthteller* intended by (6) and (7) are:

$$liar \equiv P(\,) \text{ where } \{P(\,) \simeq \neg P(\,)\}, \qquad truthteller \equiv P(\,) \text{ where } \{P(\,) \simeq P(\,)\}. \tag{15}$$

The more complex

$$liar' \equiv (\exists x)P(x) \text{ where } \{P(x) \simeq \neg P(x)\} \tag{16}$$

asserts that "the relation which is always equivalent to its own negation holds of some argument," and it too will have no truth value.

In the definition of (denotational) semantics for LPCR which follows, we will adopt a general, somewhat abstract view of formal language semantics which is unnecessarily complex for predicate logic and just barely useful for LPCR. The extra effort is worth it though, because this formulation of semantics generalizes immediately to a large class of languages (including rich fragments of natural language) and is indispensable for the description of referential intensions in the next section, even for the case of LPC.

2.3. Partial functions. Let us first agree to view an m-ary relation on a set X as a function

$$R : X^m \to TV, \qquad (TV = \{\mathbf{t}, \mathbf{f}\}) \tag{17}$$

which assigns a truth value to each m-tuple in X. A partial m-ary relation is then a *partial function*

$$P : X^m \rightharpoonup TV, \tag{18}$$

i,e., a function defined on some of the m-tuples in X with range TV—notice the symbol "$\rightharpoonup$" which suggests that for some $\vec{x}$, $P(\vec{x})$ may be undefined. In dealing with partial functions, we will use the standard notations

$$\begin{aligned} f(\vec{x})\downarrow &\iff \vec{x} \text{ is in the domain of } f, \\ f(\vec{x}) \simeq w &\iff f(\vec{x})\downarrow \text{ and } f(\vec{x}) = w, \\ f \subseteq g &\iff (\forall \vec{x}, w)[f(\vec{x}) \simeq w \implies g(\vec{x}) \simeq w], \end{aligned}$$

and we assume the usual (strict) composition rule so that (for example)

$$f(g(\vec{x}), h(\vec{x}))\downarrow \implies g(\vec{x})\downarrow \text{ and } h(\vec{x})\downarrow. \tag{19}$$

We let

$$\emptyset = \text{the totally undefined partial function.} \tag{20}$$

In set theoretic terms this is the empty set of ordered pairs and it is the least partial function in the partial ordering $\subseteq$.

2.4. Structures for LPCR. We interpret LPCR in the usual structures of predicate logic, i.e., tuples of the form $\mathbf{A} = (A, R_1, \ldots, R_l)$, where A is some non-empty set and each relation R_j interprets the corresponding formal relation symbol R_j of the signature. With the convention just described, $\mathbf{A}$ is simply a sequence of functions on its universe A, with values in TV. Of course we need more than these functions to understand the meaning of formulas in **A**, we also need the meaning of $=$, $\neg$, $\forall$, etc. These are taken for granted, but if we include them and the set of truth values TV for completeness, we get the tuple

$$\mathbf{A} = (TV, A, R_1, \ldots, R_l, eq_A, \neg, \&, \vee, \supset, \exists_A, \forall_A). \tag{21}$$

Now

$$eq_A(x,y) = \begin{cases} \mathbf{t}, \text{ if } x = y, \\ \mathbf{f}, \text{ if } x \neq y, \end{cases} \qquad \neg(a) = \begin{cases} \mathbf{f}, \text{ if } a = \mathbf{t}, \\ \mathbf{t}, \text{ if } a = \mathbf{f} \end{cases} \tag{22}$$

are clearly (total) functions on A and the set of truth values TV. The objects $\exists_A$, $\forall_A$ are *functionals*, partial functions which take unary, partial relations as arguments:

$$\exists_A(P) \simeq \begin{cases} \mathbf{t}, \text{ if for some } x \in A,\ P(x) \simeq \mathbf{t}, \\ \mathbf{f}, \text{ if for all } x \in A,\ P(x) \simeq \mathbf{f}, \end{cases} \tag{23}$$

$$\forall_A(P) \simeq \begin{cases} \mathbf{t}, \text{ if for all } x \in A,\ P(x) \simeq \mathbf{t}, \\ \mathbf{f}, \text{ if for some } x \in A,\ P(x) \simeq \mathbf{f}. \end{cases} \tag{24}$$

These functionals represent the quantifiers because for every unary partial relation P,

$$(\exists x \in A)[P(x) \simeq w] \iff \exists_A(P) \simeq w,$$

and more generally, on partial relations of any arity,

$$(\exists x \in A)[P(x,\vec{y}) \simeq w] \iff \exists_A(\lambda(x)P(x,\vec{y})) \simeq w,$$

with the obvious meaning for the λ-operator on partial functions: the term $\exists_A(\lambda(x)P(x,\vec{y}))$ is equivalent to (takes exactly the same values as) the formula $(\exists x)P(x,\vec{y})$ on the set A.

Notice that $\forall_A(P)$ is not defined if P is a partial relation which never takes the value $\mathbf{f}$. However—and this is a basic property of these objects—both $\exists_A$ and $\forall_A$ are *monotone*, i.e.,

$$[\exists_A(P) \simeq w \ \&\ P \subseteq P'] \implies \exists_A(P') \simeq w,$$
$$[\forall_A(P) \simeq w \ \&\ P \subseteq P'] \implies \forall_A(P') \simeq w.$$

This is obvious from their definition.

One is tempted to interpret $\&$, $\vee$ and $\supset$ by the usual, total functions on TV, but this will not work if we want to apply them to partial relations; because if $\vee$ were a binary function, then by the rules of strict composition we would have $(P \vee Q)\downarrow \implies [P\downarrow \ \&\ Q\downarrow]$, while we certainly want $P \simeq \mathbf{t} \implies (P \vee Q) \simeq \mathbf{t}$, even when Q is undefined. We set instead:

$$P \vee Q \simeq \begin{cases} \mathbf{t}, \text{ if } P() \simeq \mathbf{t} \text{ or } Q() \simeq \mathbf{t}, \\ \mathbf{f}, \text{ if } P() \simeq Q() \simeq \mathbf{f}, \end{cases} \quad P \,\&\, Q \simeq \begin{cases} \mathbf{t}, \text{ if } P() \simeq Q() \simeq \mathbf{t}, \\ \mathbf{f}, \text{ if } P() \simeq \mathbf{f} \text{ or } Q() \simeq \mathbf{f}, \end{cases} \tag{25}$$

and similarly for $\supset$. Here P and Q are partial, propositional variables and clearly

$$[P \subseteq P',\ Q \subseteq Q',\ P \,\&\, Q \simeq w] \implies P' \,\&\, Q' \simeq w$$

and similarly with $\vee$, i.e., the propositional connectives are also monotone.

We now define precisely (partial, monotone) *functionals* and *functional structures.* The complication is due to the unavoidable fact that these are *typed objects*, never too simple to deal with.

2.5. Partial, monotone functionals. A (many sorted) *universe* is any collection of sets

$$\mathcal{U} = \{\mathcal{U}(\bar{u}) \mid \bar{u} \in B\}$$

indexed by some (non-empty) set B of *basic types.* Members of $\mathcal{U}(\bar{u})$ are of (basic) *type* $\bar{u}$ in $\mathcal{U}$. For each Cartesian product

$$U = \mathcal{U}(\bar{u}_1) \times \cdots \times \mathcal{U}(\bar{u}_n), \quad (\bar{u}_1, \ldots, \bar{u}_n \in B)$$

and each $W = \mathcal{U}(\bar{w})$, $\bar{w} \in B$, we let

$$P(U, W) = \{p \mid p : U \rightharpoonup W\}$$

be the set of all partial functions on U to W; the objects in $P(U, W)$ are of *partial function* or *pf type* $((\bar{u}_1, \ldots, \bar{u}_n) \rightharpoonup \bar{w})$ in $\mathcal{U}$. We allow U to be *the Cartesian product of no factors* I, in which case the objects of pf type $(() \rightharpoonup \bar{w})$ are the *partial constants* of type $\bar{w}$. A *point type* in B is any tuple $\bar{x} = (\bar{x}_1, \ldots, \bar{x}_n)$, where each $\bar{x}_i$ is either a basic or a pf type in B and a *point of type* $\bar{x}$ is any tuple $x = (x_1, \ldots, x_n)$ where each object x_i has type $\bar{x}_i$ in $\mathcal{U}$. The (product) space of all points of type $\bar{x}$ is naturally partially ordered by

$$(x_1, \ldots, x_n) \leq (y_1, \ldots, y_n) \iff x_1 \leq_1 y_1 \ \& \ \ldots \ \& \ x_n \leq_n y_n,$$

where $\leq_i$ is just the identity $=$ when $\bar{x}_i$ is a basic type and for a pf type $\bar{x}_i$, $\leq_i$ is the standard partial ordering on partial functions. Finally, a (partial, monotone) *functional* of type $(\bar{x} \rightharpoonup \bar{w})$ is any partial function $f : X \rightharpoonup W$ which is monotone in this partial ordering, i.e.,

$$[f(x) \simeq w \ \& \ x \leq x'] \Longrightarrow f(x') \simeq w.$$

Notice that pf types are also functional types and every partial function in $\mathcal{U}$ is a (degenerate) functional.

2.6. Functional structures. A *finite* (functional) *signature* is a triple $\tau = (B, \{\mathrm{f}_1, \ldots, \mathrm{f}_n\}, d)$, where $B = \{\mathrm{b}_1, \ldots, \mathrm{b}_k\}$ is a finite set of basic types, $\mathrm{f}_1, \ldots, \mathrm{f}_n$ are arbitrary function symbols and d is a mapping which assigns a functional type to each f_i; a *functional structure* of signature τ is a tuple

$$\mathbf{A} = (\mathcal{U}, f_1, \ldots, f_n) = (U_1, \ldots, U_k, f_1, \ldots, f_n) \tag{26}$$

where $\mathcal{U}$ is a universe over B with the basic sets $U_i = \mathcal{U}(\mathrm{b}_i)$, $i = 1, \ldots, k$, and each f_j is a functional on $\mathcal{U}$ of type $d(\mathrm{f}_j)$.

For example, this representation of a standard first-order structure $\mathbf{A}$ in (21) above has just two basic sets TV and A of respective basic types (say) bool and a. An n-ary partial relation P on A is a functional of type $((\mathsf{a}^n) \rightharpoonup \mathsf{bool})$ and similarly, an n-ary partial function $p : A^n \rightharpoonup A$ is of pf and functional type $((\mathsf{a}^n) \rightharpoonup \mathsf{a})$. The functional $\&$ if of type $((((\,) \rightharpoonup \mathsf{bool}), ((\,) \rightharpoonup \mathsf{bool})) \rightharpoonup \mathsf{bool})$ and $\exists_A$ is of type $(((\mathsf{a}) \rightharpoonup \mathsf{bool}) \rightharpoonup \mathsf{bool})$, as is $\forall_A$. Not much is gained by computing these complex types, but it is important to notice that we are dealing with typed objects.[6] We will concentrate on this concrete example for the rest of this section, but everything we say generalizes directly to arbitrary functional structures which can be used to model languages with additional propositional connectives, generalized quantifiers, modal operators and the like.

2.7. The denotational semantics of LPCR. Fix a structure **A** as in (21). With each formula ϕ and each sequence of (individual and partial relation) variables $\vec{\mathrm{x}} = \mathrm{x}_1, \ldots, \mathrm{x}_n$ which includes all the free variables of ϕ, we associate a (partial, monotone) functional

$$f_{\vec{\mathrm{x}},\phi} = den_{\mathbf{A}}(\vec{\mathrm{x}}, \phi) : X \rightharpoonup TV,$$

where X is the space of all n-tuples $(x_1, \ldots, x_n)$ such that the type of x_i matches that of the formal variable x_i. The definition is by induction on the length of ϕ and it is quite trivial, except for the case of the new primitive where of LPCR. For example, skipping the subscripts,

$$\left.\begin{aligned} den(\vec{\mathrm{x}}, \mathrm{x}_i = \mathrm{x}_j)(\vec{x}) &= eq_A(x_i, x_j), \\ den(\vec{\mathrm{x}}, \mathrm{R}_m(\mathrm{x}_i, \mathrm{x}_j, \mathrm{x}_k))(\vec{x}) &= R_m(x_i, x_j, x_k), \\ den(\vec{\mathrm{x}}, \phi \vee \psi)(\vec{x}) &\simeq \lambda(\,)den(\vec{\mathrm{x}}, \phi)(\vec{x}) \vee \lambda(\,)den(\vec{\mathrm{x}}, \psi)(\vec{x}), \\ den(\vec{\mathrm{x}}, (\exists \mathrm{y})\phi(\vec{\mathrm{x}}, \mathrm{y}))(\vec{x}) &\simeq \exists_A(\lambda(y)den(\vec{\mathrm{x}}, \mathrm{y}, \phi(\vec{\mathrm{x}}, \mathrm{y}))(y, \vec{x})). \end{aligned}\right\} \quad (27)$$

Notice the use of the dummy λ operation in the case of $\vee$, which is needed to turn an element in TV (or "the undefined") into a partial, propositional constant. Proof that the operation defined in each of these cases is a (monotone) functional is routine, depending only on the trivial fact that the composition of monotone, partial functions is also monotone.

For the less trivial case of the where construct, suppose

$$\phi \equiv \phi_0 \text{ where } \{P_1(\vec{u}^1) \simeq \phi_1,\ \ldots,\ P_n(\vec{u}^n) \simeq \phi_n\}$$

and for each $\vec{x}$, use the induction hypothesis to define

$$f_{\vec{x},i}(\vec{u}^i, P_1, \ldots, P_n) \simeq den(\vec{u}^i, \mathrm{P}_1, \ldots, \mathrm{P}_n, \phi_i)(\vec{u}^i, P_1, \ldots, P_n), \quad (i = 0, \ldots, n).$$

Now the functionals $f_{\vec{x},i}$ are all monotone and it follows from basic results in the theory of least-fixed-point recursion that the system of equations

$$P_i(\vec{u}^i) \simeq f_{\vec{x},i}(\vec{u}^i, P_1, \ldots, P_n), \quad (i = 1, \ldots, n) \tag{28}$$

has a (unique) sequence of simultaneous *least* (as partial functions) solutions $\overline{P}_{\vec{x},i}$, $i = 1, \ldots, n$. We set

$$den(\vec{\mathrm{x}}, \phi)(\vec{x}) \simeq f_{\vec{x},0}(\overline{P}_{\vec{x},1}, \ldots, \overline{P}_{\vec{x},n}). \tag{29}$$

We need to verify that this is a functional again, i.e., monotone, but the proof is quite direct.

The denotation of a sentence (with no free variables) is independent of its arguments and we set

$$value(\phi) \simeq den(\phi)(\,). \tag{30}$$

[6]If both A and $P(A, A)$ are among the basic sets in a universe with respective basic types a and paa, then a partial function $f : P(A, A) \rightharpoonup A$ may be viewed as either of pf type (paa $\rightharpoonup$ a) or (if it is monotone), as a functional, of type ((a $\rightharpoonup$ a) $\rightharpoonup$ a),

There is a lot of notation here and an appeal to results which are not as widely known as they might be, but the underlying idea is very simple. Consider the *liar* defined by (6). The definition gives

$$value(liar) \simeq \overline{P}(),$$

where $\overline{P}$ is the least solution of the equation

$$P() \simeq \neg(P());$$

since the totally undefined $P = \emptyset$ satisfies this equation, the *liar* has no truth value and neither does the *truthteller*. For *liar'* we need the least solution of the equation

$$P(x) \simeq \neg P(x)$$

which is again $\emptyset$, so that it too receives no truth value.

Consider also the following two versions of the *ancestral*, in which we will use the notation

$$x \prec y \iff x \text{ is a child of } y$$

and we take I and *Euclid* to be individual constants with fixed references:

$$\phi \equiv P(I) \text{ where } \{P(x) \simeq x \prec Euclid \vee (\exists y)[x \prec y \,\&\, P(y)]\}, \tag{31}$$

$$\psi \equiv Q(Euclid) \text{ where } \{Q(z) \simeq I \prec z \vee (\exists y)[y \prec z \,\&\, P(y)]\}. \tag{32}$$

Both ϕ and ψ express the intended assertion

I am a descendant of Euclid,

but in different ways, intuitively

ϕ : I am a child of Euclid or the child of a descendant of Euclid,

ψ : Euclid is my father or the father of an ancestor of mine.

It is easy to verify that both ϕ and ψ take the value **t** if, indeed, I am a descendant of Euclid. In the more likely, opposite case, it can be verified that ϕ will take the value **f** only if there is an upper bound to the length of ascending ancestral chains of the form

$$I \prec y_1 \prec y_2 \prec \cdots \prec y_n,$$

while ψ will be defined and false only when there is an upper bound to the length of descending chains of parenthood of the form

$$Euclid \succ z_1 \succ z_2 \succ \cdots \succ z_n.$$

Both of these conditions are evidently true, but the first one undoubtedly claims a bit more about our world (and the meaning of the relationship "child") than the relatively innocuous second one.

2.8. Definability in LPCR. A (possibly partial) relation $R \subseteq A^n$ is *weakly representable* in LPCR if there is a formula ϕ with free variables $\mathrm{x}_1, \ldots, \mathrm{x}_n$ such that

$$R(\vec{x}) \iff den(\vec{\mathrm{x}}, \phi)(\vec{x}) \simeq \mathbf{t},$$

and *strongly representable* or just *definable* in LPCR if we have

$$R(\vec{x}) \simeq den(\vec{\mathrm{x}}, \phi)(\vec{x}).$$

Notice that strongly representable total relations are defined by *total formulas*, with no truth gaps. Those familiar with the theory of *inductive definability* [22] can verify without much difficulty that *a total relation on A is* LPCR-*weakly representable just in case it is absolutely inductive in* **A** and it is LPCR-*strongly representable just in case it is absolutely hyperelementary in* **A**. We will not pursue here this connection or the relation of LPCR with the several languages with fixed-point operators studied in logic and theoretical computer science.

Suppose now that the structure **A** is infinite and *acceptable* in the sense of [21], so that we can code tuples from A by single elements of A and we can manipulate these tuples in LPCR.[7] We can then define a copy of the integers in A and we can assign *Gödel numbers* to formulas of LPCR and *Gödel codes* to sentences in the expanded language, where we add a formal name for each element of A. Finally, we can define the *partial truth predicate* of the structure,

$$T_{\mathbf{A}}(s) \simeq \begin{cases} value(\theta_s), & \text{if } s \text{ is the Gödel code of a sentence } \theta_s, \\ \mathbf{f}, & \text{otherwise.} \end{cases} \tag{33}$$

2.9. THEOREM. *If* **A** *is acceptable, then its partial truth predicate $T_{\mathbf{A}}$ for* LPCR *is definable in* LPCR.

The proof of this is neither difficult nor surprising: we simply express the inductive definition of $T_{\mathbf{A}}(s)$ directly by a where formula of LPCR.

2.10. Kripke's language. Kripke [19] considers the formal extension of the language LPC of predicate logic by a single unary relation symbol T which is meant to denote (on acceptable structures) the truth predicate of the extended language containing it. To define the interpretation T of T (on Gödel codes of sentences), he considers the natural inductive conditions that T must satisfy and then takes the least[8] partial relation which satisfies these conditions; the truth value of a sentence θ then (when it has one) is the value $T(s)$ of T on the Gödel code s of θ. It is quite routine to define a recursive mapping

$$\phi \mapsto \tau\rho(\phi) \tag{34}$$

which assigns to each formula ϕ of the language of Kripke a formula $\tau\rho(\phi)$ of LPCR which "means the same thing," and in particular has the same denotation: basically, we replace each occurrence of T by the LPCR formula which defines $T_{\mathbf{A}}$ on the appropriate argument. In particular, the usual versions of the *liar* and other "paradoxical" sentences which employ indirect self reference are expressible in LPCR and are assigned denotations in accordance with the grounded semantics of Kripke.

---

[7] The precise definition of *acceptability* is not important. Kripke [19] points out that much weaker hypotheses will do, actually all we need is the existence of a first-order-definable, one-to-one function $\pi : A \times A \to A$.

[8] These are Kripke's *grounded semantics*, which are most relevant to what we are doing here. We will discuss briefly alternative fixed-point semantics in the next section.

2.11. Descriptions. Suppose we now expand LPCR by a construct of the form

$$\phi \mapsto (\text{the x})\phi(\text{x}) \tag{35}$$

which assigns a term to every formula $\phi(\text{x})$, where the variable x may or may not occur free in $\phi(\text{x})$. We understand (35) as a new clause in the recursive definition of terms and formulas, by which terms can occur wherever free variables can occur.

Notice first that the where construct gives an easy mechanism to control the *scoping* of descriptions. For example, Russell's "bald King of France" example in [26] is best formalized by

$$BKF \equiv \text{The King of France is bald} \equiv P((\text{the x})KF(\text{x})) \text{ where } \{P(\text{x}) \simeq B(\text{x})\}, \tag{36}$$

whose reformulation according to the theory of descriptions

$$BKF \iff (\exists!\text{x})KF(x) \;\&\; (\exists\text{x})[KF(\text{x}) \;\&\; B(\text{x})] \tag{37}$$

is false, as Russell would wish it. If $B(\text{x})$ is a complex expression, as it most likely is, Russell's analysis of the simpler $B((\text{the x})KF(\text{x}))$ might turn out to be true or false depending on the particular form of the formula $B(\text{x})$.

We naturally wish to set

$$den(\vec{y}, (\text{the x})\phi(\text{x}))(\vec{y}) \simeq the_A(\lambda(x)den(\vec{y}, \text{x}, \phi(\text{x}))(\vec{y}, x) \tag{38}$$

with a suitable functional

$$the_A : P(A, TV) \rightharpoonup A, \tag{39}$$

and it is clear that the_A should satisfy

$$[(\forall x)P(x)\downarrow \;\&\; (\exists!x)P(x) = \mathbf{t}] \Longrightarrow the_A(P) \simeq \text{ the unique } x \text{ such that } P(x) = \mathbf{t}. \tag{40}$$

It is also clear that (to insure monotonicity) $the_A(P)$ cannot be always defined, e.g., $the_A(\emptyset)$ will take no value. The question is how to set $the_A(P)$ when we know from the values of P that no $P' \supseteq P$ defines a single object, i.e., when the following holds:

$$Error(P) \iff (\forall x)[P(x) \simeq \mathbf{f}] \vee (\exists x)(\exists y)[x \neq y \;\&\; P(x) \simeq P(y) \simeq \mathbf{t}]. \tag{41}$$

There are three plausible ways to proceed and we can label them by the denotational semantics for descriptive phrases which they generate.

2.12. Frege's solution. If $Error(P)$, we let $the_A(P)$ stay undefined. Russell argues eloquently against this solution which would also leave BKF without a truth value. Now Russell's—and everybody else's theory—would certainly leave undefined the term

$$(\text{the x})[[\text{x} = 1 \;\&\; liar] \vee [\text{x} = 0 \;\&\; \neg liar]], \tag{42}$$

which we do not know how to handle, but it does not seem right to have $the_A(P)$ undefined when $Error(P)$ holds and we know exactly what is wrong.

2.13. Russell's solution. We add to the basic set A a new element e_A, we decree that

$$Error(P) \Longrightarrow the_A(P) \simeq e_A, \tag{43}$$

and we extend all partial relations so that

$$x_i = e_A \Longrightarrow P(x_1,\ldots,x_n) \simeq \mathbf{f}, \quad (i = 1,\ldots,n). \tag{44}$$

Notice that this makes $the_A(P)$ always defined on total functions P, and (easily) it assigns the same truth values to sentences as Russell's theory of descriptions.

2.14. The programmer's solution. In an effort to test Russell's solution by the experimental method, I posed directly to four friends the hot question: *do you think that the King of France is bald?* The one who knew too much responded "*Russell thought not,*" but the other three gave me exactly the same, predictable answer: "*there is no King of France.*" To make sense out of this thinking, we add a new element **e** (for "error") to the set of truth values TV, we decree (43) as above and we extend all partial relations so that

$$x_i = e_A \Longrightarrow P(x_1,\ldots,x_n) \simeq \mathbf{e}. \quad (i = 1,\ldots,n). \tag{45}$$

We must also extend the connectives and quantifiers in the obvious way, to take account of the new truth value, e.g.,

$$\mathbf{t} \,\&\, \mathbf{e} = \mathbf{e}, \quad \mathbf{t} \vee \mathbf{e} = \mathbf{t}, \quad \neg\mathbf{e} = \mathbf{e},$$

$$\exists_A(P) \simeq \begin{cases} \mathbf{t}, \text{ if for some } x \in A,\ P(x) \simeq \mathbf{t}, \\ \mathbf{f}, \text{ if for all } x \in A,\ P(x) \simeq \mathbf{f}, \\ \mathbf{e}, \text{ if } P \text{ is total },\ (\forall x \in A)[P(x) \neq \mathbf{t}],\ (\exists x \in A)[P(x) = \mathbf{e}]. \end{cases}$$

This is a refinement of the Russell solution: when we know that there is no good way to interpret a descriptive phrase, we record an "error" and we let the Boolean operations propagate the error when they receive it from an argument they need. We might in fact enrich the set TV with a whole lot of distinct "error messages" which explain just what went wrong with the descriptions we attempted to compute, one of these error messages presumably being "there is no King of France".

These functional interpretations of **the** yield denotational semantics for descriptive phrases which make it possible to view (**the** x)ϕ(x) as a constituent of every sentence θ in which it occurs while assigning to θ the correct (expected) denotation. Of course, denotations alone cannot account for the difference in meaning between "*Scott is the author of Waverly*" and "*Scott is Scott*", both of them being true by any of the proposed solutions. These sentences mean different things because they have distinct referential intensions, the first calling for us to check whether Scott indeed was Scott (a trivial algorithm) while the second requires that we verify whether Scott wrote the Waverly novels. We will define the referential intensions of descriptive phrases in the next section.

2.15. The Formal Language of Recursion FLR. In [23] we associated with each functional signature τ the *Formal Language of Recursion* FLR $=$ FLR(τ), a language of terms which in addition to the function symbols $\mathrm{f}_1,\ldots,\mathrm{f}_n$ of τ has basic and partial function variables over the types of τ and the additional type bool (standardly interpreted), constants **t** and **f** naming themselves, a *conditional construct* and a general where construct which allows recursive definitions in all basic types. Quite obviously, we can view LPC, LPCR and its natural extensions as fragments of FLR, with special signatures and some "sugaring" of the notation; and the denotational semantics we defined are special cases of the denotational semantics for FLR, which are treated in [23]. The mathematical theory of self-referential semantics, referential intension and sense is best developed for FLR and then applied to the specific cases of interest, but its usefulness and justification rest on these applications.

§3. Recursive algorithms and referential intensions.

The distinction between an *algorithm* f and *the object* $\overline{f}$ *computed by that algorithm*—typically a *function*—is well understood. Consider, for example the need to alphabetize (sort) a list of words in some alphabet which comes up in a myriad of computing applications, from compilers of programming languages to sophisticated word processing programs. There are hundreds of known *sorting algorithms* and the study of their properties is a sizable cottage industry straddling theoretical and applied computer science: they all compute the same *sorting function* which assigns to every list of words its ordered version.

Equally clear is the difference between an algorithm and its various *implementations* in specific machines or the *programs* which express it in specific programming languages. It is true that this distinction is sometimes denied by computer scientists who take a formalist or nominalist position towards the foundations of computing and recognize nothing but programs, their syntactic properties and the "symbolic computations" they define. Yet I have never known a programmer who did not know the difference between "programming the mergesort algorithm in LISP" (a relatively trivial task) and programming the mergesort in some assembly language, a notoriously difficult exercise in introductory programming courses. What can the last quoted sentence possibly mean if there is no such thing as "the mergesort algorithm"? The question is surely more serious than this flippant remark would suggest, as serious as the general question of the formalist and nominalist approaches to mathematics, but this is not the place to address it seriously. Suffice it to say that I am adopting a classical, realist view which takes it for granted that algorithms "exist" independently of the programs which express them, much as real numbers "exist" independently of the definitions which can be given for some of them in specific formal languages.[9]

Although the concept of algorithm is universally recognized as fundamental for computer science, it is traditional to treat it as a "premathematical" notion and avoid defining it rigorously. Where there is need for a precise definition, for example in complexity theory, it is generally assumed that algorithms are faithfully represented by *models of computation*, automata, Turing machines, random access machines and the like. We record here the most general definition of this type.

[9]A sequence of recent court decisions in patent law may possibly lead to a legal distinction between a program and the algorithm it expresses! Euclid could get a copyright on the *Elements*, which includes a definition of his famous algorithm for finding the greatest common divisor of two integers, but could he also patent the algorithm itself so that no one could use it without paying him a fee? Amazingly, the question came up in the courts in connection with the *fast multiplication algorithm*, which solves in a novel way a problem even more elementary than Euclid's.

3.1. An **iterator** (or *abstract sequential machine*) on the set X to W is a tuple

$$f = (S, T, \pi, in, out), \tag{46}$$

where the following conditions hold.

1. S is a non-empty set, the *states* of f, and $T \subseteq S$ is a designated subset of *terminal states.*
2. $\pi : S \to S$ is a total function, the *transition function* of f.
3. $in : X \to S$ and $out : T \to W$ are the *input* and *output* functions of f.

To compute with f on a given $x \in X$, we first find the initial state $in(x)$ associated with x and then we iterate the transition function beginning with this,

$$\pi^0(x) = in(x),\ \pi^1(x) = \pi(\pi^0(x)),\ \ldots, \pi^{n+1}(x) = \pi(\pi^n(x)),\ \ldots \tag{47}$$

until (and if) we find some n such that $\pi^n(x)$ is a terminal state; if such an integer exists, we set

$$N(x) \simeq (\text{the least } n)[\pi^n(x) \in T] \tag{48}$$

and we give as output the value of *out* on the terminal state $N(x)$,

$$\overline{f}(x) \simeq out(\pi^{N(x)}(x)). \tag{49}$$

Equations (47), (48), (49) taken together define the partial function $\overline{f} : X \rightharpoonup W$ *computed by the iterator* f.

In the most familiar example of a Turing machine with a one-way tape set up to compute a unary, partial function on the integers, states are triples (s, τ, i) where s is an internal state of the machine, τ is a description of the tape with finitely many (non-blank) symbols on it and i is the scanned square; the transition function π is determined by the finite table of the machine; (s, τ, i) is terminal if the machine halts in the situation described by it; for each integer n, $in(n) = (s_0, \tau_n, 0)$, where s_0 is the initial, internal state and τ_n is the tape with a string of $n+1$ consecutive 1's on it; and *out* counts the consecutive 1's on the tape and subtracts 1 from them.

Turing's compelling analysis of mechanical computation in [28] establishes beyond doubt that *every partial function on the integers which can be computed by a finitary, mechanical (deterministic) device can also be computed by a Turing machine*, what we now call *the Church-Turing Thesis.* His arguments also make it quite obvious that the more general iterators can represent faithfully all the intensional aspects of mechanical computation, and I take "the orthodox view" to be that algorithms can be faithfully represented by these mechanical procedures.[10] The theory developed in [22, 23, 24] starts from a radically different perspective which takes *recursion* rather than *step-by-step computation* as the essence of an

[10]For example, the *computational methods* of Knuth [16] are essentially the same as "finitary" iterators. Knuth reserves the term "algorithm" for computational methods which produce a value on each input, so that most programs do not express algorithms until we have proved that they, in fact converge. It would seem that whatever name we choose for it, the basic notion is that of an effective procedure which might or might not converge.

algorithm. We will review it briefly and discuss just two arguments in its favor, those most relevant to the purpose at hand.

3.2. The meaning of a recursive definition. A typical *mutual* or *simultaneous* recursive definition has the form

$$\left.\begin{array}{rcl} p_1(u_1) & \simeq & f_1(u_1, p_1, \dots, p_n, x), \\ & \cdots & \\ p_n(u_n) & \simeq & f_n(u_n, p_1, \dots, p_n, x), \end{array}\right\} \tag{50}$$

where $f_1, \dots, f_n$ are (monotone) functionals on the individual arguments $u_1, \dots, u_n, x$ and the partial function arguments $p_1, \dots, p_n$ and we understand the definition to determine for each value x of *the parameter* the simultaneous *least fixed points* $\overline{p}_1^x, \dots, \overline{p}_n^x$, the least partial functions which satisfy its n equations in (50) with the fixed x. We may choose to call the first equation *principal* and say that the recursion defines the partial function $\lambda(u_1, x)\overline{p}_1^x(u_1)$, or (more conveniently, following the syntax of LPCR) say that the recursion defines the partial function

$$\overline{\mathsf{f}}(x) \simeq f_0(\overline{p}_1^x, \dots, \overline{p}_n^x, x), \tag{51}$$

where f_0 is one more functional associated with the recursion, its *head*.

It follows from early results of Kleene in the theory of recursion, that if we have for each $i = 0, \dots, n$ a Turing machine T_i which computes $f_i(u_i, p_1, \dots, p_n)$ (with a suitable coding of the argument u_i and access to "oracles" that yield on demand requested values for each of the p_j's), then we can construct a new Turing machine $\overline{T}$ which computes $\overline{f}$. The argument generalizes easily to arbitrary iterators. Thus, it is argued, since *recursion can be reduced to iteration*, the notion of iterator is rich enough to represent algorithms expressed by recursive definitions.

A serious problem with this analysis is that the claimed reduction of recursion to iteration is by no means unique or natural. If we apply any of the popular known reductions to the specific, recursive definition of the *mergesort* mentioned above, we get a Turing machine which spends most of its time building "stacks," implementing "calls" and copying parts of its tape, i.e., implementing recursion rather than sorting; and if we use another notion of computational model like the assembly language of some actual computing machine (especially one capable of substantial parallel computation), then we would likely use a different reduction procedure and execution of the resulting program will bear little discernible resemblance to the computation of the Turing machine allegedly expressing the same algorithm. At the same time, the most important properties of recursive algorithms are typically proved directly from the recursive equations which define them, without any reference to details of the specific method by which we aim to reduce the recursion to iteration in order to implement it on a mechanical device.[11]

[11] This is discussed in some detail in [22] for the mergesort, whose basic property is that it will alphabetize a list of n words "using" no more than $n \cdot \log(n)$ comparisons. The result follows almost trivially from the recursive definition of the algorithm and has nothing to do with its many and diverse implementations.

On this account, it seems natural to look for a more abstract notion of *recursive algorithm* which can be read directly from the equations (50), (51), just as its fundamental, implementation-free properties can be proved directly from (50), (51).

3.3. A **recursor** on a set X to W is any tuple of (partial, monotone) functionals $[f_0, f_1, \dots, f_n]$ whose types are such that the equations (50), (51) make sense when $x \in X$, and where the image of f_0 is W. We write

$$\mathbf{f} = [f_0, f_1, \dots, f_n] : X \hookrightarrow W, \tag{52}$$

we call $f_0, \dots, f_n$ *the parts* of $\mathbf{f}$ and we say that $\mathbf{f}$ *computes* the functional $\overline{\mathbf{f}} : X \rightharpoonup W$ defined by (51), its *denotation*. A functional f may be identified with the trivial recursor $[f]$ which computes it. Two recursors $\mathbf{f}$ and $\mathbf{g}$ are *equal* if they have the same number of parts n and for some permutation σ of $\{0, \dots, n\}$ with $\sigma(0) = 0$ and inverse ρ,

$$f_i(u_i, p_1, \dots, p_n, x) \simeq g_{\rho(i)}(u_i, p_{\sigma(1)}, \dots, p_{\sigma(n)}, x), \quad (i = 0, \dots, n). \tag{53}$$

The definition of recursor identity seems complex, but it simply identifies recursors which determine the same recursive definitions, except for the order in which the equations are listed. For example, if $\mathbf{f} = [f_0, f_1, f_2]$ has three parts and we set

$$\begin{aligned} g_0(p_2, p_1, x) &\simeq f_0(p_1, p_2, x), \\ g_1(u_1, p_2, p_1, x) &\simeq f_1(u_1, p_1, p_2, x), \\ g_2(u_2, p_2, p_1, x) &\simeq f_2(u_2, p_1, p_2, x), \end{aligned}$$

then $\mathbf{g} = [g_0, g_2, g_1] = [f_0, f_1, f_2] = \mathbf{f}$, because $\mathbf{g}$ determines the recursion

$$\begin{aligned} p_2(u_2) &\simeq g_2(u_2, p_2, p_1, x) \simeq f_2(u_2, p_1, p_2, x), \\ p_1(u_1) &\simeq g_1(u_1, p_2, p_1, x) \simeq f_1(u_1, p_1, p_2, x), \\ \overline{g}(x) &\simeq g_0(\overline{p}_2^x, \overline{p}_1^x, x) \simeq f_0(\overline{p}_1^x, \overline{p}_2^x, x), \end{aligned}$$

which is exactly that determined by $\mathbf{f}$, with the first two equations interchanged.

The basic presupposition of the theory of recursive algorithms is that the fundamental mathematical and implementation-free properties of an algorithm can be coded by a recursor: put another way, to define an algorithm which computes a functional $\overline{f}$, it is enough to specify a recursive definition whose head least fixed point is $\overline{f}$. Notice that recursors, like iterators, need not be effective in any way. To separate the *iterative algorithms* from the iterators we need to add a restriction of effectivity, e.g., by insisting that all the objects of the iterator are finitely described, as in the case of Turing machines. In the theory of recursive algorithms we can only talk about *relatively effective algorithms*, as follows.

3.4. Intensional semantics. Fix a functional structure $\mathbf{A}$ of signature τ as in **2.6.** The theory of recursive algorithms associates with each term ϕ of FLR(τ) and each sequence of variables $\vec{x}$ which includes all the free variables of ϕ, a recursor

$$int_{\mathbf{A}}(\vec{x}, \phi) : X \hookrightarrow W,$$

where X is the space of all n-tuples $(x_1, \ldots, x_n)$ such that the type of x_i matches that of the formal variable x_i and W is the basic set with type that of the term ϕ. This *referential intension* $int_{\mathbf{A}}(\vec{x}, \phi)$ of ϕ in $\mathbf{A}$ computes the (functional) denotation $den_{\mathbf{A}}(\vec{x}, \phi) : X \rightharpoonup W$ of ϕ in $\mathbf{A}$. The *recursive algorithms* of $\mathbf{A}$ are the recursors which occur as intensions of FLR-terms on $\mathbf{A}$, and the *recursive functionals* of $\mathbf{A}$ are their denotations.

The precise definition of intensions given in [24] is a bit complex,[12] it depends on several technical results of [23] and we will not attempt to reproduce it here. The main task is to explain how recursive definitions can be combined directly, without the introduction of extraneous, implementation dependent data structures, i.e., to interpret the formulas (50), (51) when $f_0, \ldots, f_n$ are recursors. The examples of referential intensions below will illustrate how the notions work.

3.5. Ideal, infinitary recursive algorithms. Consider the recursive definition of the partial truth predicate $T_{\mathbf{A}}$ for the language of LPCR on an acceptable structure $\mathbf{A}$ as in (21). It can be written out in the form (50), (51), but two of the clauses will involve the "infinitary" operations $\forall_A$, $\exists_A$. This means that in the associated recursor

$$\mathbf{T_A} = [T_0, T_1, \ldots, T_n] : A \hookrightarrow TV \tag{54}$$

two of the functionals are of the form

$$\begin{aligned} T_i(u, P_1, \ldots, P_n) &\simeq \forall_A(\lambda(y) P_k(u, y)), \\ T_j(u, P_1, \ldots, P_n) &\simeq \exists_A(\lambda(y) P_k(u, y)). \end{aligned} \tag{55}$$

On the account we are giving, $\mathbf{T_A}$ is a recursive algorithm of $\mathbf{A}$ which computes the partial truth predicate $T_{\mathbf{A}}$. There is little doubt that on its own, $\mathbf{T_A}$ is an interesting mathematical object, it is the semantic version of the definition of truth. On the other hand, as an "algorithm," it cannot be implemented when $\mathbf{A}$ is infinite, however we define "implementations." So there is a real question whether it is useful to adopt the terminology of algorithms in connection with such infinitary objects: do our intuitions about algorithms garnered from constructing and analyzing finitary, step-by-step mechanical procedures carry over to help us understand and study such infinitary objects?

One point to consider is the standard and highly developed *fixed point theory of programs* which (in effect) interprets ordinary programs by recursive equations. Almost all the concepts and methods of this theory extend to arbitrary recursive algorithms, sometimes trivially, often with some extra effort. Recursion in higher types [14, 15], [25], [13], recursion on admissible ordinals and set recursion [27], positive elementary induction [21] and other *generalized recursion theories*

[12] In fact we have oversimplified a bit, because only the so-called *special terms* define algorithms of a structure. The subtleties about special terms are not relevant to the modeling of Frege's sense by the referential intension and we will disregard them in this paper.

[11, 10] have been developed and found fruitful applications in many parts of logic, set theory and computer science based on this idea. The results in this subject are typically stated in the form of extensional properties of various collections of "computable" or "recursive" functions, but the proofs always depend on intensional properties of the recursions by which these functions are defined. For example, Kripke [19] states at the end of his paper that (on an acceptable structure) the relations weakly definable in his language are exactly the inductive relations, and those strongly definable (by formulas with no truth gaps) are the hyperelementary relations. This is an extensional statement, but its proof depends on the *Stage Comparison Theorem* 2A.2 of [21], which is most naturally understood as a statement about the lengths of the "abstract computations" by the recursor $\mathbf{T_A}$.[13]

The intension of the *liar* $\equiv P()$ where $\{P() \simeq \neg P()\}$ is the recursor

$$\mathrm{l} = [l_0, l_1], \text{ where } l_0(P) \simeq P(),\ l_1(P) \simeq \neg P(), \tag{56}$$

and that of the *truthteller* $\equiv P()$ where $\{P() \simeq P()\}$ is

$$\mathrm{tt} = [l_0, l_0]. \tag{57}$$

We can read these directly from the sentences involved, because they are so simple. In general we need to *reduce* a sentence by the *reduction calculus* of [23] which preserves intensions, until we bring it to an irreducible *normal form*, from which we can read off its intension.

Suppose R, Q are interpreted by converse relations R, Q in a structure and a, b are constants interpreted by elements a, b. The reduction calculus gives the normal forms

$$\begin{aligned} \mathrm{R(a,b)} &\sim_{\mathbf{A}} \mathrm{R}(p_1(), p_2()) \text{ where } \{p_1() \simeq \mathrm{a},\ p_2() \simeq \mathrm{b}\}, \\ \mathrm{Q(b,a)} &\sim_{\mathbf{A}} \mathrm{Q}(p_2(), p_1()) \text{ where } \{p_2() \simeq \mathrm{b},\ p_1() \simeq \mathrm{a}\}, \end{aligned}$$

and the three-part intensions that we can read off these normal forms are equal (via the permutation $\sigma(1) = 2$, $\sigma(2) = 1$) precisely because R and Q are converse relations. A similar computation shows that in general,

$$\phi \,\&\, \psi \sim_{\mathbf{A}} \psi \,\&\, \phi, \qquad \phi \vee \psi \sim_{\mathbf{A}} \psi \vee \phi.$$

As an additional example from LPC, consider the sentence χ of (4). We exhibit some of the steps of its reduction to normal form, to illustrate the reduction calculus of [23]. The six-part intension of χ which can be read off this normal form is the recursor representation of the algorithm described by Steps (1)–(4) in Section 1.

[13]Kripke also associates with each sentence ϕ which has a truth value its *level*, the least ordinal at which his truth definition yields that value. In algorithmic terms, this is precisely the length of the computation by the recursor $\mathbf{T_A}$ on the Gödel code of ϕ. A more natural level function assigns to each ϕ the length of the computation of the recursor $int(\phi)$, with no appeal to Gödel codes.

$$
\begin{aligned}
\chi \equiv\ & (\forall x)(\exists y)R(x,y) \vee (\exists z)Q(z) \\
\sim_{\mathbf{A}}\ & \lambda()P_1() \vee \lambda()P_2() \\
& \quad \text{where } \{P_1() \simeq (\forall x)(\exists y)R(x,y), \\
& \qquad\qquad P_2() \simeq (\exists z)Q(z)\} \\
\sim_{\mathbf{A}}\ & \lambda()P_1() \vee \lambda()P_2() \\
& \quad \text{where } \{P_1() \simeq \forall_A(P_{11}) \text{ where } \{P_{11}(x) \simeq (\exists y)R(x,y)\}, \\
& \qquad\qquad P_2() \simeq \exists_A(P_{22}) \text{ where } \{P_{22}(z) \simeq Q(z)\} \\
& \qquad\quad \} \\
\sim_{\mathbf{A}}\ & \lambda()P_1() \vee \lambda()P_2() \\
& \quad \text{where } \{P_1() \simeq \forall_A(P_{11}),\ P_{11}(x) \simeq (\exists y)R(x,y), \\
& \qquad\qquad P_2() \simeq \exists_A(P_{22}),\ P_{22}(z) \simeq Q(z) \\
& \qquad\quad \} \\
\sim_{\mathbf{A}}\ & \cdots \\
\sim_{\mathbf{A}}\ & \lambda()P_1() \vee \lambda()P_2() \\
& \quad \text{where } \{P_1() \simeq \forall_A(P_{11}),\ P_{11}(x) \simeq \exists_A(\lambda(y)P_{111}(x,y)), \\
& \qquad\qquad P_{111}(x,y) \simeq R(x,y),\ P_2() \simeq \exists_A(P_{22}),\ P_{22}(z) \simeq Q(z) \\
& \qquad\quad \}.
\end{aligned}
$$

Finally, the "bald King of France" sentence (36) has the following normal form, assuming that both $KF(x)$ and $B(x)$ are prime formulas:

$$BKF \sim_{\mathbf{A}} P(q()) \text{ where } \{P(x) \simeq B(x),\ q() \simeq \text{the } (R),\ R(x) \simeq KF(x)\}.$$

Its intension has four parts, the head being

$$bkf_0(P, q, R) \simeq P(q()).$$

This and the preceding two examples illustrate the simple form of the intensions of sentences in LPC, even when we extend it with description operators or (quite easily) generalized connectives and quantifiers, modal operators and the like. They describe recursions which, however non-constructive (when the functionals in the structure are non-constructive), they are trivial as recursions, they "close" in a finite number of steps. Except for the imagery of algorithms, the theory on this level is really not that much different from Church's (Alternative 0 version) in [1] or Cresswell's, which could cover the same ground with the introduction of a natural equivalence relation on proposition surrogates or structures. It is, of course, the algorithm imagery which justifies the transformations by which we reduce sentences to normal form, compute their intensions and then use the natural identity of these recursors to get precisely the equivalence relation which is needed. When we add self-reference to the language we obtain the less trivial intensions of formulas such as the truth predicate, which describe more interesting recursions. We also make considerably more complex the relation of intensional identity.

3.6. THEOREM. *The problem of graph isomorphism for finite (unordered) graphs is polynomially reducible to the problem of intensional identity, for sentences of* LPCR *on any structure.*

PROOF (Kit Fine).[14] Given a graph with n nodes $P_1, \dots, P_n$ and $k \le n^2$ edges $E_1, \dots, E_k$, we construct a sentence $\phi(G)$ using $P_1, \dots, P_n$, $E_1, \dots, E_k$ and an additional Q as partial propositional variables. The sentence $\phi(G)$ is a **where** sentence with $n + k + 1$ parts, the head being $Q()$ and the other parts of the form

$$P_i() \simeq P_i(), \quad (i = 1, \dots, n),$$
$$E_j() \simeq P_{b(j)}() \vee P_{e(j)}(), \quad (j = 1, \dots, k),$$

where for each j, $P_{b(j)}$ and $P_{e(j)}$ are respectively the beginning and the end nodes of the edge E_j. The construction is obviously polynomial and it is quite trivial to verify that two graphs are isomorphic exactly when the corresponding sentences have the same intension, directly from the definition of recursor identity. ⊣

All the sentences constructed in the proof receive no truth value, they are silly and complex versions of the truthteller. One can build more interesting examples of complex intensions, but this simple construction exhibits the complexity of deciding intensional identity, the graph isomorphism problem being notoriously difficult.

3.7. A recursor structure of signature $\tau = (B, \{\mathrm{f}_1, \dots, \mathrm{f}_n\}, d)$ is an interpretation of FLR(τ) where the function symbols are interpreted by recursors, i.e., a tuple $\mathbf{A} = (\mathcal{U}, \mathrm{f}_1, \dots, \mathrm{f}_n)$, where each f_i is a recursor with head fixed point a functional $\bar{\mathrm{f}}_i$ of type $d(\mathrm{f}_i)$. We can think of every functional structure as a recursor structure, identifying its functionals with the associated trivial recursors which compute them, as in **3.3**. Recursor structures arise naturally when we enrich a language by adding a new symbol to express a complex notion, and then we want the new symbol to be interpreted by the complex meaning of the expression it replaces. Kripke's language is naturally interpreted in recursor structures, since (at the least) we want the ordinals attached to sentences to be part of the interpretation of T. The theory of intensions generalizes directly to the category of recursor structures, which is the natural context for it.

3.8. Non-minimal fixed points. In [19], Kripke discusses several possibilities of interpreting self-referential sentences by choosing fixed points other than the least one. He says that "the smallest fixed point is probably the most natural model for the intuitive concept of truth," and the algorithmic analogy supports this. On the other hand, the referential intension of a sentence determines not only the smallest fixed point but all of them; thus, if we identify *sense* with *referential intension*, we do not lose those aspects of meaning which may be represented by the non-minimal fixed points. For example, Kripke calls the *liar* "paradoxical" because it cannot have a truth value in any fixed point of $\mathbf{T_A}$. The paradoxical nature of the *liar* is coded by its referential intension, a recursor with the property that none of its simultaneous fixed points yields a truth value.

[14] This argument of Fine's (included with his permission) is simpler than my original proof and shows that in fact *the problem of intensional identity for the propositional part of* LPCR is at least as hard as the graph isomorphism problem.

§4 Sense identity and indirect reference.

Van Heijenoort [29] quotes an extensive passage from a 1906 letter from Frege to Husserl which begins with the following sentence:

> "It seems to me that we must have an objective criterion for recognizing a thought as the same thought, since without such a criterion a logical analysis is not possible."

This could be read as asserting the existence of a decision procedure for sense identity, but unfortunately, the letter goes on to suggest that logically equivalent sentences have the same sense, a position which is contrary to the whole spirit of [12]. It is apparently not clear what Frege thought of this question or if he seriously considered it at all. Kreisel and Takeuti [17] raise explicitly the question of *synonymity* of sentences which may be the same as that of identity of sense. If we identify sense with referential intension, the matter is happily settled by a theorem.

4.1. THEOREM. *For each recursor structure* $\mathbf{A} = (U_1, \ldots, U_k, f_1, \ldots, f_n)$ *of finite signature, the relation* $\sim_{\mathbf{A}}$ *of intensional identity on the terms of* FLR *interpreted on* $\mathbf{A}$ *is decidable.*

For each structure $\mathbf{A}$ and arbitrary integers n, m, let

$$S_{\mathbf{A}}(n,m) \iff n, m \text{ are Gödel numbers of sentences or terms } \theta_n, \theta_m \text{ of FLR and } \theta_n \sim_{\mathbf{A}} \theta_m. \tag{58}$$

The rigorous meaning of **4.1** is that this relation $S_{\mathbf{A}}$ is decidable, i.e., computable by a Turing machine. By the usual coding methods then, we get immediately:

4.2. COROLLARY. *The relation* $S_{\mathbf{A}}$ *of intensional identity on Gödel numbers of expressions of* FLR *is elementary (definable in* LPC*), over each acceptable structure* $\mathbf{A}$.

The Corollary is useful because it makes it possible to talk indirectly about FLR intensions within FLR. In general, we cannot do this directly because the intensions of a structure $\mathbf{A}$ are higher type objects over $\mathbf{A}$ which are not ordinarily[15] members of any basic set of the universe of $\mathbf{A}$. One reason we might want to discuss FLR intensions within FLR is to express *indirect reference*, where Frege's treatment deviates from his general doctrine of separate compositionality principles for sense and denotation. Frege argued that "the indirect denotation of a word is ...its customary sense," so that in

$$\text{Othello believed that Cassio and Desdemona were lovers,} \tag{59}$$

[15] If the universe of $\mathbf{A}$ contains the powerset of every basic set in it and the Cartesian product of every two basic sets, then of course it contains all recursors over basic sets and with suitably rich primitives we can develop the theory of intensions of $\mathbf{A}$ within LPCR. These are typed structures, however, of infinite signature, which lack a natural universe, a largest basic set. More interesting would be the structure of the universe of sets, whose only basic "set" is the class of all sets. The intensions of this structure are certainly not sets.

the object of Othello's belief would be the sense of the sentence 'Cassio and Desdemona were lovers'. Since we cannot refer to sense directly in any fragment of natural language formalizable within FLR (if we identify it with intension), we might attempt to make belief an attribute of sentences, or (equivalently) their Gödel numbers. This means that (59) is expressed by

$$\text{Othello believed 'Cassio and Desdemona were lovers'}, \tag{60}$$

where 'Cassio and Desdemona were lovers' is the Gödel number of the sentence within the quotes. But then we would certainly expect (60) to imply

$$\text{Othello believed 'Desdemona and Cassio were lovers'}, \tag{61}$$

and we would like to express in the language the general assertion that belief depends only on the sense, not the syntactic form of a sentence, i.e.,

$$[\text{Othello believes } m \ \&\ S_{\mathbf{A}}(m,n)] \Longrightarrow \text{Othello believes } n. \tag{62}$$

The Corollary says precisely that (62) is expressible already in LPC. The method is evidently quite general: if we view propositional attitudes as attributes of Gödel codes, we can express in the language that they respect sense identity, those indeed which should respect it.

Proof of the Main Theorem **4.1**. The intension of a term in a recursor structure is computed in the associated functional expansion by Def. **3.6** of [24], so we may assume that the interpretations $f_1, \ldots, f_n$ of the function symbols of the language in $\mathbf{A}$ are functionals. We fix such a functional structure $\mathbf{A}$ then and we assume (for the time being) that *all basic sets in* $\mathbf{A}$ *are infinite.* We will discuss the interesting case of finite basic sets at the end.

Since intensions are preserved by passing to the normal form, the problem of intensional identity on $\mathbf{A}$ comes down to this: given two irreducible, recursive terms

$$\begin{aligned} \phi &\equiv \phi_0 \text{ where } \{p_1(u_1) \simeq \phi_1, \ldots, \ p_n(u_n) \simeq \phi_n\}, \\ \psi &\equiv \psi_0 \text{ where } \{q_1(v_1) \simeq \psi_1, \ldots, \ q_m(v_m) \simeq \psi_m\}, \end{aligned}$$

is $n = m$ and can we match the terms so that they define the same functionals? By trying all possible ways to match the parts[16] and using the form of irreducible, explicit terms (2B.4 of [23]), we can further reduce the problem to that of deciding whether an arbitrary identity in one of the following three forms holds in $\mathbf{A}$:

$$f(z_1, \ldots, z_m) \simeq g(z_{m+1}, \ldots, z_l), \tag{63}$$

$$f(z_1, \ldots, z_m) \simeq p(w_1, \ldots, w_k). \tag{64}$$

$$q(w_1, \ldots, w_m) \simeq p(w_{m+1}, \ldots, w_l). \tag{65}$$

Here the following conditions hold:

1. The functionals f and g are among the finitely many givens of $\mathbf{A}$, or the constants $\mathbf{t}$, $\mathbf{f}$ or the conditional.

[16]This trivial part of the algorithm is (on the face of it) in NP (non-deterministic, polynomial time) in the length of the given terms and the rest will be seen to be no worse. I do not know a better upper bound for the complexity of intensional identity on a fixed structure and the best lower bound I know is that of Theorem **3.6**.

2. Each z_i is an *immediate expression* (in the full set of variables) by 2B.2 of [23], i.e., either a basic variable, or $p(\vec{x})$ where the x_i's are basic variables, or $\lambda(\vec{s})p(\vec{x})$ with $p(\vec{x})$ as above.

3. Each w_j is either a basic variable or $r(\vec{x})$, where the x_i's are basic variables and where $r \equiv p$ and $r \equiv q$ are allowed.

The decision question is trivial for identities in form (65) because of the following elementary result from equational logic.

4.3. LEMMA. *An identity (65) is valid on any fixed structure with infinite basic sets only if its two sides are identical.*

We can view (64) as a special case of (63), with the following *evaluation functional* substituted for g on the right:

$$ev^k(p, x_1, \dots, x_k) \simeq p(x_1, \dots, x_k). \tag{66}$$

Notice, however, that there are infinitely many such evaluation functionals. There are also infinitely many possible identities in form (63), because although f and g are chosen from a finite set, there is an infinite number of immediate expressions from which to choose the z_i's. The proof splits into two parts. First we will show that if we expand the structure by a fixed, finite number of evaluation functionals, then every identity in form (64) is effectively equivalent to one in form (63). In the second part we will show how to decide the validity of equations in form (63).

4.4. A basic variable v is *placed* in an identity (63) or (64) if $v \equiv z_i$ for some i. For example, the placed variables of $f(v, p(x, y), u) = p(s, r(x, y))$ are v and u.

4.5. LEMMA.[17] *Suppose the identity*

$$f(z_1, \dots, z_m) \simeq p(w_1, \dots, w_k). \tag{67}$$

is valid in the structure $\mathbf{A}$ *with infinite basic sets and* $w_i \equiv r(\vec{x})$ *is one of the terms on the right. Then there exists some* z_j *on the left such that either* $w_i \equiv z_j$, *or* $z_j \equiv \lambda(s_1, \dots, s_k)r(\vec{y})$ *and* $r(\vec{x})$ *can be obtained from* $r(\vec{y})$ *by the substitution of placed variables for* $s_1, \dots, s_k$.

PROOF. To keep the notation simple we assume that there is only one basic set and we consider the special case where

$$w_2 \equiv r(x, u, x, y, v), \quad u,\ v \text{ placed},\ x,\ y \text{ not placed}. \tag{68}$$

CASE 1. $r \not\equiv p$. Choose disjoint sets D_x, D_y, D_u, D_v, W and some $\bar{c}$ outside all of them and first set all variables other than x, y, u, v to $\bar{c}$ and all partial function variables other than r, p to constant functions with value $\bar{c}$. Next set u and v to constant values $\bar{u}$, $\bar{v}$ in the corresponding sets D_u, D_v. For each arbitrary partial function

$$\omega : D_x \times D_x \times D_y \rightharpoonup W,$$

[17] I am grateful to Joan Moschovakis for a counterexample which killed a plausible simplification of this proof, before I invested too much time in it.

set r by the conditions

$$[s_1 \notin D_x \vee s_2 \notin D_u \vee s_3 \notin D_x \vee s_4 \notin D_y \vee s_5 \notin D_v] \Longrightarrow r(s_1, s_2, s_3, s_4, s_5) \simeq \bar{c},$$

$$[s_1 \in D_x \ \&\ s_2 \in D_u \ \&\ s_3 \in D_x \ \&\ s_4 \in D_y \ \&\ s_5 \in D_v] \Longrightarrow r(s_1, s_2, s_3, s_4, s_5) \simeq \omega(s_1, s_3, s_4)$$

and finally set

$$\sigma(t) = \begin{cases} t, & \text{if } t \in W, \\ \bar{c}, & \text{otherwise}, \end{cases} \qquad p(s_1, s_2, \ldots, s_k) \simeq \sigma(s_2). \tag{69}$$

Consider the result of further substituting in (67) arbitrary values $x \in D_x$, $y \in D_y$. Suppose $w_j \equiv q(\vec{t})$ is one of the terms within p on the right. If $q \not\equiv r$, then with these substitutions w_j is defined, set either to $\bar{c}$ or to $\sigma(t_2)$, if $q \equiv p$. If $q \equiv r$ and the sequence of variables $\vec{t}$ is not *exactly* x, u, x, y, v, then w_j again takes the value $\bar{c}$. Thus the only term which may possibly be undefined among the w_j's is w_2 (which may of course occur more than once) and hence the right-hand-side of (67) is defined exactly when w_2 is defined and we have a valid identity:

$$f(z_1(\omega, x, y), \ldots, z_m(\omega, x, y)) \simeq \omega(x, x, y), \tag{70}$$
$$(x \in D_x,\ y \in D_y,\ \omega : D_x \times D_x \times D_y \rightharpoonup W).$$

The typical expression $z_i(\omega, x, y)$ on the left evaluates to the constant $\bar{c}$ or some function with the constant value $\bar{c}$ if neither r nor p occurs in z_i. If $z_i \equiv \lambda(\vec{s})p(\vec{t})$, then again z_i has a value independent of ω, x, y, because of the definition of p and the fact we we set no variable equal to a member of W. Finally, if

$$z_i \equiv \lambda(\bar{s})r(t_1, t_2, t_3, t_4, t_5), \tag{71}$$

but some t_i is free or a constant and is not the ith variable or constant in the pattern $x, \bar{u}, x, y, \bar{v}$, then again the expression evaluates to $\bar{c}$, by the definition of r. Thus z_i depends on ω, x, y only when at most t_1, t_3 or t_4 are free, and those among them which are free are set to the corresponding value x, x or y. If all three are free in some such z_i, then the lemma clearly holds. In the opposite case the partial function ω satisfies an identity of the form

$$\omega(x, x, y) \simeq h(\omega, \omega(\cdot, x, y), \omega(x, \cdot, y), \omega(x, x, \cdot), \omega(\cdot, \cdot, y), \omega(\cdot, x, \cdot), \omega(x, x, \cdot)), \tag{72}$$

where h is a monotone operation on partial functions and $\cdot$ is the algebraic notation for λ-abstraction, e.g.,

$$\omega(\cdot, x, \cdot) = \lambda(s, t)\omega(s, x, t).$$

For example, suppose

$$z_i \equiv \lambda(st)r(x, \bar{u}, s, t, \bar{v}) = \beta;$$

then

$$\beta(s, t) \simeq \begin{cases} \omega(x, s, t), & \text{if } s \in D_x,\ t \in D_y, \\ \bar{c}, & \text{otherwise}, \end{cases}$$

so that $z_i = h_i(\omega(x, \cdot, \cdot))$ with a monotone h_i. A similar evaluation of z_i in terms of some *section* of ω is involved in each of the cases and the substitution of all these monotone h_i's into f yields a monotone operation.

Finally, we obtain a contradiction from the alleged validity of (72). Choose distinct points x_0, x_1, y_0, y_1 in the respective sets D_x, D_y and define two partial functions with only the indicated values, where $0, 1$ are distinct points in W.

$$\alpha(x_0, x_0, y_0) \simeq 0, \quad \gamma(x, x', y) \simeq \begin{cases} 0, & \text{if } x = x_0 \vee x' = x_0 \vee y = y_0, \\ 1, & \text{otherwise.} \end{cases}$$

From (72) applied to α,

$$h(\alpha, \alpha(\cdot, x_0, y_0), \ldots,) \simeq \alpha(x_0, x_0, y_0) \simeq 0. \tag{73}$$

But obviously $\alpha \subseteq \gamma$ and an easy computation shows that every section of γ at (x_1, x_1, y_1) extends the corresponding section of α at (x_0, x_0, y_0), for example

$$\lambda(s)\alpha(x_0, s, y_0) \subseteq \lambda(s)\gamma(x_1, s, y_1),$$

simply because $\gamma(x_1, x_0, y_1) \simeq 0$. Thus by the monotonicity of h, (73) and (72) applied to γ, we should have

$$0 \simeq g(\gamma, \gamma(\cdot, x_1, y_1), \ldots,) \simeq \gamma(x_1, x_1, y_1),$$

while by its definition $\gamma(x_1, x_1, y_1) \simeq 1$. This completes the proof of the Lemma in the first case.

CASE 2. $r \equiv p$. We consider again a typical, simple case

$$f(z_1, \ldots, z_m) \simeq p(w_1, p(x, y, u), w_2), \ u \text{ placed, } x, y \text{ not placed.}$$

As before, we restrict the variables to disjoint sets D_x, D_y, D_u, W and we set:

$$p(s_1, s_2, s_3) \simeq \begin{cases} \omega(s_1, s_2), & \text{if } s_1 \in D_x,\ s_2 \in D_y,\ s_3 \in D_u, \\ s_2, & \text{otherwise, if } s_2 \in W, \\ \bar{c}, & \text{otherwise.} \end{cases}$$

From this it follows that we get a valid identity of the form (72) for an arbitrary $\omega : D_x \times D_y \rightharpoonup W$, the main points being that all the terms on the right which are not identical with w_2 are defined and only the sections show up on the left, and then the proof is finished as before. ⊣

4.6. LEMMA. *An identity of the form*

$$f(z_1, \ldots, z_m) \simeq p(w_1, \ldots, w_k) \tag{74}$$

cannot be valid in a structure $\mathbf{A}$ *with infinite basic sets if the number* n *of distinct terms (not variables) on the right is greater than a fixed number* d*, which depends only on the type of* f*; if* $n \leq d$*, then we can compute from (74) an equivalent identity of the form*

$$f(z_1, \ldots, z_m) \simeq ev^n(W_0, W_1, \ldots, W_n). \tag{75}$$

PROOF. If (74) is valid, then by the preceding Lemma **4.5**, each w_i which is a term either is identical with some z_j or can be obtained by the substitution of placed variables in some z_j. If there are $q \leq m$ placed variables, and if z_j is a λ-term, it is of the form $\lambda(s_1, \ldots, s_{l(j)})z_j^*$, where the number $l(j)$ can be computed from the type of f, so it can generate by substitution of placed variables into its bound variables at most $q^{l(j)}$ distinct terms; hence the total number of distinct terms on the right cannot exceed

$$d = \sum_{j=1}^{m} q^{l(j)}. \tag{76}$$

Suppose the right-hand-side of (74) is $p(x, A, u, B, A, z)$, where distinct caps indicate distinct terms and the lower case letters are variables. We then have

$$\begin{aligned} p(x, A, u, B, A, x) &\simeq (\lambda(a, b)p(x, a, u, b, a, x))(A, B) \\ &\simeq ev^2(\lambda(a, b)p(x, a, u, b, a, x), A, B). \end{aligned}$$

The general case is similar. ⊣

This last lemma reduces the decision problem of intensional identity to equations in form (63), where there is a finite choice of f's, the functionals in the signature, and a finite choice of g's, those in the structure and the ev^k's, for k less than d computed by (76) for every functional in the structure.

4.7. Extended sets and assignments. Before describing the procedure which determines the validity of identities in form (63), we consider a simple example which illustrates one of the annoying subtleties we will need to deal with. Suppose g is a total, unary function on some set A and we define the total, binary function f by

$$f(x, y) \simeq g(x). \tag{77}$$

Clearly (77) is a valid identity involving the old g and the new f we just defined. Suppose we substitute a term in this to get

$$f(x, p(x)) \simeq g(x); \tag{78}$$

now this is not valid, because for some values of the partial function p, $p(x)$ will not be defined, so neither will $f(x, p(x))$, while the right-hand-side is defined. In dealing with partial functions and functionals as we have, *validity of identities is not preserved by substitution of terms.* One way to deal with this problem is to add an element $\perp$ to each basic set and view partial functions as total functions, which take the value $\perp$ when they should be undefined. For each set A, we set

$$A^{\perp} = A \cup \{\perp\} = \text{ the extension of } A. \tag{79}$$

We can now try to interpret identities by allowing the basic variables to range over the extended sets, so that the validity of (77) implies the validity of (78); this is fine, except that now (77) *fails for the f we originally defined,* because we still have $f(x, \perp) = \perp \neq g(x)$, when $x \in A$. Of course, some might say we defined the wrong f, but in fact it is these "strict" identities we need to decide to settle the question of intensional identity. In practice we will need to work both with strict and with extended identities and we must keep the context clear.

We will use "=" to denote equality in the extended basic sets and set

$$x\downarrow \iff x \in A \iff x \neq \perp, \quad (x \in A^{\perp}). \tag{80}$$

A *strict assignment* π in a structure **A** assigns partial functions to pf variables and members of the basic sets to basic variables, as usual. An *extended assignment* behaves exactly like a strict assignment on pf variables, but assigns members of the extended basic sets to the basic variables, i.e., it can set $\pi(v) = \bot$. An identity is ***strictly valid*** when it holds for all strict assignments, and ***extendedly valid*** if it holds for all extended assignments.

4.8. Dictionary lines. Choose once and for all fixed, *special variables* $x_1, \ldots, x_l$ of types such that

$$f(x_1, \ldots, x_m) \simeq g(x_{m+1}, \ldots, x_l) \tag{81}$$

is well formed. A *dictionary line* for f and g is an implication of the form

$$\phi_1,\ \phi_2,\ \ldots,\ \phi_n \Longrightarrow f(x_1, \ldots, x_m) \simeq g(x_{m+1}, \ldots, x_l) \tag{82}$$

where each formula ϕ_k in the *antecedent of the line* may involve additional *extra, basic variables* other than the $x_1, \ldots, x_l$ and satisfies one of the following conditions.

1. $\phi_k \equiv x_i = u$, where x_i is one of the special, basic variables. At most one formula of this type in the antecedent involves each x_i.

2. ϕ_k is $\lambda(\vec{s})x_i(\vec{u}) = \lambda(\vec{s})x_j(\vec{v})$ or $x_i(\vec{u}) = x_j(\vec{v})$. At most one formula of this type in the antecedent involves each pair x_i and x_j.

3. ϕ_k is $u \downarrow$ or $u \neq v$, where the basic variables u, v occur free in the line in formulas of type (1) or (2).

4.9. Dictionaries. A line is valid (in the given structure) if every extended assignment which satisfies its hypothesis also satisfies its conclusion. This means that the choice of specific extra variables is irrelevant to the validity of a line, and then a simple counting argument shows that the types of f and g determine an upper bound on the number of distinct (up to alphabetic change and reordering of hypotheses) lines. We fix a sufficiently large set of extra variables and list once and for all, all the lines in these variables which are valid for f and g in the given structure; this is *the dictionary for f and g*.

The dictionary of the structure $\mathbf{A}$ is the union of the dictionaries for all the pairs of functionals in $\mathbf{A}$. It is a finite list of lines, perhaps not easy to construct for specific structures with non-constructive givens, but in principle it can be written down.

We will associate (effectively) with each identity (63) a specific set of lines L such that the *strict validity* of (63) is equivalent to the *extended validity* of all the lines in L. It will be convenient to express these lines using the variables which occur in (63). To decide a specific (63), we translate the lines of L into equivalent lines in the fixed, chosen variables by an alphabetic change, and then (63) will be equivalent to the presence of these lines in the dictionary.

For example, (77) will be expressed (essentially) by the single line

$$x_1 = x_3,\ x_1\downarrow,\ x_2\downarrow,\ x_3\downarrow \Longrightarrow f(x_1, x_2) = g(x_3),$$

which is valid, while for (78) we will get

$$x_1 = x_3,\ x_1\downarrow,\ x_3\downarrow \Longrightarrow f(x_1, x_2) = g(x_3),$$

which is not. (Actually there will be some additional "fringe" on the lines produced by the formal decision procedure, which will not affect the validity of the lines.)

4.10. Bound variable unifiers. Let

$$E \equiv \lambda(u_1, \ldots, u_m)A, \quad F \equiv \lambda(v_1, \ldots, v_n)B, \tag{83}$$

be two immediate λ-expressions. A *bound variable unifier* or just *bvu* for these two expressions is a triple

$$(\tau, \sigma, \vec{s}) = (\tau, \sigma, (s_1, \ldots, s_l)),$$

where

$$\tau : \{u_1, \ldots, u_m\} \to \text{variables}, \quad \sigma : \{v_1, \ldots, v_n\} \to \text{variables}$$

are substitution maps on the bound variables, $s_1, \ldots, s_l$ are variables which do not occur in A, B but do occur in both $\tau[A]$ and $\sigma[B]$, and

$$\lambda(s_1, \ldots, s_l)\tau[A] \equiv \lambda(s_1, \ldots, s_l)\sigma[B]. \tag{84}$$

The variables $s_1, \ldots, s_l$ are the *bound variables* of the unifier.

It will be convenient to assume that such variable transformations are defined on all variables, by setting

$$\tau(w) \equiv w, \text{ if } w \text{ is not in the domain of } \tau.$$

We have already used this convention in writing $\tau[A]$, presumably meaning the term resulting by replacing every basic variable u in A by $\tau(u)$.

For example, we can unify

$$\lambda(u, u')r(u, u', a), \quad \lambda(v)r(v, b, a)$$

by setting

$$\tau(u) \equiv \tau(u') \equiv b, \quad \sigma(v) \equiv b, \quad \vec{s} = \emptyset,$$

which identifies both expressions with $\lambda(\,)r(b, b, a)$. It is obvious that this is not the best we can do, though, since we can also set

$$\tau'(u) \equiv s,\ \tau'(u') \equiv b, \quad \sigma'(v) \equiv s, \quad \vec{s} = (s),$$

which unifies the terms "further" to $\lambda(s)r(s, b, a)$. The next definition and lemma capture this simple idea of the existence of a unique such "maximal" unifier, when one exists at all.

4.11. Suppose $(\tau, \sigma, \vec{s})$ and $(\tau', \sigma', \vec{s}')$ are both bvu's for two λ-terms E and F. We say that $(\tau, \sigma, \vec{s})$ is *reducible to* $(\tau', \sigma', \vec{s}')$ if it "unifies no more," i.e., every variable in the sequence $\vec{s}$ is also in the sequence $\vec{s}'$, if $\tau'(w)$ is not bound in $(\tau', \sigma', \vec{s}')$ then $\tau(w) = \tau'(w)$, if $\tau(w)$ is bound in $(\tau, \sigma, \vec{s})$ then $\tau(w) = \tau'(w)$, and similarly with σ, σ'.

A bvu for two λ-terms is *maximal*, if there exists no other bvu for the same λ-terms with a longer sequence of bound variables.

4.12. LEMMA. *(1) Every bounded variable unifier for two λ-terms as in* (83) *can be reduced to a maximal one.*

(2) Two λ-terms have at most one maximal unifier, up to alphabetic change of the bound variables.

PROOF. The *critical number* for two λ-terms as in (83) is the number of distinct variables among the $u_1, \ldots, u_m, v_1, \ldots, v_n$ which actually occur in A and B. We will show by induction on the critical number, simultaneously, that if a bvu exists, then a unique maximal one exists and the given one is reducible to the maximal one. Notice that if there is any unifier at all, we must have

$$A \equiv r(a_1, \ldots, a_k), \quad B \equiv r(b_1, \ldots, b_k), \tag{85}$$

i.e., A and B must be terms involving the same pf variable.

At the basis, none of the variables occur, so there is only one possible unifier, the empty τ, σ, which reassigns no variables; it is a unifier exactly when the two expressions are identical and it is trivially maximal. We take this case of expressions of the form $\lambda()A$, $\lambda()B$ to cover also (by convention) the case of terms A, B.

For the induction step, suppose u_i actually occurs in A. We distinguish cases.

CASE 1. For some j, $a_j \equiv u_i$ but b_j is none of the v_t's.

In this case every unifier must set $\tau(u_i) \equiv b_j$, we make this replacement on the expressions and we get the result by applying the induction hypothesis to these new expressions which have smaller critical number since u_i does not occur in them.

CASE 2. For some j, $a_j \equiv u_i$, and for every j, if $a_j \equiv u_i$ then b_j is one of the v_t's, but there exist distinct $j \neq j'$ such that $a_j \equiv a_{j'} \equiv u_i$ and $b_j \not\equiv b_{j'}$.

In this case every unifier must satisfy

$$\tau(u_i) \equiv \sigma(b_j) \equiv \sigma(b_{j'}),$$

and we can apply the ind. hyp. to the expressions resulting by identifying b_j with $b_{j'}$ on the right, which have smaller critical number.

CASE 3. u_i occurs in A, and for all j and some t,

$$a_j \equiv u_i \Longrightarrow b_j \equiv v_t.$$

In this case we look at the symmetric argument, taking Cases 1, 2 and 3 on v_t. In the first of these two we can use the induction hypothesis and the last leads us to the following:

CASE 4. u_i occurs in A and there exists some v_t which occurs in B at precisely all the same places where u_i occurs in A.

In this case every unifier must set $\tau(u_i) \equiv \sigma(v_t)$, and this variable may be one of the bound or the free ones in the final expressions. We choose a variable s not occurring anywhere, we replace u_i in A and v_t in B by s throughout and we apply the ind. hyp. to these new expressions with lower critical number to obtain some maximal bvu (if it exists) $(\tau, \sigma, (s_1, \ldots, s_k))$; the maximal bvu for the original expressions is obtained by adding s to the bound variables and extending τ and σ in the obvious way,

$$\tau(u_i) \equiv \sigma(v_t) \equiv s. \qquad \dashv$$

Suppose now we are given an identity (63). We first show how to construct a single dictionary line from it, which will determine the strict truth of (63) when all the free, basic variables in it are interpreted by distinct values in the domain. The complete set of lines for (63) will contain the lines we get in this way from all the identities which result from (63) by the identification of some of its free, basic variables.

We assume that the special variables $x_1, \ldots, x_m$ we will use for the lines do not occur in the given identity.

STEP 1. For each z_i which is a basic variable, we put in the antecedent of the line the equality $x_i = z_i$ and the condition $x_i\downarrow$.

STEP 2. Consider any pair z_i, z_j $(i \neq j)$ of expressions which are not basic variables; we will view these as λ-expressions by identifying temporarily a term $r(\vec{x})$ with the λ-expression $\lambda(\,)r(\vec{x})$. If z_i, z_j cannot be unified, we add nothing to the line. In the opposite case, suppose

$$z_i \equiv \lambda(u_1, \ldots, u_{arity(i)})r(x_1, \ldots, x_k), \quad z_j \equiv \lambda(v_1, \ldots, v_{arity(j)})r(y_1, \ldots, y_k),$$

and $(\tau, \sigma, (s_1, \ldots, s_l))$ is a maximal bvu for these expressions. We add to the antecedent of the line the equation

$$\lambda(s_1, \ldots, s_l)x_i(\tau(u_1), \ldots, \tau(u_{arity(i)})) \simeq \lambda(s_1, \ldots, s_l)x_j(\sigma(v_1), \ldots, \sigma(v_{arity(j)})). \tag{86}$$

STEP 3. After Steps 1 and 2 are completed, we add to the antecedent of the line the inequality $u \neq v$ and the conditions $u\downarrow$, $v\downarrow$, for every two free, basic variables which occur free on the line and are not among the standard $x_1, \ldots, x_n$. (If only one such u occurs, we simply add $u\downarrow$.)

To recapitulate what we said above, the complete set of lines associated with an identity is obtained by applying this procedure to every identity obtained by identifying some or all of the free basic variables of the identity.

To illustrate the procedure, consider again (77). There are no terms to unify here, only Steps 1 and 3 come up and we get the following two lines:

$$x_1 = x,\ x_2 = y,\ x_3 = x,\ x_1\downarrow,\ x_2\downarrow,\ x_3\downarrow,\ x \neq y,\ x\downarrow,\ y\downarrow \implies f(x_1, x_2) = g(x_3),$$
$$x_1 = x,\ x_2 = x,\ x_3 = x,\ x_1\downarrow,\ x_2\downarrow,\ x_3\downarrow,\ x\downarrow \implies f(x_1, x_2) = g(x_3).$$

The procedure generates only one line for (78):

$$x_1 = x,\ x_3 = x,\ x_1\downarrow,\ x_3\downarrow \implies f(x_1, x_2) = g(x_3).$$

Consider also the example

$$f(\lambda(ts)p(x, y, t, s), q(x), x) \simeq g(p(x, y, x, y), q(y), x). \tag{87}$$

There are two free variables, so we will get two lines. First from the identity as it is, Step 2 will come into play with the most general bvu for z_1 and z_3 obviously being

$$\tau : t := x,\ s := y,$$

with no bound variables left, so the line produced is

$$x_3 = x,\ x_3\downarrow,\ x_6 = x,\ x_6\downarrow,\ x_1(x,\ y) = x_4,\ x \neq y,\ x\downarrow,\ y\downarrow \Longrightarrow f(x_1, x_2, x_3) = g(x_4,\ x_5, x_6).$$

If we identify $x \equiv y$, we get the identity

$$f(\lambda(ts)p(x,x,t,s), q(x), x) \simeq g(p(x,x,x,x), q(x), x)$$

which has an additional (trivial) unification and generates the line

$$x_3 = x,\ x_3\downarrow,\ x_6 = x,\ x_6\downarrow,\ x_1(x,x) = x_4,\ x_2 = x_5,\ x\downarrow \Longrightarrow f(x_1, x_2, x_3) = g(x_4,\ x_5, x_6).$$

It is quite easy to verify directly that the extended validity of these two lines is equivalent to the strict validity of the identity.

4.13. LEMMA. *The conjunction of all the lines constructed for* (63) *implies* (63).

PROOF. We will verify that if the line produced by an identity holds, then every strict assignment which assigns distinct values to distinct basic variables satisfies the identity, from which the Lemma follows by applying it to all the substitution instances of the identity.

Suppose we are given a strict assignment π to the variables of the identity and extend it to the special variables which occur on the line by setting

$$\pi(x_i) = \pi(z_i). \tag{88}$$

It will be enough to verify that this assignment satisfies the antecedent of the line, so consider how the clauses were introduced to it by the three steps of the construction.

STEP 1. We put in $x_i = z_i$ and $x_i\downarrow$ if z_i is basic, and π clearly makes these conditions true, by definition and because it is strict on the variables which occur free in the identity.

STEP 2. If the clause (86) is added to the antecedent of the line in this step, we must show that it is validated by π, or equivalently that π validates the term identity

$$x_i(\tau(u_1), \ldots, \tau(u_{arity(i)})) \simeq x_j(\sigma(v_1), \ldots, \sigma(v_{arity(j)})). \tag{89}$$

We compute:

$$\begin{aligned}\pi(x_i(\tau(u_1), \ldots, \tau(u_{arity(i)}))) &\simeq \pi(z_i)(\tau(u_1), \ldots, \tau(u_{arity(i)}))\\ &\simeq \pi((\lambda(u_1, \ldots, u_m)r(a_1, \ldots, a_k))(\tau(u_1), \ldots, \tau(u_{arity(i)})))\\ &\simeq \pi(r(\tau(a_1), \ldots, \tau(a_k)))\\ &\simeq \pi(r(\sigma(b_1), \ldots, \sigma(b_k))),\end{aligned}$$

where the last line follows from the fact that τ, σ are parts of a unifier. From this point we proceed with the opposite computation for z_j, to complete the proof of (89).

STEP 3. The clauses introduced by Step 3 are obviously validated by π, which is assumed to assign distinct, strict values to distinct basic variables. ⊣

4.14. LEMMA. *An identity implies the validity of every line it generates.*

PROOF. We now assume that we are given an extended assignment π to the variables of the line which satisfies the antecedent, and we must show that it also satisfies the consequent.

Notice that π assigns distinct values other than $\perp$ to all the basic variables which occur free both in the line and in the identity, because of the clauses we introduced in Step 3. We extend π to the remaining free basic variables in the equation by giving a distinct, new value to each of them. We want to extend π to all the pf variables also, so that we get

$$\pi(z_i) \simeq \pi(x_i), \tag{90}$$

for every $i = 1, \ldots, m$. This is already true when z_i is a basic variable, because of the equations put in the line in Step 1.

To effect (90), for

$$z_i \equiv \lambda(u_1, \ldots, u_{arity(i)})r(a_1, \ldots, a_k),$$

we want to define $\pi(r)$ so that

$$\pi(r)(\pi^*(a_1), \ldots, \pi^*(a_k)) \simeq \pi(x_i)(\pi^*(u_1), \ldots, \pi^*(u_{arity(i)})) \tag{91}$$

for every π^* which agrees with π on all the variables except (perhaps) $u_1, \ldots, u_{arity(i)}$. This will be possible, unless there is another expression

$$z_j \equiv \lambda(v_1, \ldots, v_{arity(j)})r(b_1, \ldots, b_k)$$

which then demanded

$$\pi(r)(\pi^*(b_1), \ldots, \pi^*(b_k)) \simeq \pi(x_j)(\pi^*(v_1), \ldots, \pi^*(v_{arity(j)})), \tag{92}$$

but for some assignment π^* as above we have the conflict

$$\pi^*(a_1), \ldots, \pi^*(a_k) \;\simeq\; \pi^*(b_1), \ldots, \pi^*(b_k), \tag{93}$$

$$\pi(x_i)(\pi^*(u_1), \ldots, \pi^*(u_{arity(i)})) \;\not\simeq\; \pi(x_j)(\pi^*(v_1), \ldots, \pi^*(v_{arity(j)})). \tag{94}$$

First we argue that if (93) holds, then the terms z_i and z_j can be unified. Let $w(1), \ldots, w(l)$ be the distinct values that π^* assigns to the variables $u_1, \ldots, u_{arity(i)}$, $v_1, \ldots, v_{arity(j)}$ (it maybe that $l = 0$), choose distinct, fresh variables $s(1), \ldots, s(l)$ and set

$$\tau(u_\alpha) \equiv \begin{cases} a_\beta, \text{ if } \pi^*(u_\alpha) = \pi(a_\beta), \\ s(w(\pi^*(u_\alpha))), \text{ otherwise,} \end{cases}$$

where a_β is meant to be a variable not among $u_1, \ldots, u_{arity(i)}$. Notice that if the first case of the definition applies, then the variable a_β is uniquely determined, because π^* agrees with π on these variables and assigns a distinct value to each of them. We give a similar definition of another variable transformation σ on the v_α's and then observe that because of (93), these transformations define a unifier of z_i with z_j with the bound variables $s_1, \ldots, s_l$.

Using Lemma **4.12**, we can find a maximal unifier of z_i, z_j from which the one just constructed can be obtained by further specifying or identifying some of the bound variables. It follows that in Step 2 of the construction of the line we put in a clause which expresses the maximal unification of z_i with z_j and this easily contradicts (94). ⊣

4.15. Structures with (some) finite basic sets. There is an obvious, trivial way by which we can reduce the problem of intensional identity for any structure $\mathbf{A}$ of finite signature to that of another such structure $\mathbf{A}'$ in which all basic sets are infinite. If, for example, $U = \{u_1, \ldots, u_n\}$ is finite and $f : U \times V \rightharpoonup W$ is a partial function among the givens, we replace f in $\mathbf{A}$ by n new partial functions

$$f_i : V \rightharpoonup W, \quad f_i(v) \simeq f(u_i, v), \quad i = 1, \ldots, n.$$

The translation is a bit messier for functionals but still trivial in principle. This is, in fact what we will do if U is a *small* finite set, e.g., if $U = TV = \{\mathbf{t}, \mathbf{f}\}$ is the set of truth values. If, however, n is immense, then this reduction leads to a structure with impossibly many primitives whose dictionary is totally unmanageable. Suppose, for example, that the language is a small (in theory formalized) fragment of the basic English currently in use as the common language of business in continental Europe. There are few basic sets in the intended interpretation, the citizens of France, the German cars, the Greek raisins, etc., but they are all immense. We may also assume few primitives, we are only interested in making simple assertions like

there are enough Greek raisins to satisfy the needs of all Frenchmen.

The problem of sense identity for sentences of such a language appears to be quite manageable, and in fact, the actual dictionaries we would use to translate this language into the national European languages are quite small. In contrast, the formal dictionary of the expanded language suggested by the trivial procedure of eliminating all the finite basic sets is absurdly large and involves specific entries detailing separately the relation between every Frenchman with every Greek raisin. The decision procedure we described allows a better solution.

4.16. COROLLARY (to the proof). *Suppose* $\mathbf{A} = (U_1, \ldots, U_k, f_1, \ldots, f_n)$ *is a recursor structure of finite signature, such that every basic set* U_i *has at least* d *members. Then the decision procedure for intensional identity defined in this section will decide correctly every identity on* $\mathbf{A}$ *with* n *(free and bound) basic variables, provided that* $2n + 4 \leq d$.

The Corollary suggests a method of constructing a reasonably sized "dictionary of meanings" for a structure in which some basic sets are very small—and we eliminate these—and the others are very large. The formal decision procedure for intensional identity on the basis of this dictionary is not that far from the way we would decide such questions in practice: we understand quantification over small sets by considering individual cases, while for large sets we appeal to fundamental identities relating the meanings of the primitives, including the quantifiers. The procedure will fail to resolve questions of identity of meaning which involve more quantifiers over large sets than (roughly) half the size of the structure. The proof of the Corollary follows from a careful examination of the arguments of this section, which basically require the existence of enough possible values for variables to make certain distinctions. For example, it is not hard to check that Lemma **4.3** holds, provided all the basic sets of the structure have at least 4 elements. We will omit the details.

REFERENCES

[1] A. CHURCH, *Intensionality and the paradox of the name relation.* To appear in the Proceedings of the meeting at SUNY, Buffalo.

[2] ——, *A formulation of the logic of sense and denotation, abstract*, ***Journal of Symbolic Logic***, **11** (1946), p. 31.

[3] ——, *A formulation of the logic of sense and denotation*, in ***Structure, Method and Meaning***, P. Henle, H. M. Kallen, and S.K. Langer, eds., Liberal Arts Press, New York, 1951, pp. 3–24.

[4] ——, *Outline of a revised formulation of the logic of sense and denotation, part I*, ***Noûs***, **7** (1973), pp. 24–33.

[5] ——, *Outline of a revised formulation of the logic of sense and denotation, part II*, ***Noûs***, **8** (1974), pp. 135–156.

[6] M. J. CRESSWELL, ***Structured Meanings: The Semantics of Propositional Attitudes***, The MIT Press, Cambridge, Mass, 1985.

[7] K. DONNELLAN, *Reference and definite description*, ***Philosophical Review***, **75** (1966), pp. 281–304.

[8] M. A. DUMMETT, *Frege's distinction between sense and reference*, in ***Truth and Other Enigmas***, Harvard Univ. Press, Cambridge, 1978, pp. 116–144.

[9] G. EVANS, ***The Varieties of Reference***, Clarendon Press, Oxford, 1982. Edited by J. N. McDowell.

[10] J. E. FENSTAD, R. O. GANDY, and G. E. SACKS, eds., ***Generalized Recursion Theory, II***, Studies in Logic, No. 94, North Holland, Amsterdam, 1978.

[11] J. E. FENSTAD and P. G. HINMAN, eds., ***Generalized Recursion Theory***, Studies in Logic, No. 79, North Holland/American Elsevier, Amsterdam, 1974.

[12] G. FREGE, *On sense and denotation*, in ***Translations from the Philosophical Writings of Gottlob Frege***, P. Geach and M. Black, eds., Basil Blackwell, Oxford, 1952. Translated by Max Black under the title *Sense and meaning*. In quotations from Black's translation, I have consistently changed the rendering of *Bedeutung* as *meaning* to the more current *denotation* or *reference*.

[13] A. S. KECHRIS and Y. N. MOSCHOVAKIS, *Recursion in higher types*, in ***Handbook of Mathematical Logic***, J. Barwise, ed., Studies in Logic, No. 90, North Holland, Amsterdam, 1977, pp. 681–737.

[14] S. C. KLEENE, *Recursive functionals of finite type, I*, ***Transactions of the American Mathematical Society***, **91** (1959), pp. 1–52.

[15] ——, *Recursive functionals of finite type, II*, ***Transactions of the American Mathematical Society***, **108** (1963), pp. 106–142.

[16] D. E. KNUTH, ***Fundamental Algorithms***, vol. 1 of The Art of Computer Programming, Addison-Wesley, 1968.

[17] G. KREISEL and G. TAKEUTI, *Formally self-referential propositions for cut-free classical analysis and related systems*, ***Dissertationes Mathematicae***, **118** (1974).

[18] S. A. KRIPKE, ***Naming and Necessity***, Harvard Univ. Press, Cambridge, 1972.

[19] ——, *Outline of a theory of truth*, ***Journal of Philosophy***, **72** (1975), pp. 690–716.

[20] R. MONTAGUE, *Universal grammar*, ***Theoria***, **36** (1970), pp. 373–398.

[21] Y. N. MOSCHOVAKIS, ***Elementary Induction on Abstract Structures***, Studies in Logic, No. 77, North Holland, Amsterdam, 1974.

[22] ——, *Abstract recursion as a foundation of the theory of recursive algorithms*, in ***Computation and Proof Theory***, M. M. Richter et al., eds., Lecture Notes in Mathematics, No. 1104, Springer-Verlag, Berlin, 1984, pp. 289–364.

[23] ——, *The formal language of recursion*, ***Journal of Symbolic Logic***, **54** (1989), pp. 1216–1252.

[24] ——, *A mathematical modeling of pure recursive algorithms*, in ***Logic at Botik '89***, A. R. Meyer and M. A. Taitslin, eds., Lecture Notes in Computer Science, No. 363, Springer-Verlag, Berlin, 1989, pp. 208–229.

[25] R. PLATEK, ***Foundations of Recursion Theory***, Ph.D. thesis, Stanford University, 1966.

[26] B. RUSSELL, *On denoting*, ***Mind***, **14** (1905), pp. 479–493.

[27] G. E. SACKS, ***Higher Recursion Theory***, Perspectives in Mathematical Logic, Springer-Verlag, Berlin, 1990.

[28] A. M. TURING, *On computable numbers, with an application to the Entscheidungsproblem*, ***Procedings of the London Mathematical Society***, Series II, **42** (1936–37), pp. 230–265.

[29] J. VAN HEIJENOORT, *Frege on sense identity*, in ***Selected Essays***, Bibliopolis, Napoli, 1985, pp. 65–70.

[30] ——, *Sense in Frege*, in ***Selected Essays*** [29], pp. 55–64.

Department of Mathematics
University of California, Los Angeles
ynm@math.ucla.edu

A TRANSFINITE VERSION OF PUISEUX'S THEOREM, WITH APPLICATIONS TO REAL CLOSED FIELDS

M. H. MOURGUES and J.-P. RESSAYRE

Abstract. Extending the effective version of Puiseux's theorem, we compute the roots of polynomials inside all fields of the form $k((x))^{\Gamma}$, where k is real closed and Γ divisible. We use this computation to prove that every real closed field has an integer part, that is, a discrete subring which plays for the field the same role as $\mathbf{Z}$ plays for $\mathbf{R}$.

Puiseux's theorem in its detailed form allows one to compute the roots of a polynomial, inside the field of "Puiseux series"; Lemma 3.6 and Remark 3.7 below generalize this computation to the case of a field $k((x))^{\Gamma}$, where k is a real closed field and Γ a divisible ordered abelian group, even if Γ is non-archimedian.

Two applications are given: the truncation lemma of F. Delon (see 3.5 below), and the existence of an "integer part" in every real closed field (see 1.4 below). Another proof of these results appears in [MR], but without the generalization of Puiseux's theorem which is one of the chief interests of this paper. The first author has developed extensions and other applications of such computations in [M].

§1. Definitions, remarks.

1.1. Let us denote by $\widetilde{K}$ the real closure of a totally ordered field K.

1.2. We say that a subring Z of a ring A is an *integer part* of A if it is discrete and if for any $x \in A$, there is $z \in Z$ such that $z \leq x < z + 1$. We call this unique element z the *integer part* of x and write $z = [x]$.

1.3. S. Boughattas showed in [B] that on the one hand, every totally ordered field has an ultrapower endowed with an integer part, and on the other hand, there are ordered fields without integer parts. In fact, he has for every integer p, a p-real closed field with no integer part. We show that these examples are optimal, in the sense that every real closed field has an integer part.

1.4. In fact, we will prove a stronger result:

Let A be a convex subfield of $K = \widetilde{K}$. Then any integer part Z of A can be extended to an integer part Z_K of K.

§2. Convex valuation in a totally ordered field.

The following results are classical results of valuation theory (cf. [K], [KW], [R]). Let K be a (totally) ordered field and v a convex valuation on K. We denote by k the residue field and by $\Gamma = v(K)$ the abelian totally ordered group of valuations. When $y \in K$ and $v(y) = 0$, $\overline{y}$ will be the residue image of y in k.

2.1. PROPOSITION. *If K is real closed, then k is real closed and Γ is divisible. Moreover, k can be embedded in K, and there exists a cross section, i.e., a family $\{x^\gamma : \gamma \in \Gamma\} \subset K^+$ that satisfies*

$$\forall \gamma, \gamma' \in \Gamma\ [v(x^\gamma) = \gamma \text{ and } x^\gamma \cdot x^{\gamma'} = x^{\gamma+\gamma'}].$$

2.2. The field of formal series $k((x))^\Gamma$: Given an ordered field k and an ordered abelian group Γ, we let $k((x))^\Gamma = \{0\} \cup \{\sum_{i<\mu} a_i x^{\gamma_i} : \mu$ is an ordinal, $(\gamma_i)_{i<\mu}$ a strictly increasing family of Γ, $(a_i)_{i<\mu}$ a family of elements from $k^*\}$.

Then $k((x))^\Gamma$ is a field with the usual sum and with the product induced by $x^\gamma \cdot x^{\gamma'} = x^{\gamma+\gamma'}$. Moreover, the order on k can be extended to an order on $k((x))^\Gamma$ thus: $x^\gamma > x^{\gamma'} > k$ iff $\gamma < \gamma' < 0$. In addition, if k is real closed and Γ is divisible then $k((x))^\Gamma$ is real closed. The map $d : k((x))^\Gamma \to \Gamma \cup \{\infty\}$, defined by $d(s) = \gamma_0$, is a convex valuation on $k((x))^\Gamma$ with valuation group Γ and residue field isomorphic to k.

The following result is a version of the theorem: "any henselian subfield of $k((x))^\Gamma$ is real closed" [R]. In the following L will be a subfield of $k((x))^\Gamma$ such that $d(L) = \Gamma$ and $k \subset L$.

2.3. DEFINITION. *Let y be algebraic over L with $d(y) = 0$. We say that y satisfies condition (H) if there is a polynomial $P(X) = \sum_{k=0}^{n} A_k X^k \in L[X]$ such that*

(i) $P(y) = 0$ and P is primitive (i.e., $\min d(A_k) = 0$),
(ii) $\overline{P}'(\overline{y}) \neq 0$ (where $\overline{P}$ is the image of P in $k[X]$).

2.4. PROPOSITION. *Let $y \in \widetilde{L}$ with $d(y) = 0$. Then there is $y_1, \ldots, y_k$ belonging to $\widetilde{L}$ such that $y = y_1 + \cdots + y_k$ and $y_i / x^{d(y_i)}$ satisfies the condition (H) over $L(y_1, \ldots, y_{i-1})$ for all i.*

Proof. See [MR].

§3. Integer part in subfields of $k((x))^\Gamma$, closure under truncation.

In the following k will be a real closed field and Γ a totally ordered abelian group.

3.1 DEFINITION. Let $s = \sum_{i<\mu} a_i x^{\gamma_i} \in k((x))^\Gamma$. An *initial segment* of s (abbreviated by I.S.) is any element $\sum_{i<\lambda} a_i x^{\gamma_i}$ with $\lambda < \mu$. We use $(s)_{<\gamma}$ to denote the I.S. $\sum_{i<\lambda} a_i x^{\gamma_i}$ of s where $i < \lambda$, $\gamma_i < \gamma$, and $\gamma_\lambda \geq \gamma$. Note that $(s)_\gamma = 0$ when $\gamma \leq \gamma_0$. We say that the subfield L of $k((x))^\Gamma$ is *closed under truncation* if every I.S. of any element of L also belongs to L.

3.2. LEMMA. *Assume that k has an integer part Z. Then every subfield $L \supset k$ of $k((x))^\Gamma$ which is closed under truncation, has an integer part $Z_L \supset Z$.*

Proof. Let $Z_L = \{\sum_{i\le\mu} a_i x^{\Gamma_i} \in L : \gamma_i \le 0 \text{ and } (\gamma_i = 0 \Rightarrow a_i \in Z)\}$. It is easy to see that Z_L is a discrete subring of L. Let $y \in L$ and let y' be the initial segment of y such that every exponent of y' is strictly negative. Let a_{i_0} be the term of y with exponent 0. Then if we let $[y] = y' + [a_{i_0}] - 1$ or $y = y' + [a_{i_0}]$ (depending on y), we get $[y] \in Z_L$, and $[y]$ is the integer part of y. □

Next we will study preservation of closure under truncation by field extensions.

3.3. LEMMA. *Let $s = \sum_{i<\mu} a_i x^{\alpha_i}$ and $t = \sum_{j<\nu} b_j x^{\beta_j}$ be two elements of $k((x))^\Gamma$ and $(s\cdot t)_{<\delta}$ a strict I.S. of $s\cdot t$. Then there is a unique strictly increasing subsequence $(\alpha_{i_0},\dots,\alpha_{i_n})$ of $(\alpha_i)_{i<\mu}$ and a unique strictly decreasing subsequence $(\beta_{i_0},\dots,\beta_{i_n})$ of $(\beta_j)_{j<\nu}$ such that $\alpha_{i_1}+\beta_{j_1}\ge\delta$ and*

(E) $(s\cdot t)_{<\delta} = t\cdot(s)_{<\alpha_{i_0}} + (t)_{<\beta_{j_0}}\cdot((s)_{<\alpha_{i_1}} - (s)_{<\alpha_{i_0}}) + \dots + (t)_{<\beta_{j_n}}\cdot(s-(s)_{<\alpha_{i_n}})$.

Subsequently we will say that $(s\cdot t)_{<\delta}$ is written in form (E).

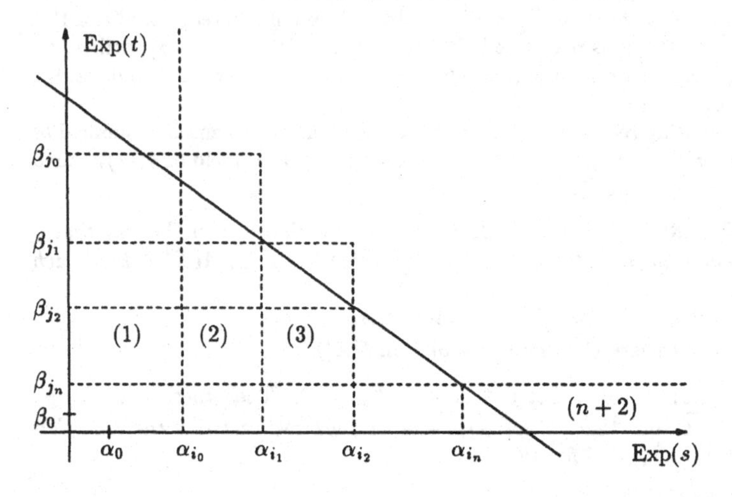

$$(s\cdot t)_{<\delta} = \overbrace{t\cdot(s)_{<\alpha_{i_0}}}^{(1)} + \overbrace{(t)_{<\beta_{j_0}}\left[(s)_{<\alpha_{i_1}} - (s)_{<\alpha_{i_2}}\right]}^{(2)} + \dots + \overbrace{(t)_{<\beta_{j_n}}\left[s-(s)_{<\alpha_{i_n}}\right]}^{(n+2)}$$

Proof. Let Exp(s) denote the set of exponents of the series s. Since $(s\cdot t)_{<\delta}$ is a strict I.S. of $s\cdot t$, there exist $\alpha\in$ Exp(s) and $\beta\in$ Exp(t) with $\alpha+\beta\ge\delta$. Let α_{i_0} be the smallest exponent of s satisfying this property, and let β_{j_0} be the smallest exponent β of t such that $\alpha_{i_0}+\beta\ge\delta$. If $\alpha+\beta<\delta$ for all $\alpha\in$ Exp(s) and for all $\beta\in$ Exp(t) such that $\alpha>\alpha_{i_0}$ and $\beta<\beta_{j_0}$, then $(s\cdot t)_{<\delta} = t\cdot(s)_{<\alpha_{i_0}} + (t)_{<\beta_{j_0}}\cdot(s-(s)_{<\alpha_{i_0}})$ and Lemma 3.3. holds. In the other case let α_{i_1} the smallest exponent α of s strictly larger that α_{i_0} such that there is $\beta\in$ Exp(t), $\beta<\beta_{j_0}$, with $\alpha+\beta\ge\delta$, and let β_{j_1} be the smallest exponent $\beta<\beta_{j_0}$ of t such that $\alpha_{i_1}+\beta\ge\delta$. If $\alpha+\beta<\delta$ for all exponents $\alpha>\alpha_{i_1}$ of s

and all $\beta > \beta_{j_1}$, then

$$(s \cdot t)_{<\delta} = t \cdot (s)_{<\alpha_{i_0}} + (t)_{<\beta_{j_0}} \cdot ((s)_{<\alpha_{i_1}} - (s)_{<\alpha_{i_0}}) + (t)_{<\beta_{j_1}} \cdot (s - (s)_{<\alpha_{i_1}})$$

and Lemma 3.3. holds again.

Going on inductively, we build two sequences (β_{j_l}) and (α_{i_l}). Since $\mathrm{Exp}(t)$ is a well-ordered set, the strictly decreasing sequence (β_{j_l}) is finite, hence so is the sequence (α_{i_l}). The uniqueness of the sequences $(\beta_{j_l})_{l=0}^{n}$ and $(\alpha_{i_l})_{l=0}^{n}$ is easy to prove. So Lemma 3.3. is proved. □

NOTES. (1) $(s \cdot t)_{<\delta}$ is the sum of the terms $a_i b_j x^{\alpha_i+\beta_j}$ of $s \cdot t$ whose exponents (α_i, β_j) are in the area $\alpha + \beta < \delta$ (see the picture).

(2) We can see from the picture that (E) is in fact the expression of $(s \cdot t)_{<\delta}$ as an exact and finite Riemann sum over this area.

(3) Note that in (E), β_{j_n} is the smallest exponent of t such that there is an $\alpha \in \mathrm{Exp}(s)$ with $\beta + \alpha \geq \delta$.

3.4. LEMMA. *Let L be a subfield of $k((x))^\Gamma$ closed under truncation and let $y \in k((x))^\Gamma$ be such that every I.S. of y belongs to L. Then $L(y)$ remains closed under truncation.*

Proof. In the following, H will be any subfield of $k((x))^\Gamma$, and $s = \sum_{i<\mu} a_i x^{\alpha_i}$, $t = \sum_{j<\nu} b_j x^{\beta_j}$ will be any two elements of H.

FACT 1. If every I.S. of s and t belongs to H, then every I.S. of $s \cdot t$ again belongs to H. Fact 1 follows from Lemma 3.3.

FACT 2. Let $t' = (t)_{<\beta_\lambda}$ be a strict I.S. of t. Then there is a unique exponent $\gamma \leq \alpha_0 + \beta_\lambda$ of $\mathrm{Exp}(s) + \mathrm{Exp}(t)$ and two finite sequences as in Lemma 3.3 with

(a) $(s \cdot t)_{<\gamma} = t' \cdot (s)_{<\alpha_{i_1}} + (t)_{<\beta_{j_1}} \cdot ((s)_{<\alpha_{i_2}} - (s)_{<\alpha_{i_1}}) + \cdots + (t)_{<\beta_{j_n}} \cdot (s - (s)_{<\alpha_{i_n}})$.

(b) For any exponent γ' of $\mathrm{Exp}(s) + \mathrm{Exp}(t)$ the form (E) of $(s \cdot t)_{<\gamma'}$ uses only *strict* I.S. of t'.

Proof. Let γ be the smallest element of the set $\{\delta \in \mathrm{Exp}(s \cdot t) : \forall j < \lambda\ (\alpha_0 + \beta_j < \delta)\}$. Since $\mathrm{Exp}(s) + \mathrm{Exp}(t)$ is a well-ordered set, γ exists and $\gamma \leq \alpha_0 + \beta_\lambda$. Let $(\alpha_{i_l})_{l=0}^{n}$ and $(\beta_{j_l})_{l=0}^{n}$ the two sequences for $(s \cdot t)_{<\gamma}$ given by Lemma 3.3. By the definition of γ, we get $\forall j < \lambda\ (\alpha_0 + \beta_j < \gamma \leq \alpha_0 + \beta_\lambda)$. Thus by the definition of α_{i_0} and β_{j_0}, $\alpha_{i_0} = \alpha_0$ and $\beta_{j_0} = \beta_\lambda$. Hence $(s)_{<\alpha_{i_0}} = (s)_{<\alpha_0} = 0$ and $(t)_{<\beta_{j_0}} = (t)_{<\beta_\lambda}$, whence (a) holds. Moreover, let $\gamma' < \gamma$ be an element of $\mathrm{Exp}(s) + \mathrm{Exp}(t)$. Then by definition of γ, there exists $j < \lambda$ such that $\gamma' < \alpha_0 + \beta_j$. Hence if we denote by $(\alpha'_{i_l})_{l=0}^{m}$ and $(\beta'_{j_l})_{l=0}^{m}$ the two sequences given by Lemma 3.2 for $(s \cdot t)_{<\gamma'}$, we get $\alpha'_{i_0} = \alpha_0$, $\beta'_{j_0} = \beta_j < \beta_\lambda$; and since the sequence $(\beta'_{j_l})_{l=0}^{m}$ is decreasing, $(t)_{<\beta_{j_l}}$ is a strict initial segment of t' and (b) holds.

FACT 3. If every I.S. of s belongs to H then every I.S. of $1/s$ again belongs to H.

Proof. If not, let $t' = (1/s)_{<\beta_\lambda}$ be the shortest I.S. of $1/s$ which does not belong to H. Then by Fact 2 applied to $t = 1/s$

$$(\exists \gamma \leq \alpha_0 + \beta_\lambda)[1 = (s \cdot 1/s)_{<\gamma} = t' \cdot (s)_{<\alpha_{i_1}} + C]$$

where C is a finite sum of finite products of I.S. of s and strict I.S. of t'. Thus $C \in H$ and $t' \in H$, contradicting the definition of t'.

Lemma 3.4 follows obviously from Facts 1 and 3. □

3.5 LEMMA (F. Delon). *Let L be a subfield of $k((x))^{\Gamma}$ such that $k \subset L$ and $d(L) = \Gamma$. If L is closed under truncation then so is $\widetilde{L}$.*

Remark. This lemma has for us an eventful history. We gave a proof in which D. Marker pointed out an important error. In order to correct it, we proved Facts 4 and 5 below, from which the lemma follows if you know Ribenboim's theorem as stated in 2.3. But we did not know it, and it is only after F. Delon outlined for us a proof of 3.5 based on her work [D] that we completed the proof presented below, which is different from Delon's proof, since it provides an explicit computation that does not follow from Delon's work; see Remark 3.7 below.

Proof of 3.5: It is a straightforward consequence of Proposition 2.4 and Lemma 3.6 below. □

3.6. LEMMA. *Let L be a subfield of $k((x))^{\Gamma}$ closed under truncation with $k \subset L$ and $d(L) = \Gamma$. Let y be any element of $\widetilde{L}$ satisfying $d(y) = 0$ and the statement (H). Then any I.S. of y belongs again to $\widetilde{L}$.*

Proof. We shall prove Lemma 3.6 by way of contradiction: if it fails, there is an I.S. of y that does not belong to L. Let $y' = (y)_{<\beta_\lambda}$ be the shortest I.S. of t satisfying this property. Let $P(X) = X^n + A_{n-1}X^{n-1} + \cdots + A_0$ be given by hypothesis and let κ be the set of exponents $k \neq 0$ of P such that $d(A_k) = 0$ (note that κ has at least one element). We have to make a distinction between two cases and get a contradiction in each of them.

First case. Assume that there exists a convex subgroup Γ_0 of Γ such that $\mathrm{Exp}(y')$ is cofinal in Γ_0 (in other words $\forall \alpha, \beta \in \mathrm{Exp}(y')\ \exists \gamma \in \mathrm{Exp}(y')\ (\alpha + \beta \leq \gamma)$). Let $(A_k)_{<\Gamma_0}$ denote the largest initial segment of A_k such that $\mathrm{Exp}((A_k)_{<\Gamma_0}) \subset \Gamma_0$ (this definition is available since Γ_0 is a convex subgroup) and let $Q(X) = \sum_{k=0}^{n}(A_k)_{<\Gamma_0}X^k \in L[X]$ (recall that L is closed under truncation). Q is not constant since $\forall k \in \kappa\ d(A_k) = 0$. So $P(X) = Q(X) + R(X)$ where $R(X) = \sum_{k=0}^{n}(A_k - (A_k)_{<\Gamma_0})X^k = \sum_{k=0}^{n} B_k X^k$ with $d(B_k) > \Gamma_0$. Hence $d(R(y)) \geq \min(d(B_k)) > \Gamma_0$. Moreover, Taylor's formula gives $Q(y) = Q(y' + y'') = Q(y') + T$, where $d(T) > \Gamma_0$. So $0 = P(y) = Q(y') + T + R(y) = Q(y') + S$ with $\mathrm{Exp}(Q(y')) \subset \Gamma_0$ and $d(S) > \Gamma_0$. Therefore $Q(y') = 0$, contradicting the choice of y'.

Second case. Assume the negation for the hypothesis of the first case: i.e., $\exists \alpha, \beta \in \mathrm{Exp}(y')\ \forall \beta_j \in \mathrm{Exp}(y')\ (\beta_j < \alpha + \beta)$.

We prove first two facts:

FACT 4. Let $\oplus_n \mathrm{Exp}(y)$ denote the set $\{\alpha_0 + \alpha_1 + \cdots + \alpha_n\colon \forall i, \alpha_i \in \mathrm{Exp}(y)\}$. Let γ be the smallest element of the set $\Delta = \{\delta \in \oplus_n \mathrm{Exp}(y) : \forall j < \lambda, \beta_j < \delta\}$. Then

(a) $$\forall k \geq 1, (y^k)_{<\gamma} = y' \cdot B_k + C_k$$

where B_k and C_k belong to the ring generated by the strict I.S. of y'. Moreover $d(B_k) = 0$ and $\overline{B}_k = k\overline{y}^{k-1}$.

(b) If $\gamma' < \gamma$ belongs to $\oplus_n \operatorname{Exp}(y)$, then $(y^k)_{<\gamma'}$ belongs to the ring generated by the strict I.S. of y'.

Proof of Fact 4. First, we can see that Δ is not an empty set; indeed by hypothesis there are α and β belonging to $\operatorname{Exp}(y')$ such that $\forall j < \lambda$, $\beta_j < \alpha + \beta$, hence Δ contains an element of $\oplus_2 \operatorname{Exp}(y)$ (since $d(y) = 0 \Rightarrow \oplus_2 \operatorname{Exp}(y) \subset \oplus_n \operatorname{Exp}(y)$). So, since $\oplus_n \operatorname{Exp}(y)$ is a well-ordered set, γ is well defined. Since $\forall k \leq n$, $\operatorname{Exp}(y^{k-1}) \subset \oplus_n \operatorname{Exp}(y)$, by the same argument as in Fact 2(a), we can easily prove that there are a sequence $\gamma > \alpha_1 > \cdots > \alpha_m > 0$ of $\operatorname{Exp}(y)$ and a sequence $0 < \gamma_1 < \cdots < \gamma_m < \gamma$ of $\oplus_n \operatorname{Exp}(y)$ such that, $\forall k$, $1 \leq k \leq n$:

(1)

$$(y^k)_{<\gamma} = (y \cdot y^{k-1})_{<\gamma} = (y)_{<\gamma} \cdot (y^{k-1})_{<\gamma_1} + (y)_{<\alpha_1} \cdot [(y^{k-1})_{<\gamma_2} - (y^{k_1})_{<\gamma_1}] + \cdots + (y)_{<\alpha_m} \cdot [(y^{k-1})_{<\gamma} - (y^{k-1})_{<\gamma_m}]$$

By definition of γ, $(y)_{<\gamma} = y'$, moreover $\forall i$, $(y)_{<\alpha_i}$ is a strict initial segment of y'. Hence we get, $\forall k \leq n$:

$$(y^k)_{<\gamma} = y' \cdot (y^{k-1})_{<\gamma_1} + y_1 \cdot [(y^{k-1})_{<\gamma_2} - (y^{k-1})_{<\gamma_1}] + \cdots + y_m \cdot [(y^{k-1})_{<\gamma} - (y^{k-1})_{<\gamma_m}] \tag{2}$$

where $y_1, \ldots, y_m$ are strict I.S. of y'. Now we are able to end the proof of Fact 4 by induction on k:

(i) $k = 1$. By definition of γ, $(y)_{<\gamma} = y'$, then Fact 4(a) holds with $C_1 = 0$ and $B_1 = 1$; moreover for any γ' in $\oplus_n \operatorname{Exp}(y)$, strictly less than γ, there is β_j in $\operatorname{Exp}(y')$ such that $\gamma' \leq \beta_j$, so $(y)_{<\gamma'}$ is a strict I.S. of y', then Fact 4(b) holds.

(ii) Assume that Fact 4 holds for $k - 1$. Recall (2) above. By induction hypothesis $(y^{k-1})_{<\gamma} = y' \cdot B_{k-1} + C_{k-1}$, and $\forall i < m$, $\gamma_i < \gamma$, then $(y^{k-1})_{<\gamma_i}$ belongs to the ring generated by the strict I.S. of y'. Let $B_k = (y^{k-1})_{<\gamma_1} - y_m \cdot B_{k-1}$ and

$$C_k = y_1 \cdot [(y^{k-1})_{<\gamma_2} - (y^{k-1})_{<\gamma_1}] + \cdots + [C_{k-1} - (y^{k-1})_{<\gamma_m}],$$

then (2) becomes

$$(y^k)_{<\gamma} = y' \cdot B_k + C_k$$

where B_k and C_k belong to the ring generated by the strict I.S. of y'. So Fact 4(a) holds. Fact 4(b) holds using the same argument as in Fact 2(b) and induction hypothesis.

FACT 5. Consider the truncation of $P(y)$ at the exponent γ given by Fact 4. Then $0 = (P(y))_{<\gamma} = y'A + B$, where $A, B \in L$. Moreover $d(A) \geq 0$ and we have $\overline{A} = \sum_{k \in \kappa} k \overline{A}_k (\overline{y})^{k-1}$.

Proof of Fact 5.

(i) First suppose that $k \in \kappa$. Lemma 3.3 with $s = A_k$ and $t = y^k$ gives

$$(A_k \cdot y^k)_{<\gamma} = y^k \cdot (A_k)_{<\alpha_{i_0}} + (y^k)_{<\beta_{j_0}} \cdot ((A_k)_{<\alpha_{i_1}} - (A_k)_{<\alpha_{i_0}}) + (y^k)_{<\beta_{j_n}} \cdot (A_k - (A_k)_{<\alpha_{i_n}}). \tag{1}$$

By the definition of $\alpha_{i_0}, \beta_{j_0}$ and since $d(A_k) = 0$, we have $\alpha_{i_0} = 0$ and $(y^k)_{<\beta_{j_0}} = (y^k)_{<\gamma}$. So (1) becomes

(2) $(A_k \cdot y^k)_{<\gamma} = (y^k)_{<\gamma} \cdot (A_k)_{<\alpha_{i_1}} + \cdots + (y^k)_{<\beta_{j_n}} \cdot (A_k - (A_k)_{<\alpha_{i_n}})$.

Since the sequence (β_{j_l}) is decreasing, all I.S. of y^k in (2) are I.S. of $(y^k)_{<\gamma}$. By Fact 4(b) they belong to the ring generated by the strict I.S. of y'. Hence they belong to L. Therefore we can write:

(3) $$(A_k \cdot y^k)_{<\gamma} = (y^k)_{<\gamma} \cdot (A_k)_{<\alpha_{i_1}} + C$$

where $C \in L$. Now Fact 4 gives

(4) $(A_k \cdot y^k)_{<\gamma} = (y'B_k + C_k) \cdot (A_k)_{<\alpha_{i_1}} + C = y' \cdot B_k(A_k)_{<\alpha_{i_1}} + S'_k$,

with $S'_k \in L$. So $(A_k \cdot y^k)_{<\gamma} = y' \cdot S_k + S'_k$, where $S_k, S'_k \in L$. Moreover $d(S_k) = 0$ and $\overline{S}_k = \overline{B}_k \cdot \overline{A}_k = \overline{A}_k \cdot k(\overline{y})^{k-1}$.

(ii) We now consider $k \neq 0$ and $k \notin \kappa$. By Lemma 3.3 we get

(1) $$(A_k \cdot y^k)_{<\gamma} = y^k \cdot (A_k)_{<\alpha_{i_0}} + (y^k)_{<\beta_{j_0}} \cdot ((A_k)_{<\alpha_{i_1}} - (A_k)_{<\alpha_{i_0}}) + \cdots + (y^k)_{<\beta_{j_n}} \cdot (A_k - (A_k)_{<\alpha_{i_n}}).$$

We have $d(A_k) + \gamma > \gamma$, and in this case, $\alpha_{i_0} = d(A_k)$ and $\beta_{j_0} \leq \gamma$.

Therefore exactly as in (i) we get $(A_k \cdot y^k)_{<\gamma} = y' \cdot S_k + S'_k$, where $S_k, S'_k \in L$. But then $d(S_k) = d(B_k) + d(A_k) > 0$, or $S_k = 0$. Note that $(A_0)_{<\gamma} \in L$.

(iii) Let $A = \sum_{k=0}^n S_k$ and $B = \sum_{k=0}^n S'_k$. Then (i) and (ii) give us $(P(y))_{<\gamma} = \sum_{k=0}^n (A_k y^k)_{<\gamma} = y' \cdot A + B$, where $A, B \in L$, and $d(A) \geq \min(d(S_k)) \geq 0$. Moreover, (i) and (ii) yield $\overline{A} = \sum_{k \in \kappa} k\overline{A}_k(\overline{y})^{k-1}$. So Fact 5 holds.

We next prove that $A \neq 0$. Observe that $\sum_{k \in \kappa} k\overline{A}_k(\overline{y})^{k-1} = \overline{P}'(\overline{y})$. By hypothesis, $\overline{P}'(\overline{y}) \neq 0$. Thus $\overline{A} \neq 0$, hence $A \neq 0$. Finally, $y' = -B/A \in L$, contradicting the choice of y'. So Lemma 3.6 is proved in this case too. □

3.7. *Remark*. The above proof computed y' from its strict initial segments; this computation actually holds for every I.S. y' of y, and together with Proposition 2.4 and Lemma 3.4, it provides a computation, as effective as can be in the most general case, of any roots of a polynomial inside $k((x))^\Gamma$. This is developed in [M].

§4. Every real closed field has an integer part.

Let K be closed, A a convex subring, and Z an integer part of A. Let v be the convex valuation given by A. Then the residue field k can be embedded in K in such a way that Z becomes an integer part of k. Let $\Gamma = v(K)$, $\{x^\gamma : \gamma \in \Gamma\}$ as in 2.2, and denote by H_0 the subfield generated by k and $\{x^\gamma : \gamma \in \Gamma\}$.

4.1. LEMMA. *Let H be a real closed subfield of K, $H_0 \subset H$, and assume $f : H \to k((x))^\Gamma$ with the following properties:*

(a) the restriction of f to H_0 is Id

(b) f is an injective homomorphism of ordered fields

(c) if $y \in H$, then $v(y) = d(f(y))$

(d) $f(H)$ is closed under truncation.
Then for any $y \in K - H$, f can be extended to $f' : \widetilde{H}(y) \to k((x))^\Gamma$ which satisfied the conditions (a), (b), (c), (d).

Proof. By hypothesis we may identify H with $f(H)$ in $k((X))^\Gamma$. Let us say that the series $\sum_{i<\alpha} a_i x^{\gamma_i} \in H$ is a ***development at order α of y with respect to H***, if $v(y - \sum_{i<\alpha} a_i x^{\gamma_i}) > \gamma_i$ for all $i < \alpha$. By 2.3, since $H_0 \subset H$, y has a development at any finite order with respect to H. Let $S(y)$ denote the set of all the developments of y with respect to H. $S(y)$ is totally ordered by the relation "initial segment of." Let $f'(y) \in k((x))^\Gamma$ be the least upper bound of $S(y)$ for the relation "I.S. of". We can see that $f'(y) \notin H$. Moreover, we have:

FACT. $\forall z \in H \ (z < y \Leftrightarrow f(z) < f'(y))$.

Proof. $z < y \Rightarrow v(z) \geq v(y)$. If $v(z) > v(y)$, then $d(z) > d(f'(y))$ and thus $f(z) < f'(y)$. In the other case $v(z) = v(y)$. Then $y - z = ax^\alpha + y'$, where $a > 0$ and $v(y') > \alpha = v(y - z) \geq v(y) = v(y)$. Let $z' = (z)_{<\alpha+1} + ax^\alpha$, then since H is closed under truncation, $z' \in H$, and z' is the development of y at order α with respect to H. We get

$$z < (z)_{\alpha+1} + (a/2)x^\alpha < (z)_{<\alpha+1} + ax^\alpha + s$$

for any s such that $v(s) > \alpha$. Hence z is strictly smaller than every element of $S(y)$, and finally $z < f'(y)$.

Thus we can extend f' to an isomorphism of $\widetilde{H}(y)$ onto $\widetilde{L}$ where L is the field generated over H by $f'(y)$ in $k((x))^\Gamma$. In addition, by the construction of $f'(y)$ all strict I.S. of $f'(y)$ belong to H. Thus L and $\widetilde{L}$ are closed under truncation, by Lemmas 3.4 and 3.5. □

4.2. COROLLARY. *Under the same hypotheses as in 4.1, there is $f' : K \to k((x))^\Gamma$ with the properties (a), (b), (c), (d).*

Proof. Transfinite iteration of 4.1, beginning with $H = \widetilde{H}_0$. □

From Lemma 3.2 applied to the image of the application f given by 4.2, we have finally:

4.3. THEOREM. *K has an integer part $Z_K \supset Z$.*

4.4. COROLLARY. *Every real closed field admits an integer part.*

Proof: Apply the preceding theorem with $Z = \mathbb{Z} \subset K$. □

Acknowledgments: Thanks to D. Marker who pointed out an error, and to F. Delon who allowed us to correct it.

REFERENCES

[B] S. BOUGHATTAS. *Résultats optimaux sur l'existence d'une partie entière dans les corps ordonnés.* ***The journal of symbolic logic***, vol. 58 (1993), pp. 326–333.

[D] F. DELON. *Indécidabilité de la théorie des paires immédiates de corps valués henseliens.* ***The journal of symbolic logic***, vol. 56 (1991), pp. 1236–1242.

[K] I. KAPLANSKY. *Maximal fields with valuations.* ***Duke mathematical journal***, vol. 9 (1942), pp. 303–321.

[KW] M. KNEBUSCH and M. J. WRIGHT. *Bewertungen mit reeller Henselisierung.* ***Journal für die reine und angewandte Mathematik***, n° 286/287 (1976), pp. 314–321.

[M] M. H. MOURGUES. ***Thèse***, Paris 7 (1992).

[MR] M. H. MOURGUES and J.-P. RESSAYRE. *Every real closed field has an integer part.* ***The journal of symbolic logic***, vol. 58 (1993), pp. 641–647.

[R] P. RIBENBOIM. ***Théorie des valuations.*** Les Presses de l'Université de Montréal, 2e éd., 1968.

Equipe de logique U. A. 753
couloir 45–55. 5° etage
Université de Paris 7
2, Place Jussieu
75251 Paris cedex 05

ON SIMILARITIES OF COMPLETE THEORIES

T. G. MUSTAFIN

In classical model theory two objects of different nature correspond to every signature σ:

L — the first-order language of σ and
K — the class of all structures of σ.

But there is a well-known one-to-one correspondence between maximal consistent sets of sentences of L and minimal axiomatizable classes in L of structures from K. When we say that we study the complete theory T we usually mean the pair $\langle T, \mathrm{Mod}(T)\rangle$, where $\mathrm{Mod}(T)$ is the class of all models of T. In connection with this duality of the nature of complete theories I want to introduce two notions of similarity which play the role of isomorphisms and two notions of nearness of theories.

§1. Syntactical similarity.

Let $F_n(T)$, $n < \omega$, be the Boolean algebras of formulas of T with exactly n free variables $v_1, \ldots, v_n$, and $F(T) = \bigcup_n F_n(T)$.

DEFINITION 1. Complete theories T_1 and T_2 are *syntactically similar* if and only if there exists a bijection $f : F(T_1) \to F(T_2)$ such that

(i) $f \restriction F_n(T_1)$ is an isomorphism of the Boolean algebras $F_n(T_1)$ and $F_n(T_2)$, $n < \omega$;
(ii) $f(\exists v_{n+1}\varphi) = \exists v_{n+1} f(\varphi)$, $\varphi \in F_{n+1}(T)$, $n < \omega$;
(iii) $f(v_1 = v_2) = (v_1 = v_2)$.

EXAMPLE 1. The following theories T_1 and T_2 of the signature $\sigma = \langle \varphi, \psi\rangle$ are syntactically similar, where φ, ψ are binary functions:

$$T_1 = \mathrm{Th}(\langle \mathbb{Z}; +, \cdot\rangle), \qquad T_2 = \mathrm{Th}(\langle \mathbb{Z}; \cdot, +\rangle).$$

§2. Semantic similarity.

From the point of view of a model-theoretician, the object $\langle \mathrm{Mod}(T); \simeq, \preccurlyeq\rangle$ is important for the study of the class $\mathrm{Mod}(T)$. Properties of this object are more completely characterized by the triple $\langle \mathfrak{C}, \mathrm{Aut}(\mathfrak{C}), \mathcal{N}(\mathfrak{C})\rangle$, where $\mathfrak{C}$ is the monster-model of T, $\mathrm{Aut}(\mathfrak{C})$ is the group of all automorphisms of $\mathfrak{C}$ and $\mathcal{N}(\mathfrak{C})$ is the class of all elementary substructures of $\mathfrak{C}$. Therefore the following definition of semantic similarity is justified.

I shall begin with some preliminary notions.

DEFINITION 2. (1) By a *pure triple* we mean $\langle A, \Gamma, \mathcal{M}\rangle$, where $A \neq \emptyset$, Γ is a permutation group on A, and $\mathcal{M}$ is a family of subsets of A such that

$$M \in \mathcal{M} \Rightarrow g(M) \in \mathcal{M} \quad \text{for every } g \in \Gamma.$$

(2) If $\langle A_1, \Gamma_1, \mathcal{M}_1\rangle$ and $\langle A_2, \Gamma_2, \mathcal{M}_2\rangle$ are pure triples, and $\psi : A_1 \to A_2$ is a bijection, then ψ is an *isomorphism*, if

(i) $\Gamma_2 = \{\, \psi g \psi^{-1} : g \in \Gamma_1 \,\}$;

(ii) $\mathcal{M}_2 = \{\, \psi(E) : E \in \mathcal{M}_1 \,\}$.

DEFINITION 3. The pure triple $\langle |\mathfrak{C}|, G, \mathcal{N}\rangle$ is called the *semantic triple* of T (abbreviated s.t.), where $|\mathfrak{C}|$ is the universe of $\mathfrak{C}$, $G = \mathrm{Aut}(\mathfrak{C})$, and $\mathcal{N}$ is the class of all subsets of $|\mathfrak{C}|$ which are universes of suitable elementary submodels of $\mathfrak{C}$.

DEFINITION 4. Complete theories T_1 and T_2 are *semantically similar* if and only if their semantic triples are isomorphic.

EXAMPLE 2. The following theories T_1 and T_2 are semantically similar, where

$$\begin{aligned} T_1 &= \mathrm{Th}(\langle \mathcal{M}_1; P_n, n < \omega; a_{nm}, n, m < \omega\rangle), \\ \mathcal{M}_1 &= \{\, a_{nm} : n, m < \omega \,\}, \\ P_n(\mathcal{M}_1) &= \{\, a_{nm} : m < \omega \,\}, \end{aligned}$$

and

$$\begin{aligned} T_2 &= \mathrm{Th}(\langle \mathcal{M}_2; Q_n, n < \omega; Q_{nm}, n, m < \omega; b_{nmk}, n, m, k < \omega\rangle), \\ \mathcal{M}_2 &= \{\, b_{nmk} : n, m, k < \omega \,\}, \\ Q_n(\mathcal{M}_2) &= \{\, b_{nmk} : m, k < \omega \,\}, \\ Q_{nm}(\mathcal{M}_2) &= \{\, b_{nmk} : k < \omega \,\}. \end{aligned}$$

§3. Criteria of syntactical and semantical similarities.

It turns out that the notions of syntactical and semantical similarity may be defined in a common language, namely, in terms of so-called semisystems.

DEFINITION 5. (1) By a *semisystem* we mean a pair $\langle A, \mathcal{F}\rangle$, $A \neq \emptyset$, $\mathcal{F} \subseteq \bigcup_n \mathcal{P}(A^n)$, where $\mathcal{P}(x)$ denotes the set of all subsets of X.

(2) If $\langle A_1, \mathcal{F}_1\rangle$ and $\langle A_2, \mathcal{F}_2\rangle$ are semisystems, $\psi : A_1 \to A_2$ is a bijection, then ψ is called an *isomorphism* if and only if $\mathcal{F}_2 = \{\, \psi(E) : E \in \mathcal{F}_1 \,\}$, where $\psi(E) = \{\, \langle \psi(e_1), \dots, \psi(e_n)\rangle : \langle e_1, \dots, e_n\rangle \in E \,\}$.

DEFINITION 6. $X \in \mathsf{F}(\mathfrak{C}) \iff \exists n < \omega,\ \varphi \in F_n(T)$ such that

$$X = \{\, \langle a_1, \dots, a_n\rangle \in |\mathfrak{C}^n| : \mathfrak{C} \models \varphi(a_1, \dots, a_n) \,\}.$$

THEOREM 1. *The following are equivalent:*

(i) T_1 and T_2 are syntactically similar.

(ii) The semisystems $\langle |\mathfrak{C}_1|, \mathsf{F}(\mathfrak{C}_1)\rangle$ and $\langle |\mathfrak{C}_2|, \mathsf{F}(\mathfrak{C}_2)\rangle$ are isomorphic.

DEFINITION 7. $X \in \mathrm{TV}(\mathfrak{C})$ (i.e., X is a Tarski–Vaught set) if and only if there is $n < \omega$ such that
(i) $X \subseteq |\mathfrak{C}|^n$;
(ii) $\forall g \in \mathrm{Aut}(\mathfrak{C})(X = g(X))$;
(iii) $\forall M \prec \mathfrak{C}$, $\forall m, 1 \leq m < n$, $\forall b_1, \dots, b_m, b_{m+2}, \dots, b_n \in M$

$$\exists y \in |\mathfrak{C}|(\langle b_1, \dots, b_m, y, b_{m+2}, \dots, b_n\rangle \in X) \Rightarrow$$
$$\exists y \in M(\langle b_1, \dots, b_m, y, b_{m+2}, \dots, b_n\rangle \in X).$$

THEOREM 2. *The following are equivalent:*
(i) T_1 and T_2 are semantically similar.
(ii) The semisystems $\langle |\mathfrak{C}_1|, TV(\mathfrak{C}_1)\rangle$ and $\langle |\mathfrak{C}_2|, TV(\mathfrak{C}_2)\rangle$ are isomorphic.

PROPOSITION 1. *If T_1 and T_2 are syntactically similar, then T_1 and T_2 are semantically similar. The converse implication fails.*

Proof. "$\Rightarrow$" is easy; "$\not\Leftarrow$" follows from Example 2.

§4. A list of semantic properties of theories.

DEFINITION 8. A property (or a notion) of theories (or models, or elements of models) is called *semantic* if and only if it is invariant relative to semantic similarity.

PROPOSITION 2. *The following properties and notions are semantic:*
(1) type,
(2) forking,
(3) λ-stability,
(4) Lascar rank,
(5) strong type,
(6) Morley sequence,
(7) orthogonality, regularity of types,
(8) $I(\aleph_\alpha, T)$—the spectrum function.

§5. Quasisimilarity of theories.

DEFINITION 9. Let $\langle A, \Gamma, \mathcal{M}\rangle$ be an arbitrary pure triple, $\sim$ an equivalence relation on A. Then $\sim$ is *congruence* if and only if
(i) $a_1 \sim a_2 \Rightarrow g(a_1) \sim g(a_2)$, $\forall g \in \Gamma$, $\forall a_1, a_2 \in A$;
(ii) $a_1 \in M \;\&\; M \in \mathcal{M} \;\&\; a_1 \sim a_2 \Rightarrow a_2 \in M$, $\forall a_1, a_2 \in A$.

Remark. (1) If $\sim$ is a congruence, then $\sim$ induces a group congruence $\approx$ on the group Γ in the following way:

$$g_1 \approx g_2 \iff \forall a \in A(g_1(a) \sim g_2(a)), \text{ where } g_1, g_2 \in \Gamma.$$

(2) If $\sim$ is a congruence, then the triple $\langle A/{\sim}, \Gamma/{\sim}, \mathcal{M}/{\sim}\rangle$ will be a pure triple too which is called the *quotient pure triple*. Here

$$\mathcal{M}/{\sim} = \{\, M/{\sim} : M \in \mathcal{M} \,\}.$$

(3) The following relation ε on the semantic triple of any theory is a congruence:

$$a \,\varepsilon\, b \iff \begin{cases} a = b, & \text{if } a, b \in \operatorname{acl}(\emptyset); \\ \operatorname{acl}(a) = \operatorname{acl}(b) & \text{in the other case.} \end{cases}$$

DEFINITION 10. (1) T_1 and T_2 are *ε-similar* if and only if the semantic quotient triples $\langle |\mathfrak{C}_1|/\varepsilon, G_1/\varepsilon, \mathcal{N}_1/\varepsilon \rangle$ and $\langle |\mathfrak{C}_2|/\varepsilon, G_2/\varepsilon, \mathcal{N}_2/\varepsilon \rangle$ are isomorphic.

(2) T_1 and T_2 are *quasisimilar* if and only if there are $M_1 \models T_1$, $M_2 \models T_2$ such that $\mathrm{Th}((M_1, m)_{m \in M_1})$ and $\mathrm{Th}((M_2, m)_{m \in M_2})$ are ε-similar.

EXAMPLE 3. $\mathrm{Th}(\langle \mathbb{Z}; ' \rangle)$, where $x' = x + 1$, is quasisimilar to the theory of infinite sets without any structure.

The following question is natural: What kind of theories are quasisimilar to theories with only unary predicates? To answer this question we need some notions.

DEFINITION 11. We say that a theory T *admits* a closure operator J if and only if J is a closure operator on $|\mathfrak{C}|$ which satisfies the condition

$$J(g(B)) = g(J(B)),\ \forall g \in \operatorname{Aut}(\mathfrak{C}),\ \forall B \subset |\mathfrak{C}|.$$

Notations.

(1) If $a, b \in |\mathfrak{C}| \setminus B$, then $b \in C^J_B(a)$ if and only if there are $n < \omega$ and $b_0, \dots, b_n \in |\mathfrak{C}| \setminus B$ such that $b_0 = a$, $b_n = b$ and $b_i \in J(b_{i+1})$ or $b_{i+1} \in J(b_i)$, for every $i < n$.

(2) $C^J_B(A) = \bigcup \{ C^J_B(a) : a \in A \}$.

(3) $\hat{B} = \bigcup \{ J(b) : b \in B \}$.

(4) If $M \models T$, and $a, b \in |\mathfrak{C}| \setminus M$, then $a E^J_M b \iff (M \cap J(a) = M \cap J(b))$.

(5) $\chi^J_M(a) = \left| C^J_M(a) / E^J_M \right|$.

(6) $\chi^J(T) = \min\{ \mu : \chi^J_M(a) < \mu,\ \forall M \models T,\ \forall a \in |\mathfrak{C}| \setminus M \}$, if such a cardinal exists. Otherwise $\chi^J(T) = \infty$.

THEOREM 3. *The following are equivalent:*

(i) T is quasisimilar to some theory of unary predicates.

(ii) T is a bounded dimensional superstable theory which admits a closure operator J for which the following conditions are satisfied:

(1) $M \models T \Rightarrow \hat{M} = M$,

(2) $B = \hat{B}$ & $C^J_B(\bar{a}) = C^J_B(\bar{b}) \Rightarrow \bar{a} \downarrow_B \bar{b}$,

(3) $\chi^J(T) < \infty$.

As an application of this theorem we have

PROPOSITION 3. *$\mathrm{Th}(\langle M; f \rangle)$, where f is unary function (i.e., the theory of so called unars), is ω_1-categorical if and only if it is quasisimilar to the theory of infinite sets without any structure.*

Remark. In connection with the last proposition it is necessary to note the following. For the description of some classes of concrete algebraic systems defined in model theory language there may not exist a characterisation in an appropriate algebraic language. For example, Shishmarev [5] gave in 1972 a criterion of ω_1-categoricity of unars in a complex mixed language (model theoretical language with algebraic one). But until now no appropriate criterion in a purely algebraic language was found. In these cases, as the above proposition shows, the language of quasisimilarity can become useful and understandable.

§6. Envelope and almost envelope.

If $Q \in F_1(T)$ and $M \models T$, then $M^Q = \langle Q(M); R_\varphi \rangle_{\varphi \in F(T)}$, where $R_\varphi = (Q(M))^n \cap \varphi(M)$, for $\varphi \in F_n(T)$. $T^Q = \mathrm{Th}(M^Q)$.

DEFINITION 12. A theory T_1 is the *envelope* of T_2 if and only if there is $Q \in F_1(T_1)$ such that
(i) T_2 is syntactically similar to T_1^Q;
(ii) For all $N \models T_1^Q$ there is $M \models T_1$ such that $N = M^Q$;
(iii) $M = \mathrm{dcl}(Q(M))$, $\forall M \models T_1$.

EXAMPLE 4. Let T_2 be the theory of infinite sets without any structure, $T_1 = \mathrm{Th}(\langle M_1; Q, f \rangle)$, where Q is a unary relation, f is a unary function such that (i) $f^2 = \mathrm{id}$, (ii) $f \restriction Q$ is a bijection between Q and $M \setminus Q$. (See Figure 1.)

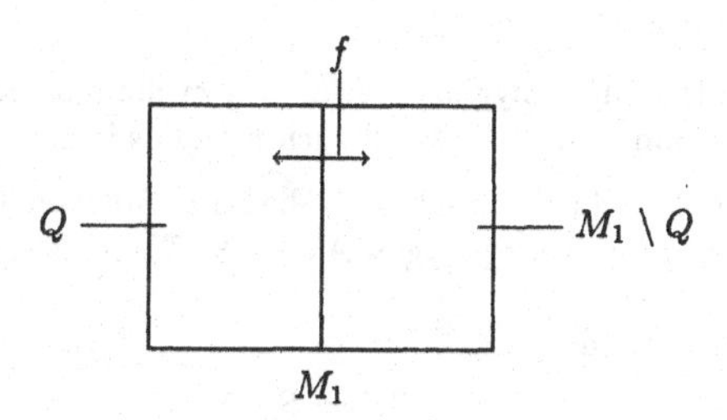

Figure 1.

Then T_2 is syntactically similar to T_1^Q and T_1 is an envelope of T_2.

DEFINITION 13. By a *polygon* over a monoid S we mean a structure with only unary functions $\langle A; f_\alpha : \alpha \in S \rangle$ such that
(i) $f_e(a) = a$, $\forall a \in A$, where e is the unit of S;
(ii) $f_{\alpha\beta}(a) = f_\alpha(f_\beta(a))$, $\forall \alpha, \beta \in S$, $\forall a \in A$.

THEOREM 4. *For every theory T_2 in a finite signature there is a theory T_1 of polygons such that some inessential extension of T_1 is an envelope of T_2.*

For the case when the signature is infinite we have a weak variant of this theorem. I shall introduce new notions for the formulation of the next theorem.

DEFINITION 14. A type $p \in S_1(T)$ is called *neutral* if and only if
(i) $M \models T \Rightarrow M_p \models T$, where $M_p = M \setminus p(M)$;
(ii) $p(M) \setminus A$ is indiscernable over A, $\forall M \models T$, $\forall A \subset M$.

DEFINITION 15. T_1 is an *almost envelope* of T_2 if and only if there are $Q \in F_1(T)$ and a neutral type $p \in S_1(T)$ such that
(i) T_2 is syntactically similar to T_1^Q;
(ii) $\forall N \models T_1^Q \; \exists M \models T_1 \; (N = M^Q)$;
(iii) $M_p = \text{dcl}(Q(M))$, $\forall M \models T_1$.

THEOREM 5. *For every theory T_2 in an infinite signature there is a theory T_1 of polygons such that some inessential extension of T_1 is an almost envelope of T_2.*

Remark. The notions of envelope and almost envelope express the close connection between theories. The following shows it:

PROPOSITION 4. *If T_1 is an envelope (or almost envelope) of T_2, then*
(i) T_1 is ω-stable $\iff$ T_2 is λ-stable, $\forall \lambda$;
(ii) $I(\aleph_\alpha, T_1) = I(\aleph_\alpha, T_2)$ (or $I(\aleph_\alpha, T_1) = I(\aleph_\alpha, T_2) + |\aleph_0 + \alpha|$), $\forall \alpha$.

From this it is clear that many problems of model theory can be reduced to the analogous problems of polygon theory in an exact way. In particular, this is true for Vaught's conjecture about the number of countable models.

§7. Polygons.

It is known that the polygons over an any cyclic monoid have a superstable theory. The question of describing all such monoids is natural.

DEFINITION 16. (1) Monoid S is called a *stabilisator* (or *superstabilisator*, or *ω-stabilisator*) if for every polygon A over S, Th(A) is stable (or superstable, or ω-stable).

(2) If $\alpha, \beta \in S$ then

$$\alpha \trianglelefteq \beta \iff S_\alpha \supseteq S_\beta,$$
$$\alpha \sim \beta \iff S_\alpha = S_\beta,$$
$$I_S = |S/{\sim}|.$$

(3) If $\langle S/{\sim}; \trianglelefteq \rangle$ is linearly ordered (or well ordered) then S is called *LO-monoid* (or *WO-monoid*).

THEOREM 6. *(i) S is a stabilisator if and only if S is LO-monoid.*
(ii) S is a superstabilisator if and only if S is WO-monoid.

Problem: When is S is an ω-stabilisator?

We have the following information on the problem.

THEOREM 7. *(i) If S is a stabilisator then $I_S \leq 2$.*
(ii) If $I_S = 1$ (i.e., S is a group), then S is an ω-stabilisator if and only if S has at most countably many subgroups.
(iii) If $I_S = 2$ (in which case S may be represented by $S = G \cup J$, where J is the unique proper left ideal, $G = S \setminus J$ and G is a group) and S is an ω-stabilisator, then
(1) G has at most countably many subgroups;
(2) $|G| < \omega \Rightarrow |S| < \omega$.

Conjecture. If $I_S = 2$ then S is an ω-stabilisator $\Longleftrightarrow$ $|S| < \omega$.

The proofs of the results are given in [1]–[4].

REFERENCES

[1] T. G. MUSTAFIN. *O podobiiah iblizosti polnyh teorii*, ***Algebra i logika***, vol. 29, no. 2 (1990), pp. 179–191.

[2] T. G. MUSTAFIN. *Ob operatorah zamykaniia na modeliah stabilnyh teorii*, ***Teoretiko-modelnaia algebra***, Alma-Ata 1989, pp. 68–90.

[3] T. G. MUSTAFIN. *O stabilnostnoi teorii poligonov*, ***Teoriia modelei i ee primeneniia, Trudy im soan SSSR***, vol. 8, 1988, pp. 92–108.

[4] T. G. MUSTAFIN, T. A. NURMAGAMBETOV. ***Vvedente v prikladnuiu teoriiu modelei***, Karaganda 1987.

[5] IU. E. SHISHMAREV. *O kategorichnyh teoriiah odnoi funkcii*, ***Matematicheskie zametki***, vol. 11, no. 1 (1972), pp. 89–98.

Karaganda University
Karaganda 470074
Kazakhstan

DECIDABILITY QUESTIONS FOR THEORIES OF MODULES

F. POINT

§0. Introduction.

Which finite rings with identity have a decidable theory of unitary left modules? This question has been raised by S. Burris and R. McKenzie in their paper on decidable varieties with modular congruence lattices. They showed that if a locally finite variety with modular congruence lattice does not decompose as a product of a discriminator variety and an affine variety, then it interprets the theory of all finite graphs. Then, they reduced the problem of classifying the decidable locally finite affine varieties to the problem of classifying the finite rings which have a decidable theory of modules.

First, we will see how this question arises in the context of decidable locally finite varieties. Then, we will restrict our attention to the decidability question for theories of modules. We will establish a connection between the decidability of the theory of modules over a finite-dimensional algebra and the representation type of that algebra.

This leads to the following questions: what are the relationships

- between the theory of R-modules and the theory of finitely generated R-modules?
- between theories of modules which are Morita equivalent?

§1. Locally finite varieties.

A *variety* is a class of L-structures, where the language L only contains function symbols, defined by some set of equations (or equivalently closed under products, substructures and homomorphisms). A variety is *locally finite* if every finitely generated algebra is finite.

S. Burris and R. McKenzie proved a decomposition theorem for decidable locally finite varieties with modular congruence lattice. They show that it decomposes as the product of a discriminator variety and an affine variety. (See [B,M]).

R. McKenzie and M. Valeriote generalized their decomposition theorem for decidable locally finite varieties. Before stating the result of McKenzie and Valeriote, we make this notion of decomposition precise.

DEFINITION. *$\mathcal{V} = \mathcal{A} \otimes \mathcal{B}$ means that $\mathcal{V}$ is generated by $\mathcal{A}$ and $\mathcal{B}$ and that there exists a term $\tau(x,y)$ such that $\mathcal{A} \models \tau(x,y) = x$ and $\mathcal{B} \models \tau(x,y) = y$. If $M \in \mathcal{V}$, there exist unique up to isomorphism $A \in \mathcal{A}$, $B \in \mathcal{B}$ such that $M \cong A \times B$.*
If $\mathcal{V} = \mathcal{A} \otimes \mathcal{B}$, then $\mathcal{V}$ is decidable iff $\mathcal{A}$ and $\mathcal{B}$ are decidable. (See [B,M].)

THEOREM 1. (See [M,V].) *If $\mathcal{V}$ is a locally finite decidable variety, then there are three subvarieties of $\mathcal{V}$, $\mathcal{A}$, $\mathcal{S}$ and $\mathcal{D}$ such that $\mathcal{V} = \mathcal{A} \otimes \mathcal{S} \otimes \mathcal{D}$ where $\mathcal{A}$ is an affine variety, $\mathcal{S}$ is a strongly abelian variety and $\mathcal{D}$ is a discriminator variety.*

This theorem reduces the question of classifying the decidable locally finite varieties to classifying the decidable locally finite discriminator, strongly abelian and affine varieties. M. Valeriote settled the question for decidable locally finite strongly abelian varieties.

A. Strongly abelian varieties.

DEFINITION. *An algebra A is strongly abelian if for every term t and tuples $\overline{a}, \overline{b}, \overline{c}, \overline{d}, \overline{e}$ from A such that length($\overline{a}$) = length($\overline{c}$) and length($\overline{b}$) = length($\overline{d}$) = length($\overline{e}$), then*

$$(t(\overline{a},\overline{b}) = t(\overline{c},\overline{d}) \Rightarrow t(\overline{a},\overline{e}) = t(\overline{c},\overline{e})).$$

EXAMPLE. Any algebra which language only contains unary functions and constants is strongly abelian.

Valeriote in his thesis characterized the decidable locally finite strongly abelian varieties.

DEFINITION. *Let W be a multi-sorted unary variety.*
W is linear if for all non constant terms $s(x), t(y)$ where x and y are of the same sort, there exists $w(z)$ such that

$$\begin{array}{ll} \text{either} & W \models t(x) = \omega(s(x)) \\ \text{or} & W \models s(x) = \omega(t(x)). \end{array}$$

THEOREM 2. (See [V].) *A locally finite strongly abelian variety is decidable iff it is bi-interpretable with a multi-sorted linear unary variety.*

B. Discriminator varieties.

DEFINITION. *A discriminator variety $\mathcal{V}$ is a variety such that there is a class $\mathcal{K}$ included in $\mathcal{V}$ and a ternary term $t(x,y,z)$ such that $\mathcal{V}$ is generated by $\mathcal{K}$ and for every $A \in \mathcal{K}$,*

$$\begin{aligned} t^A(x,y,z) &= z \quad \text{if } x = y \\ &= x \quad \text{if } x \neq y. \end{aligned}$$

EXAMPLES. Boolean algebras ($\mathcal{K}$ consists of the Boolean algebra with 2 elements), rings satisfying $x^m = x$.

THEOREM 3. (See [W], [B,W].) *Every finitely generated discriminator variety with finite language is decidable.*
A finitely generated discriminator variety $\mathcal{V}$ is decidable iff $Th(\mathcal{K})$ is decidable where $\mathcal{K}$ is the class of simple algebras in $\mathcal{V}$.

Burris and Werner used the sheaf representation of a discriminator variety and a generalization of a technique of Ershov for bounded Boolean powers of a finite structure.

Recent progress has been made by Burris, McKenzie and Valeriote concerning the question of which are the decidable locally finite discriminator varieties. The first observation is that the class of simple elements is definable and so decidable. Therefore by theorem 1, it can be decomposed as the product of a discriminator variety, a strongly abelian variety and an affine variety. By a result of Burris on undecidability of iterated discriminator variety, no discriminator term can appear (see [B]). They settled the question for homogeneous discriminator varieties where the class of simple elements is strongly abelian. (See [B,M,V].)

Added in proof: Valeriote and Willard settled the question for $\mathcal{K}$ an affine variety. The corresponding discriminator variety $\mathcal{V}$ is decidable if and only if $\mathcal{K}$ is polynomially equivalent to a variety of left modules over a finite semi-simple ring. (See [V,W].)

C. Affine varieties.

An affine variety $\mathcal{A}$ is polynomially equivalent to a variety of unitary left R-modules and there exists a term $t(x,y,z)$ such that $t(x,y,z) = x - y + z$ (see [F,M]). If $\mathcal{A}$ is locally finite, then R is finite.

The problem of classifying the decidable locally finite affine varieties effectively reduces to the problem of determining which are the finite rings R for which the variety of unitary left R-modules is decidable. (See [B,M].)

We will examine this question in more details, but first let us come back to theorem 1.

McKenzie and Valeriote show that if the locally finite variety $\mathcal{V}$ does not decompose as $\mathcal{A} \otimes \mathcal{S} \otimes \mathcal{D}$ then it interprets the class of all (finite) graphs. (Actually they interpreted BP^1 which is a subclass of the class of Boolean pairs: $(A, S) \in BP^1$ if A is atomic and the atoms of A are included in S). Which finishes the proof by the result of Lavrov who proved that the set of sentences true in all graphs and the set of sentences which become false in some finite graph are recursively inseparable (see [L]).

Now we are going to make a digression to point out the connection between having few models and being decidable for strongly abelian varieties.

DEFINITION. *Let $\mathcal{K}$ be a class of L-structures and let $I(\mathcal{K}, \lambda)$ be the number of non isomorphic models in $\mathcal{K}$ of cardinality λ. Then $\mathcal{K}$ has few models if there exists $\lambda > |L|$ such that $I(\mathcal{K}, \lambda) < 2^\lambda$.*

A consequence of the proof of McKenzie and Valeriote is that if $\mathcal{V}$ is a locally finite variety which has few models, then $\mathcal{V}$ decomposes as $\mathcal{A} \otimes \mathcal{S} \otimes \mathcal{D}$. We may eliminate $\mathcal{D}$ since any non trivial discriminator variety contains an algebra whose complete theory is unstable.

For the strongly abelian varieties, we have the following theorem:

THEOREM 4. (See [H,V].) *Let $\mathcal{V}$ be a strongly abelian variety. Then $\mathcal{V}$ has few models iff $\mathcal{V}$ is bi-interpretable with a multi-sorted unary variety which is linear and has the ascending chain condition.*

So for a strongly abelian locally finite variety, the properties of having few models and being decidable coincide.

Let R be a countable ring and $\mathcal{M}_R$ be the variety of all unitary left R-modules. Baldwin and McKenzie showed that if $I(\mathcal{M}_R, \lambda) < 2^\lambda$ for some λ, then every left R-module is ω-stable. (See [Ba,M].) This implies that every left module is a direct sum of indecomposable submodules that are finitely generated (see [G]). If the ring R is an Artin algebra (in particular if either R is finite or if R is a finite-dimensional algebra) then this implies that there are only finitely many indecomposable modules. (See [Pr1].) (An Artin algebra is a ring which is finitely generated as a module over its center and its center is an artinian ring).

So we see that having few models has some relationships with being decidable. But in case of modules as we shall see, it is a too strong property.

From now on, we will concentrate on the following question: for which finite rings R is the theory of unitary left R-modules decidable?

§2. Locally finite affine varieties.

Let R be a ring with unity 1. Let $L = \{+, -, 0, r; r \in R\}$, r is viewed as a unary function symbol. Let T_R be the theory of unitary left R-modules.

Baur gave the first example of a finite ring $R = \mathbb{Z}/p^9\mathbb{Z}[x]/(x^2)$ for which T_R is undecidable (p is any prime number). (See [Ba1].) So the theory of $\mathbb{Z}[x]$-modules is undecidable. (The theory of $\mathbb{Z}$-modules is decidable; see [Sz].)

THEOREM 5. (See [Ba].) *The theory T of pairs of abelian groups of exponent p^9 is undecidable.*

The idea of the proof is to interpret in a finite extension of T the word problem for a finitely presented semi-group on two generators.

Let $R = \mathbb{Z}/p^9\mathbb{Z}[x]/(x^2)$.

COROLLARY. *T_R is undecidable.* (See [Ba1].)

PROOF. Let $M \models T_R$. Define $\mathcal{A}_M = (\ker x, \operatorname{im} x)$. Then $\mathcal{A}_M \models T$. Let $(\mathcal{A}, \mathcal{B}) \models T$. Let $\mathcal{B}_1$ be an isomorphic copy of $\mathcal{B}$. Set $M = \mathcal{A} \times \mathcal{B}_1$. Define $x.a = 0$ for all $a \in \mathcal{A}$, $x.b_1 = b$ where b_1 is sent to b by the isomorphism between $\mathcal{B}_1$ and $\mathcal{B}$. So we have a faithful interpretation of T in T_R.

COROLLARY. (See [Ba1].)

1. *The theory of pairs of $k[x]$-modules is undecidable, where k is a finite field.*
2. *The theory $T_{k[x,y]}$ of $k[x,y]$-modules is undecidable, where k is a finite field.*

Let $C = \{(V,W,f) : V, W$ are k-vector spaces, $V \supseteq W, f \in \text{End}(V)\}$. Then $\text{Th}(C)$ is undecidable. (This follows from the point 1 of the Corollary above.) Let k be a field.

DEFINITIONS. *Let A be a k-algebra i.e., A is a ring with unity and a left k-module satisfying:*

$$r(x.y) = (r.x).y = x.(r.y) \quad \text{for all } x, y \in A, r \in k.$$

Let J be the Jacobson radical of A i.e., $J = \bigcap\{M : M$ is a maximal left ideal of $A\}$.
A is local if $A/J \cong k$.
A is left artinian if A has the descending chain condition on left ideals.

PROPOSITION 1. (See [P1].) *Let A be a commutative artinian ring. Let $\mathcal{I}$ be the set of maximal ideals of A and A_I the localization of A by $I, I \in \mathcal{I}$. Suppose that A_I is a k-algebra over some field k. Then T_A is decidable iff each T_{A_I} is decidable, for all $I \in \mathcal{I}$.*

PROPOSITION 2. (See [P1].) *Let A be a local artinian commutative k-algebra. Suppose that characteristic of $k \neq 2$ or that k is finite.*
Then, either

1. *A has a residue ring isomorphic to $k\langle x,y,z\rangle/I_1$, where I_1 is the two-sided ideal generated by all monomials of degree 2, and T_A is undecidable.*

2. *A has a residue ring isomorphic to $k\langle x,y\rangle/(x^2, xy - yx, y^2x, y^3)$ and T_A is undecidable.*

3. *A is isomorphic to $B_{n,m} = k\langle x,y\rangle/I$ where I is the two-sided ideal generated by $x.y$, $y.x$, x^n, y^m with $n + m \geq 5$.*

4. *Let $\overline{k}$ be a quadratic extension of k and $\overline{A} = A \otimes \overline{k}$. Then $\overline{A}$ is isomorphic to $\overline{B}_{n,m} = \overline{k}\langle x,y\rangle/I$.*

5. *A is isomorphic to $B_2 = k\langle x,y\rangle/I_2$ where I_2 is generated by x^2, y^2, $xy - yx$.*

6. *A is isomorphic to $B_{2,2} = k\langle x,y\rangle/I_{2,2}$ where $I_{2,2}$ is generated by $x^2, y^2, x.y, y.x$.*

7. *A is isomorphic to $B_1(n) = k\langle x\rangle/I_3$ where I_3 is the two-sided ideal generated by x^n.*

In cases 5, 6 and 7, T_A is decidable.

The idea of the proof of cases 1 and 2 is the following one:
Let T be the theory of pairs (V,W) of $k[x]$-modules with $V \supseteq W$. We show that there exists a finite extension T^* of T_A such that in any model M of T^* we can define a pair (V,W) of $k[x]$-modules with $V \supseteq W$, where the action of x is definable and every model of T is of that form.

This proof is inspired by the proof of Drozd of wildness of these algebras (see [D1]). He showed that there is an exact and faithful embedding of the finite-dimensional elements of $\mathcal{C}$ into the class of finitely generated modules over each of these algebras.

Since there is a link between the representation type of a finite-dimensional algebra and the decidability of the theory of modules over it, for sake of completeness, we shall give the definition of the various kinds of representation types for a finite-dimensional k-algebra. (If k is not algebraically closed, tensor up with the algebraic closure of k.)

DEFINITION. (See [R1].) *Let R be a finite-dimensional k-algebra and let R-mod be the class of finitely generated left R-modules. Assume k is algebraically closed.*

1. *R is of finite representation type if there are only finitely many indecomposable elements (up to isomorphism) in R-mod.*
2. *R is of tame representation type provided R is not of finite representation type and for any dimension d there is a finite number of embedding functors $F_i : k[x]$-mod $\to R$-mod such that all but a finite number of indecomposable finitely generated R-modules of dimension d are of the form $F_i(M)$ for some i and for some indecomposable finitely generated $k[x]$-module M.*
   In case there exists (independently of d) a finite number of such embedding functors F_i, then R is domestic.
   If R is tame, not domestic but there is a finite bound on the number of functors needed, then R is of finite growth.
   If R is tame, not domestic and not of finite growth, then R is infinite growth.
3. *R is of wild representation type if there is a functor from $k\langle x, y\rangle$- mod which preserves and reflects indecomposability and isomorphism.*

(See also [Pr1]. This definition can be phrased for an arbitrary k-algebra.)

REMARKS.

1. Let R be a finite-dimensional algebra over an algebraically closed field k. Then either R is of finite representation type or else there are infinitely many d_i's such that there are infinitely many pairwise non isomorphic indecomposable R-modules of dimension d_i, for all i. (Those indecomposable are indexed by the elements of k). (See [N,R].)
2. Every finite-dimensional k-algebra of infinite representation type is either tame or wild but not both. (See [D2].)

EXAMPLES. Let $\tilde{k}$ be the algebraic closure of k.

1. Then $\tilde{k} \otimes B_1(n)$ is of finite representation type.
(The indecomposable finitely generated modules are isomorphic to $\tilde{k}[x]/(x)^m$, $m \leq n$.
2. Then $\tilde{k} \otimes B_2$ and $\tilde{k} \otimes B_{2,2}$ are of tame domestic representation type. (See [G,P2].)
3. Then $\tilde{k} \otimes B_{n,m}$, $n + m \geq 5$, are of tame, infinite growth representation type. (See [G,P1].)

Using the classification of Ringel of the representation type of local k-algebras, k an algebraically closed field, one shows:

PROPOSITION 3. (See [P1].) *Let R be a local, complete k-algebra with k an algebraically closed field. If R is of wild representation type and R is not a residue ring of the group algebra of the generalized quaternion algebra, then T_R is undecidable.*

Recent work by A. Marcja, M. Prest and C. Toffalori showed undecidability results for wild classes of modules over group rings of the form $\mathbb{Z}/p^k\mathbb{Z}[G]$, where G is a p-group. (See [M,P,T].) Their notion of a wild class is the following: for some field K one can interpret in it a class of $K\langle x,y\rangle$-modules including the finite-dimensional $K\langle x,y\rangle$-modules in such a way that finite-dimensional $K\langle x,y\rangle$-modules are interpreted in structures of the class which are "finite-dimensional" in some sense (e.g. finitely generated).

Now we come to the decidability proofs which go deeper in the structure of the models.

First we are going to recall some notions of module theory.

DEFINITIONS. *A module M is pure-injective (p.i.) if every system of equations with parameters in M which is finitely satisfiable in M, has a solution in M. A p.p.-formula $\varphi(y,\bar{b})$ is a formula of the form*

$$\exists \bar{z} \begin{pmatrix} & A & \end{pmatrix} \begin{pmatrix} \bar{z} \\ \\ y \end{pmatrix} = \begin{pmatrix} \\ \bar{b} \\ \\ \end{pmatrix},$$

where A is a matrix with coefficients in R and $\bar{b}$ are parameters from a module M. If $\bar{b} = \bar{0}$, the set $\{y \in M : M \models \varphi(y,\bar{0})\}$ is a (p.p. definable) subgroup of M; $\varphi(y,\bar{0})$ asserts the solvability of a finite system of linear equations with parameter y.
An invariant sentence is of the form $(\varphi,\psi) \geq k$, where $k \in \mathbb{N} - \{0\}$, φ,ψ are p.p. formulas without parameters, $\psi \to \varphi$ and $(\varphi,\psi) \geq k$ means that the subgroup defined by ψ has index $\geq k$ in the subgroup defined by φ.

To prove decidability of T_R, you may expect to use the result of Ziegler which roughly says that if you know the space of pure-injective indecomposable R-modules with its topology, then T_R is decidable.

THEOREM 6. (See [Z].) *Let R be a ring which is finitely presented as an R-module with decidable word problem. Suppose that there is a recursive enumeration of all those conditions of the form $\bigwedge_i(\varphi_i, \varphi_j) \in [m, n)$ which are satisfied by some pure-injective indecomposable R-module, where $n, m \in \mathbb{N} \cup \{+\infty\}, m \neq +\infty$ and (φ_i) a recursive enumeration of all the p.p. formulas. Then T_R is decidable.*

Going back to Proposition 2, one has for cases (5) and (6):

(a) $T_{B_{2,2}}$ is decidable iff T_{B_2} is decidable.

(b) $T_{B_{2,2}}$ is decidable since it is interpretable in the theory of quadruples (i.e., the theory of a vector space and four subspaces).

THEOREM 7. (See [Ba2].) *The theory of k-vector spaces with four subspaces is decidable, whenever the theory of $k[x]$-modules is decidable.*

THEOREM 8. (See [E,F].) *Let k be a recursive field with decidable word problem and a splitting algorithm (i.e., an algorithm which determines for any element of $k[x]$ its irreducible factors). Then the theory of $k[x]$-modules is decidable.*

In his proof of decidability of the theory of quadruples, Baur described the $\aleph_1$-saturated models and used the description of finitely generated indecomposable models given by Gelfand and Ponomarev (see [G,P2]).

This case is the cutting line. The theory of a vector space and five subspaces is undecidable, and the theory of a vector space and three subspaces is of course decidable (the proof in this case is much simpler). (See [Ba3].)

Extending the result of Baur on quadruples, Prest proved:

THEOREM 9. (See [Pr2].) *Let Δ be a quiver without relations, the underlying diagram of which is extended Dynkin. Then the theory of modules over the path algebra $k[\Delta]$ is decidable.*

DEFINITIONS.

1. *A quiver Δ is a finite directed graph with no oriented cycles.*

2. *The basis of $k[\Delta]$ as a k-vector space is the set of all oriented paths of Δ which includes the path of length zero at each vertex. The product of two basis elements is composition of paths when defined and zero otherwise.*
   The paths of length zero compose as "local identities".

The theory of quadruples over the field k can be translated into the theory of modules over the path algebra of the quiver:

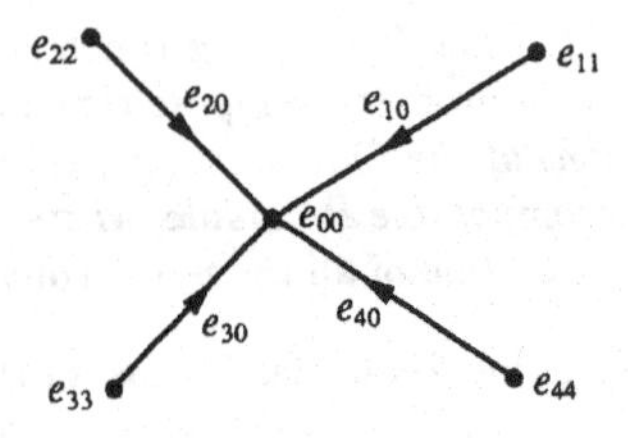

or over the algebra

$$\begin{pmatrix} k & k & k & k & k \\ & k & \cdot & 0 & \cdot \\ & & k & \cdot & \cdot \\ & 0 & & k & \cdot \\ & & & & k \end{pmatrix}$$

and vice-versa.

Roughly, the decidability proof of Prest for $k[\Delta]$ is done taking the proof that $k[\Delta]$ is of tame domestic representation type, replacing indecomposable finitely generated modules by pure-injective indecomposable modules (to prove that he obtains all of them he uses the density of the set of regular indecomposable modules) and showing that the functors respects p.p. definable subgroups. So the fact that the topology of the space of indecomposable pure-injective modules over $k[x]$ is given explicitly transfers to the space of indecomposable pure-injective modules over $k[\Delta]$.

Gabriel established the connection between finite-dimensional algebras and path algebras with relations (a relation is a linear combination of sums of paths with same starting point and same end point which is declared to be zero).

THEOREM 10. (See [Ga].) *Over an algebraically closed field, any finite-dimensional algebra is Morita equivalent to a path algebra over a quiver with relations.*

THEOREM 11. (See [P,Pr].) *Over sufficiently decidable rings, decidability of the theory of modules is a Morita invariant.*

Do there exist a transfer theorem linking the representation type of a finite-dimensional algebra R and the decidability of T_R?

A more modest question is: since the representation type concerns the class of finitely generated modules or of finitely presented modules, do we have $T_R = T_R^{f.p.}$, where $T_R^{f.p}$ is the theory of finitely presented R-modules? (If R is artinian, then $T_R^{f.p.} = T_R^{f.g.}$, where $T_R^{f.g.}$ is the theory of finitely generated R-modules.

PROPOSITION 4. (See [P,Pr].) *If R is of finite representation type, then* $T_R = T_R^{f.p.}$.

THEOREM 12. (See [P,Pr].) *If R is the path algebra over a quiver without relations, of infinite representation type, then $T_R \neq T_R^{f.p.}$.*

THEOREM 13. (See [P,Pr].) *Let R be a noetherian local k-algebra of infinite representation type. Then $T_R \neq T_R^{f.p.}$.*

COROLLARY. *Let R be a finite-dimensional commutative k-algebra of infinite representation type. Then $T_R \neq T_R^{f.p.}$.*

The motivating idea is the following: Let $R = \mathbb{Z}$, let $\sigma =$

$$\exists v(p.v = 0 \wedge v \neq 0) \wedge \forall v \exists \omega(v = p.\omega).$$

$\mathbb{Z}_{p^\infty} \models \sigma$ and if an abelian group satisfies σ, it is infinitely generated. Thus, $T_\mathbb{Z} \neq T_\mathbb{Z}^{f.g.}$ Replacing p by $x - \alpha$, $\alpha \in k$, we have the same result for $R = k[x]$ i.e., $T_{k[x]} \neq T_{k[x]}^{f.g.}$.

Returning to Proposition 2 and the decidability question for T_R with R a commutative artinian local k-algebra, we have

(a) in cases (1), (2) of Proposition 2, both T_R and $T_R^{f.g.}$ are undecidable;

(b) in cases (5), (6), both T_R and $T_R^{f.g.}$ are decidable. (See [P2].)

For (a), we use a theorem of Slobodskoi (see [S]) that the word problem for the class of finite groups is undecidable. This implies that the theory of finite-dimensional vector-spaces with two endomorphisms is undecidable. This implies that the theory of pairs of finite-dimensional k-vector spaces with an endomorphism is undecidable (see [Pr1], Corollary 17.7).

For (b), we use the fact that the theory of finitely generated $k[x]$-modules is decidable and the proof of Prest of Baur's result on the theory of quadruples.

To finish, we would like to make some comments about case (3) of Proposition 2. In that case, there are $2^{\aleph_0}$ pure-injective indecomposable R-modules (see [Pr1]).

Let $R = k\langle a, b\rangle/(a^n, b^m, ab, ba)$ with $n + m \geq 5$. The finitely generated indecomposable modules have been described by Gelfand and Ponomarev (see [G,P1]). They are of two types: string and band.

Let us give an example of each type (suppose $n = 2, m = 3$).

A *string* module $M(C)$ where $C = b^{-2}ab^{-2}ab^{-1}$:

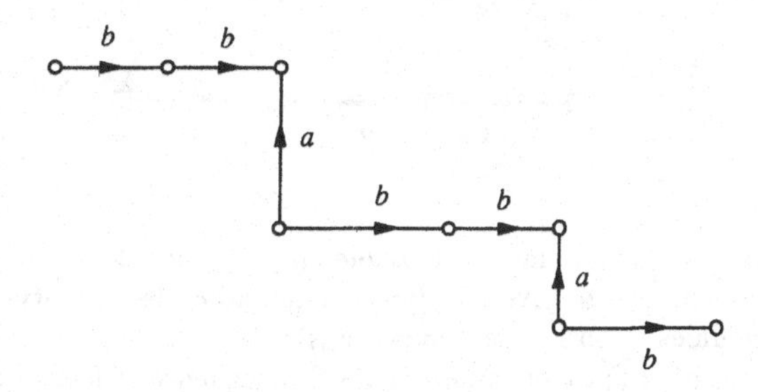

The points represent isomorphic copies of an one dimensional k-subspace V,

M is the direct sum of those and the actions of a and b are represented by the arrows.

A *band* module $M(C)$ where $C = b^{-1}ab^{-1}ab^{-1}$:

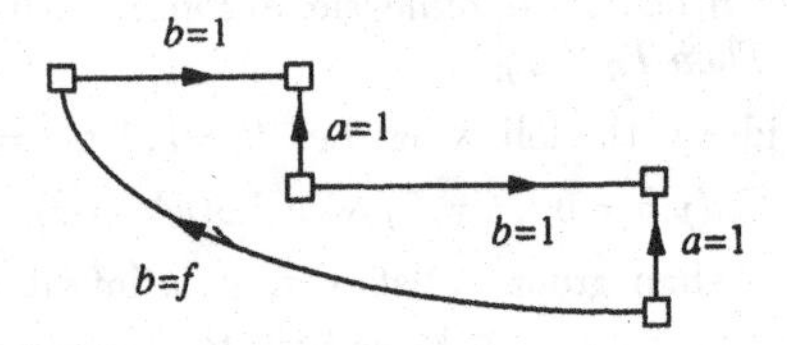

The boxes represent isomorphic copies of a q-dimensional k-subspace V and M is the direct sum of those; f is an automorphism of V such that (V, f) is an indecomposable $k[x, x^{-1}]$-module. The actions of a and b are represented by the arrows.

An analogous case is the case of the dihedral algebra:

$$R = k\langle a, b\rangle/(a^2, b^2, (ab)^n, (ba)^m).$$

The finitely generated indecomposable have been described by Ringel (see [R2]). They are of two types: string and band.

We give an example of each type (with the same interpretation of the diagrams).

A *string* module $M(C)$ where $C = ab^{-1}ab$:

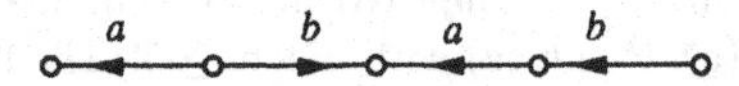

A *band* module $M(C)$ where $C = a^{-1}b^{-1}ab$:

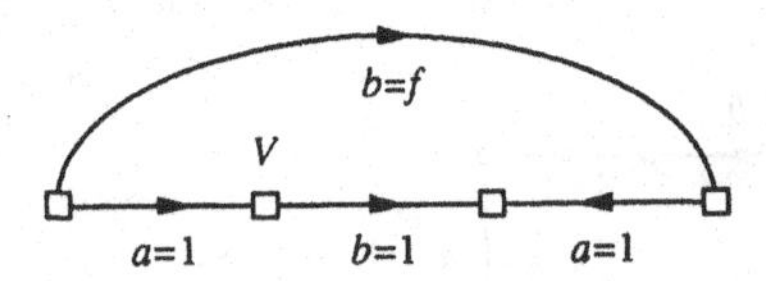

$f \in \text{Aut}(V)$ and (V, f) is an indecomposable $k[x, x^{-1}]$-module.

Now we would like to give an idea of a tentative decidability proof of the theory of R-modules in both cases. (See [P3].)

A word will be a finite or infinite one-sided sequence of letters belonging to $\{a, a^{-1}, b, b^{-1}\}$. Let W_a be the set of words beginning by an a or a^{-1}.

In any R-module, one can attach to an element x an ordered pair of two infinite one-sided words belonging to $W_b \times W_a$ (see [R2]). One can introduce a partial order on the words (see [R2]) and so on the pairs of words.

An element x is maximal in M if one cannot decompose x into a sum of 2 elements x_1, x_2 which have a strictly smaller pair of words associated to them.

NOTATION. Let $(D, C) \in W_b \times W_a$. We write D from the right to the left and we replace each letter by its inverse. We denote this new sequence by D^-. $D^-{}^\frown C$ is the concatenation of D^- and C.

The strategy of the proof in both cases is the following:

1. We show that each module is elementarily equivalent to a direct sum of pure-injective indecomposable modules containing a maximal element.
2. We give two criteria of independence:
   Let x, y be two maximal elements belonging to a pure-injective module.
   (a) Suppose that for all $(\alpha, \beta) \in k^2 - \{(0,0)\}$, the pair associated with $\alpha x + \beta y$ is (D, C) and that $\alpha x + \beta y$ is maximal. Suppose that $D^-{}^\frown C$ is not an infinite periodic two-sided word. Then x and y belong to distinct summands of M.
   (b) Let (F, E) be the pair of words associated with y and (D, C) be the pair of words associated with x.
   Suppose that $F^-{}^\frown E \neq D^-{}^\frown C$, then x and y belong to distinct summands of M.
3. We describe the pure-injective indecomposable modules containing a maximal element. There are 2 types of them, the string ones associated with a word (finite, one-sided infinite, two-sided infinite), the band ones associated with a finite word of even length and an indecomposable $k[x, x^{-1}]$-module where the automorphism x is given by the action of the word.
4. Then we show that we only need invariant sentences of a certain kind (i.e., those associated with a pair of finite words) to distinguish, up to elementary equivalence, between two pure-injective non-periodic indecomposable modules containing a maximal element.

REFERENCES

[Ba1] BAUR W., *Undecidability of the theory of abelian groups with a subgroup*, ***Proc. Amer. Math. Soc.***, **55**(1) (1976), 125–128.

[Ba2] BAUR W., *On the elementary theory of quadruples of vector spaces*, ***Ann. Math. Logic***, **19** (1980), 243–262.

[Ba3] BAUR W., *Decidability and undecidability of theories of abelian groups with predicates for subgroups*, ***Compositio Math.***, **31** (1975), 23–30.

[Ba,M] BALDWIN J. T., MCKENZIE R., *Counting models in universal Horn classes*, ***Algebra Universalis***, **15** (1982), 359–384.

[B] BURRIS S., *Iterated discriminator varieties have undecidable theories*, ***Algebra Universalis***, **21** (1985), 54–61.

[B,M] BURRIS S., MCKENZIE R., ***Decidability and Boolean Representations***, Memoirs Amer. Math. Soc., No. **246**, 1981.

[B,M,V] BURRIS S., MCKENZIE R., VALERIOTE M., *Decidable discriminator varieties from unary varieties*, ***J. Symbolic Logic***, **56**(4) (1991), 1355–1368.

[B,W] BURRIS S., WERNER H., *Sheaf constructions and their elementary properties*, ***Trans. Amer. Math. Soc.***, **248** (1979), 269–309.

[D1] DROZD J. A., *Representations of commutative algebras*, ***Funct. Analysis and its Appl.***, **6** (1972), 286–288.

[D2] DROZD J. A., *Tame and wild matrix problems in "Representations and Quadratic Forms,"* ***Akad. Nauk. Ukrain. SSR***, Inst. Mat. Kiev, 1979, pp. 39–74 and 154.

[E,F] EKLOF P. C., FISHER F., *The elementary theory of abelian groups*, ***Ann. Math. Logic***, **4**(2) (1972), 115–171.

[F,M] FREESE R., MCKENZIE R., ***Commutator theory for congruence modular varieties***, London Math. Soc. Lecture Note Series, **125**, 1987.

[Ga] GABRIEL P., *Auslander-Reiten sequences and representation-finite algebras*, in ***Representation Theory I***, Springer L.N.M., **831**, 1980, pp. 1–71.

[G] GARAVAGLIA S., *Decomposition of totally transcendental modules*, ***J. Symbolic Logic***, **45**(1) (1980), 155–164.

[G,P1] GELFAND I. M., PONOMAREV V. A., *Indecomposable representations of the Lorentz group*, ***Usp. Mat. Nauk***, **23** (1968), 3–60. Engl. Transl.: ***Russian Math. Surv.***

[G,P2] GELFAND I. M., PONOMAREV V. A., *Problems of linear algebra and classification of quadruples of subspaces in a finite-dimensional vector space*, in ***Coll. Math. Soc. Bolyai 5***, North-Holland, Amsterdam, 1972, pp. 163–237.

[H,V] HART B., VALERIOTE M., *A structure theorem for strongly abelian varieties*, ***J. Symbolic Logic***, **56**(3) (1991), 832–852.

[L] LAVROV I. A., *Effective inseparability of the sets of identically true formulas and finitely refutable formulae for certain elementary theories*, ***Algebra i Logika Sem.*** 2, vol. **1** (1963), 5–18.

[M,P,T] MARCJA A., PREST M., TOFFALORI C., *Stability theory for abelian-by-finite groups and modules over a group ring*, ***J. London Math Soc.*** (to appear).

[M,V] MCKENZIE R., VALERIOTE M., ***The structure of decidable locally finite varieties***, Progress in Mathematics, **79**, Birkhäuser, 1989.

[N,R] NAZAROVA L. A., ROJTER A. V., *Categorical matrix problems and the Brauer-Thrall conjecture*, preprint, Inst. Math. Acad. Sci., Kiev, 1973, German transl. Mitt. Math. Sem. Giesen, **115**, 1975.

[P1] POINT F., *Problèmes de décidabilité pour les théories de modules*, ***Bull. Belg. Math. Soc.***, Ser. B, **38** (1986), 58–74.

[P2] POINT F., *Relationships between decidability of a class of R-modules and the subclass of its finitely generated elements*, preprint.

[P3] POINT, F., *Decidability of the theory of modules over the dihedral algebra (respectively over the Gelfand and Ponomarev algebra)*, submitted.

[P,Pr] POINT F., PREST M., *Decidability for theories of modules*, ***J. London Math. Soc.*** (2), **38** (1988), 193–206.

[Pr1] PREST M., ***Model theory and modules***, London Math. Soc. Lecture Note Series, **130**, 1988.

[Pr2] PREST M., *Tame categories of modules and decidability*, University of Liverpool, 1985, preprint.

[R1] RINGEL C. M., *Tame algebras (on algorithms for solving vector space problems II)* in ***Representation Theory I***, Proceedings Ottawa, Carleton University, 1979, Springer L.N.M., **831**, 1980.

[R2] RINGEL C. M., *The indecomposable representations of the dihedral 2-groups*, ***Math. Ann.***, **214** (1975), 19–34.

[S] SLOBODSKOI A. M., *The unsolvability of the universal theory of finite groups*, ***Algebra i Logika***, **20**(2) (1981), 207–230.

[Sz] SZMIELEW W., *Elementary properties of abelian groups*, ***Fund. Math.***, **41** (1955), 203–271.

[V] VALERIOTE M., *On decidable locally finite varieties*, Ph.D. dissertation, University of California, Berkeley, 1986.

[V,W] VALERIOTE M., WILLARD R., *Dicriminating varieties*, ***Algebra Universalis*** (to appear).

[W] WERNER H., ***Discriminator algebras***, Studien zur Algebra und ihre Anwendungen, **6**, Akademie-Verlag, Berlin, 1978.

[Z] ZIEGLER M., *Model theory of modules*, ***Ann. Pure Appl. Logic***, **26**(2) (1984), 149–213.

Chercheur qualifié F.N.R.S.
15, Avenue Maistriau
7000 Mons
Belgium

ON $\mathrm{CH} + 2^{\aleph_1} \to (\alpha)^2_2$ FOR $\alpha < \omega_2$

SAHARON SHELAH[1]

§1. Introduction.

We prove the consistency of

$$\mathrm{CH} + 2^{\aleph_1} \text{ is arbitrarily large} + 2^{\aleph_1} \not\to (\omega_1 \times \omega)^2_2$$

(Theorem 1). In fact, we can get $2^{\aleph_1} \not\to [\omega_1 \times \omega]^2_{\aleph_0}$, see 1A. In addition to this theorem, we give generalizations to other cardinals (Theorems 2 and 3). The $\omega_1 \times \omega$ is best possible as CH implies

$$\omega_3 \to (\omega \times n)^2_2.$$

We were motivated by a question of J. Baumgartner, in his talk in the MSRI meeting on set theory, October 1989, on whether $\omega_3 \to (\alpha)^2_2$ for $\alpha < \omega_2$ (if $2^{\aleph_1} = \aleph_2$, it follows from the Erdős–Rado theorem). Baumgartner proved the consistency of a positive answer with CH and $2^{\aleph_1}$ large. He has also proved [BH] in ZFC + CH a related polarized partition relation:

$$\binom{\aleph_3}{\aleph_2} \to \binom{\aleph_1}{\aleph_1}^{1,1}_{\aleph_0}$$

Note. The main proof here is that of Theorem 1. In that proof, in the way things are set up, the main point is proving the $\aleph_2$-c.c. The main idea in the proof is using $\mathbf{P}$ (defined in the proof). It turns out that we can use as elements of $\mathcal{P}$ (see the proof) just pairs (a, b). Not much would be changed if we used $\langle (a_n, \alpha_n) : n < \omega \rangle$, a_n a good approximation of the nth part of the suspected monochromatic set of order type $\omega_1 \times \omega$. In 1A, 2, and 3 we deal with generalizations and in Theorem 4 with complementary positive results.

§2. The main result.

THEOREM 1. *Suppose*

(a) CH

(b) $\lambda^{\aleph_1} = \lambda$.

Then there is an $\aleph_2$-c.c., $\aleph_1$-complete forcing notion $\mathbf{P}$ *such that*

(i) $|\mathbf{P}| = \lambda$

(ii) $\Vdash_{\mathbf{P}}$ "$2^{\aleph_1} = \lambda,\ \lambda \not\to (\omega_1 \times \omega)^2_2$"

(iii) $\Vdash_{\mathbf{P}}$ CH

(iv) Forcing with $\mathbf{P}$ *preserves cofinalities and cardinalities.*

[1] Publication number 424. Partially supported by BSF.

Proof. By Erdős and Hajnal [EH] there is an algebra $\mathbf{B}$ with $2^{\aleph_0} = \aleph_1$ ω-place functions, closed under composition (for simplicity only), such that

$$\otimes \qquad \text{If } \alpha_n < \lambda \text{ for } n < \omega, \text{ then for some } k$$
$$\alpha_k \in \text{cl}_{\mathbf{B}}\{\, \alpha_l : k < l < \omega \,\}.$$

($\otimes$ implies that for every large enough k, for every m, $\alpha_k \in \text{cl}_{\mathbf{B}}\{\, \alpha_l : m < l < \omega \,\}$.) Let

$$\mathcal{R}_\delta = \{\, b : b \subseteq \lambda,\ \text{otp}(b) = \delta,\ \alpha \in b \Rightarrow b \subseteq \text{cl}_{\mathbf{B}}(b \setminus \alpha) \,\}.$$

So by $\otimes$ we have

$$\oplus \qquad \text{If } \alpha \text{ is a limit ordinal, } b \subseteq \lambda,\ \text{otp}(b) = \alpha,$$
$$\text{then for some } \alpha \in b,\quad b \setminus \alpha \in \textstyle\bigcup_\delta \mathcal{R}_\delta.$$

Let $\mathcal{R}_{<\omega_1} = \bigcup_{\alpha<\omega_1} \mathcal{R}_\alpha$. Let $\mathbf{P}$ be the set of forcing conditions

$$(w, c, \mathcal{P})$$

where w is a countable subset of λ, $c : [w]^2 \to \{\text{red}, \text{green}\} = \{0,1\}$ (but we write $c(\alpha,\beta)$ instead of $c(\{\alpha,\beta\})$), and $\mathcal{P}$ is a countable family of pairs (a,b) such that

(i) a, b are subsets of w

(ii) $b \in \mathcal{R}_{<\omega_1}$ and a is a finite union of members of $\mathcal{R}_{<\omega_1}$

(iii) $\sup(a) < \min(b)$

(iv) if $\sup(a) \leq \gamma < \min(b)$, $\gamma \in w$, then $c(\gamma, \cdot)$ divides a or b into two infinite sets.

We use the notation

$$p = (w^p, c^p, \mathcal{P}^p)$$

for $p \in \mathbf{P}$. The ordering of the conditions is defined as follows:

$$p \leq q \iff w^p \subseteq w^q \ \&\ c^p \subseteq c^q \ \&\ \mathcal{P}^p \subseteq \mathcal{P}^q.$$

Let

$$\underset{\sim}{c} = \bigcup\{\, c^p : p \in \underset{\sim}{G}_{\mathbf{P}} \,\}.$$

FACT A. *$\mathbf{P}$ is $\aleph_1$-complete.*

Proof. Trivial—take the union. □

FACT B. *For $\gamma < \lambda$, $\{\, q \in \mathbf{P} : \gamma \in w^q \,\}$ is open dense.*

Proof. Let $p \in \mathbf{P}$. If $\gamma \in w^p$, we are done. Otherwise we define q as follows: $w^q = w^p \cup \{\gamma\}$, $\mathcal{P}^q = \mathcal{P}^p$, $c^q \restriction w^p = c^p$ and $c^q(\gamma, \cdot)$ is defined so that if $(a,b) \in \mathcal{P}^q$, then $c^q(\gamma,\cdot)$ divides a and b into two infinite sets. □

FACT C. $\Vdash_{\mathbf{P}}$ "$2^{\aleph_1} \geq \lambda$ and $\underset{\sim}{c} : [\lambda]^2 \to \{\text{red, green}\}$."

Proof. The second phrase follows from Fact B. For the first phrase, define $\underset{\sim}{\rho}_\alpha \in {}^{\omega_1}2$, for $\alpha < \lambda$, by: $\underset{\sim}{\rho}_\alpha(i) = \underset{\sim}{c}(0, \alpha + i)$. Easily

$$\Vdash_{\mathbf{P}} \text{“}\underset{\sim}{\rho}_\alpha \in {}^{\omega_1}2 \text{ and for } \alpha < \beta < \lambda,\ \underset{\sim}{\rho}_\alpha \neq \underset{\sim}{\rho}_\beta\text{”}$$

so $\Vdash_{\mathbf{P}}$ “$2^{\aleph_1} \geq \lambda$.” □

FACT D. $\mathbf{P}$ *satisfies the* $\aleph_2$*-c.c.*

Proof. Suppose $p_i \in \mathbf{P}$ for $i < \aleph_2$. For each i choose a countable family $\mathcal{A}^i$ of subsets of w^{p_i} such that $\mathcal{A}^i \subseteq \mathcal{R}_{<\omega_1}$ and $(a, b) \in \mathcal{P}^{p_i}$ implies $b \in \mathcal{A}^i$ and a is a finite union of members of $\mathcal{A}^i$. For each $\gamma \in c \in \mathcal{A}^i$ choose a function $F^i_{\gamma,c}$ (from those in the algebra $\mathbf{B}$) such that $F^i_{\gamma,c}(c \setminus (\gamma + 1)) = \gamma$. Let v_i be the closure of w_i (in the order topology).

We may assume that $\langle v_i : i < \omega_2 \rangle$ is a Δ-system (we have CH) and that $\operatorname{otp}(v_i)$ is the same for all $i < \omega_2$. Without loss of generality (w.l.o.g.) for $i < j$ the unique order-preserving function $h_{i,j}$ from v_i onto v_j maps p_i onto p_j, $\mathcal{A}^i$ onto $\mathcal{A}^j$, $w^{p_i} \cap w^{p_j} = w^{p_0} \cap w^{p_1}$ onto itself, and

$$F^i_{\gamma,c} = F^j_{h_{i,j}(\gamma), h_{i,j}\text{“}c}$$

for $\gamma \in c \in \mathcal{A}^i$ (remember: $\mathbf{B}$ has $2^{\aleph_0} = \aleph_1$ functions only). Hence

$\otimes_1$ $\qquad h_{i,j}$ is the identity on $v_i \cap v_j$ for $i < j$.

Clearly by the definition of $\mathcal{R}_{<\omega_1}$ and the condition on $F^i_{\gamma,c}$:

$\otimes_2$ $\qquad$ If $a \in \mathcal{A}^i$, $i \neq j$ and $a \not\subseteq w^{p_i} \cap w^{p_j}$,
then $a \setminus (w^{p_i} \cap w^{p_j})$ is infinite.

We define q as follows.

$w^q = w^{p_0} \cup w^{p_1}$.

$\mathcal{P}^q = \mathcal{P}^{p_0} \cup \mathcal{P}^{p_1}$.

c^q extends c^{p_0} and c^{p_1} in such a way that, for $e \in \{0, 1\}$,

(*) for every $\gamma \in w^{p_e} \setminus w^{p_{1-e}}$ and every $a \in \mathcal{A}^{1-e}$, $c^q(\gamma, \cdot)$ divides a into two infinite parts, provided that

(**) $a \setminus w^{p_e}$ is infinite.

This is easily done and $p_0 \leq q$, $p_1 \leq q$, *provided that* $q \in \mathbf{P}$. For this the problematic part is c^q and, in particular, part (iv) of the definition of $\mathbf{P}$. So suppose $(a, b) \in \mathcal{P}^q$, e.g., $(a, b) \in \mathcal{P}^{p_0}$. Suppose also $\gamma^* \in w^q$ so that $\sup(a) \leq \gamma^* < \sup(b)$. If $\gamma^* \in w^{p_0}$, there is no problem, as $p_0 \in \mathbf{P}$. So let us assume $\gamma^* \in w^q \setminus w^{p_0} = w^{p_1} \setminus w^{p_0}$. If $a \setminus w^{p_1}$ or $b \setminus w^{p_1}$ is infinite, we are through in view of condition (*) in the definition of c^q. Let us finally assume $a \setminus w^{p_1}$ is finite. But $a \subseteq w^{p_0}$. Hence $a \setminus (w^{p_0} \cap w^{p_1})$ is finite and $\otimes_2$ implies it is empty, i.e., $a \subseteq w^{p_0} \cap w^{p_1}$. Similarly, $b \subseteq w^{p_0} \cap w^{p_1}$. So $h_{0,1} \restriction (a \cup b)$ is the identity. But $(a, b) \in \mathcal{P}^{p_0}$. But $h_{i,j}$ maps p_i onto p_j. Hence $(a, b) \in \mathcal{P}^{p_1}$. As $p_1 \in \mathbf{P}$, we get the desired conclusion. □

FACT E. $\Vdash_{\mathbf{P}}$ *"There is no $\underset{\sim}{c}$-monochromatic subset of λ of order-type $\omega_1 \times \omega$."*

Proof. Let p force the existence of a counterexample. Let G be $\mathbf{P}$-generic over V with $p \in G$. In $V[G]$ we can find $A \subseteq \lambda$ of order-type $\omega_1 \times \omega$ such that $\underset{\sim}{c}^G \restriction [A]^2$ is constant. Let $A = \bigcup_{n<\omega} A_n$ where $\mathrm{otp}(A_n) = \omega_1$ and $\sup(A_n) \leq \min(A_{n+1})$. We can replace A_n by any $A'_n \subseteq A_n$ of the same cardinality. Hence we may assume w.l.o.g.

$$(*)_1 \qquad A_n \in \mathcal{R}_{\omega_1} \quad \text{for } n < \omega.$$

Let $\delta_n = \sup(A_n)$ and

$$\beta_n = \min\{\,\beta : \delta_n \leq \beta < \lambda,\ d(\beta,\cdot) \text{ does not divide } \bigcup_{l\leq n} A_l \text{ into two infinite sets}\,\},$$

where $d = \underset{\sim}{c}^G$. Clearly $\beta_n \leq \min(A_{n+1})$. Hence $\beta_n < \beta_{n+1}$. Let $d_n \in \{0,1\}$ be such that $d(\beta_n, \gamma) = d_n$ for all but finitely many $\gamma \in \bigcup_{l\leq n} A_l$. Let u be an infinite subset of ω such that d_n is constant for $n \in u$ and $\{\beta_n : n \in u\} \in \mathcal{R}_\omega$. Let $A_l = \{\,\alpha^l_i : i < \omega_1\,\}$ in increasing order. So p forces all this on suitable names

$$\langle \underset{\sim}{\beta}_n : n < \omega \rangle,\ \langle \underset{\sim}{\alpha}^l_i : i < \omega_1 \rangle,\ \langle \underset{\sim}{\delta}_n : n < \omega \rangle.$$

As $\mathbf{P}$ is $\aleph_1$-complete, we can find $p_0 \in \mathbf{P}$ with $p \leq p_0$ so that p_0 forces $\underset{\sim}{\beta}_l = \beta_l$ and $\underset{\sim}{\delta}_n = \delta_n$ for some β_l and δ_n. We can choose inductively conditions $p_k \in \mathbf{P}$ such that $p_k \leq p_{k+1}$ and there are $i_k < j_k$ and α^l_i (for $i < j_k$) with

$$\begin{aligned} p_{k+1} \Vdash \text{“}&\alpha^l_{i_k} > \sup(w^{p_k} \cap \delta_l),\\ &\alpha^l_i \in w^{p_{k+1}} \text{ for } i < j_k,\\ &\{\,\alpha^l_i : i < i_k\,\} \subseteq \mathrm{cl}_{\mathbf{B}}\{\,\alpha^l_i : i_k < i < j_k\,\},\\ &\underset{\sim}{\alpha}^l_i = \alpha^l_i \text{ for } i < j_k,\\ &\underset{\sim}{c}(\beta_n, \alpha^l_i) = d_n \text{ for } l \leq n, i > i_0, \text{ and}\\ &\gamma \in [\delta_m, \beta_m) \cap w^{p_k} \text{ implies } \underset{\sim}{c}(\gamma,\cdot) \text{ divides}\\ &\qquad \{\,\alpha^l_i : i < j_k,\ l \leq m\,\} \text{ into two infinite sets”} \end{aligned}$$

(remember our choice of β_m). Let

$$\begin{aligned} l(*) &= \min(u)\\ a &= \{\,\alpha^l_i : l \leq l(*),\ i < \bigcup_k j_k\,\}\\ b &= \{\,\beta_l : l \in u\,\}\\ q &= (\bigcup_k w^{p_k}, \bigcup_k c^{p_k}, \bigcup_k P^{p_k} \cup \{(a,b)\}). \end{aligned}$$

Now $q \in \mathbf{P}$. To see that q satisfies condition (iv) of the definition of $\mathbf{P}$, let $\sup(a) \leq \gamma < \min(b)$. Then $\sup\{\,\alpha^{l(*)}_{i_k} : k < \omega\,\} \leq \gamma < \beta_{l(*)}$. But $\gamma \in w^q =$

$\bigcup_k w^{p_k}$, so for some k, $\gamma \in w^{p_k}$. This implies

$$\gamma \notin \left(\alpha^{l(*)}_{i_k+1}, \delta_{l(*)}\right),$$

whence $\gamma \geq \delta_{l(*)}$ and

$$\{\alpha^l_i : l \leq l(*),\ i < j_k\} \subseteq a,$$

which implies the needed conclusion.

Also $q \geq p_k \geq p$. But now, if $r \geq q$ forces a value to $\alpha^{l(*)}_{\bigcup_k j_k}$; we get a contradiction. □

Remark 1A. Note that the proof of Theorem 1 also gives the consistency of $\lambda \not\to [\omega_1 \times \omega]^2_{\aleph_0}$: replace "$c(\gamma, \cdot)$ divides a set x into two infinite parts" by "$c(\gamma, \cdot)$ gets all values on a set x."

§3. Generalizations to other cardinals.

How much does the proof of Theorem 1 depend on $\aleph_1$? Suppose we replace $\aleph_0$ by μ.

THEOREM 2. *Assume* $2^\mu = \mu^+ < \lambda = \lambda^\mu$ *and* $2 \leq \kappa \leq \mu$. *Then for some* μ^+*-complete* μ^{++}*-c.c. forcing notion* $\mathbf{P}$ *of cardinality* 2^μ*:*

$$\Vdash_{\mathbf{P}} 2^\mu = \lambda, \qquad \lambda \not\to [\mu^+ \times \mu]^2_\kappa.$$

Proof. Let $\mathbf{B}$ and $\mathcal{R}_\delta$ be defined as above (for $\delta \leq \mu^+$). Clearly

⊕ If $a \subseteq \lambda$ has no last element, then for some $\alpha \in a$, $a \setminus \alpha \in \bigcup_\delta \mathcal{R}_\delta$.

Hence, if $\delta = \mathrm{otp}(a)$ is additively indecomposable, then $a \setminus \alpha \in \mathcal{R}_\delta$ for some $\alpha \in a$.

Let $\mathbf{P}_\mu$ be the set of forcing conditions

$$(w, c, \mathcal{P})$$

where $w \subseteq \lambda$, $|w| \leq \mu$, $c : [w]^2 \to \kappa$, and $\mathcal{P}$ is a set of $\leq \mu$ pairs (a, b) such that

(i) a, b are subsets of w
(ii) $b \in \mathcal{R}_\mu$, and a is a finite union of members of $\bigcup_{\mu \leq \delta < \mu^+} \mathcal{R}_\delta$
(iii) $\sup(a) < \min(b)$
(iv) if $\sup(a) \leq \gamma < \min(b)$, $\gamma \in w$, then the function $c(\gamma, \cdot)$ gets all values ($< \kappa$) on a or on b.

With the same proof as above we get

$$\mathbf{P}_\mu \text{ satisfies the } \mu^{++}\text{-c.c.},$$
$$\mathbf{P}_\mu \text{ is } \mu^+\text{-complete},$$

(so cardinal arithmetic is clear) and

$$\Vdash_{\mathbf{P}_\mu} \lambda \not\to [\mu^+ \times \mu]^2_\kappa.$$

□

What about replacing μ^+ by an inaccessible θ? We can manage by demanding

$$\{\, a \cap (\alpha,\beta) : (a,b) \in \mathcal{P},\ \bigcup_n \mathrm{otp}(a \cap (\alpha,\beta)) \times n = \mathrm{otp}(a)$$
$$(\alpha,\beta) \text{ maximal under these conditions} \,\}$$

is free (meaning there are pairwise disjoint end segments) and by taking care in defining the order. Hence the completeness drops to θ-strategical completeness. This is carried out in Theorem 3 below.

THEOREM 3. *Assume $\theta = \theta^{<\theta} > \aleph_0$ and $\lambda = \lambda^{<\theta}$. Then for some θ^+-c.c. θ-strategically complete forcing $\mathbf{P}$, $|\mathbf{P}| = \lambda$ and*

$$\Vdash_{\mathbf{P}} 2^\theta = \lambda,\ \lambda \not\to (\theta \times \theta)^2_2.$$

Proof. For W a family of subsets of λ, each with no last element, let

$$\mathrm{Fr}(W) = \{\, f : f \text{ is a choice function on } W \text{ such that}$$
$$\{\, a \setminus f(a) : a \in W \,\} \text{ are pairwise disjoint} \,\}.$$

If $\mathrm{Fr}(W) \neq \emptyset$, W is called *free*.

Let $\mathbf{P}_{<\theta}$ be the set of forcing conditions

$$(w, c, \mathcal{P}, W)$$

where $w \subseteq \lambda$, $|w| < \theta$, $c : [w]^2 \to \{\text{red}, \text{green}\}$, W is a free family of $< \theta$ subsets of w, each of which is in $\bigcup_{\delta<\theta} \mathcal{R}_\delta$, and $\mathcal{P}$ is a set of $< \theta$ pairs (a,b) such that

(i) a, b are subsets of w

(ii) $b \in \mathcal{R}_\omega$

(iii) $\sup(a) < \min(b)$ and for some $\delta_0 < \delta_1 < \cdots < \delta_n$, $\delta_0 < \min(a)$, $\sup(a) \leq \delta_n$, $a \cap [\delta_l, \delta_{l+1}) \in W$

(iv) if $\sup(a) \leq \gamma < \min(b)$, $\gamma \in w$, then $c(\gamma, \cdot)$ divides a or b into two infinite sets.

We order $\mathbf{P}_{<\theta}$ as follows:

$$p \leq q \text{ iff } w^p \subseteq w^q,\ c^p \subseteq c^q,\ \mathcal{P}^p \subseteq \mathcal{P}^q,\ W^p \subseteq W^q \text{ and every}$$
$$f \in \mathrm{Fr}(W^p) \text{ can be extended to a member of } \mathrm{Fr}(W^q).$$

□

§4. A provable partition relation.

CLAIM 4. *Suppose $\theta > \aleph_0$, $n, r < \omega$, and $\lambda = \lambda^{<\theta}$. Then*

$$(\lambda^+)^r \times n \to (\theta \times n, \theta \times r)^2_2.$$

Proof. We prove this by induction on r. Clearly the claim holds for $r = 0, 1$. So w.l.o.g. we assume $r \geq 2$. Let c be a 2-place function from $(\lambda^+)^r \times n$ to $\{\text{red}, \text{green}\}$. Let $\chi = \beth_2(\lambda)^+$. Choose by induction on l a model N_l such that

$$N_l \prec (H(\chi), \in, <^*),$$

$|N_l| = \lambda$, $\lambda + 1 \subseteq N_l$, $N_l^{<\theta} \subseteq N_l$, $c \in N_l$ and $N_l \in N_{l+1}$. Here $<^*$ is a well-ordering of $H(\chi)$. Let

$$A_l = \left[(\lambda^+)^r \times l,\ (\lambda^+)^r \times (l+1)\right),$$

and let $\delta_l \in A_l \setminus N_l$ be such that $\delta_l \notin x$ whenever $x \in N_l$ is a subset of A_l and $\text{otp}(x) < (\lambda^+)^r$. W.l.o.g. we have $\delta_l \in N_{l+1}$. Now we shall show

(*) If $Y \in N_0$, $Y \subseteq A_m$, $|Y| = \lambda^+$ and $\delta_m \in Y$,
then we can find $\beta \in Y$ such that $c(\beta, \delta_l) = \text{red}$ for all $l < n$.

Why does (*) suffice? Assume (*) holds. We can construct by induction on $i < \theta$ and for each i by induction on $l < n$ an ordinal $\alpha_{i,l}$ such that

(a) $\alpha_{i,l} \in A_l$ and $j < i \Rightarrow \alpha_{j,l} < \alpha_{i,l}$
(b) $\alpha_{i,l} \in N_0$
(c) $c(\alpha_{i,l}, \delta_m) = \text{red}$ for $m < n$
(d) $c(\alpha_{i,l}, \alpha_{i_1,l_1}) = \text{red}$ when $i_1 < i$ or $i_1 = i \,\&\, l_1 < l$.

Accomplishing this suffices as $\alpha_{i,l} \in A_l$ and

$$l < m \Rightarrow \sup A_l \leq \min A_m.$$

Arriving in the inductive process at (i, l), let

$$Y = \{\beta \in A_l : c(\beta, \alpha_{j,m}) = \text{red} \quad \text{if } j < i,\ m < n, \text{ or } j = i,\ m < l\}.$$

Now clearly $Y \subseteq A_l$. Also $Y \in N_0$ as all parameters are from N_0, their number is $< \theta$ and $N_0^{<\theta} \subseteq N_0$. Also $\delta_l \in Y$ by the induction hypothesis (and $\delta_l \in A_l$). So by (*) we can find $\alpha_{i,l}$ as required.

Proof of (*). $Y \nsubseteq N_0$, because $\delta_m \in Y$ and $Y \in N_0$. As $|Y| = \lambda^+$, we have $\text{otp}(Y) \geq \lambda^+$. But $\lambda^+ \to (\lambda^+, \theta)^2$, so there is $B \subseteq Y$ such that $|B| = \lambda^+$ and $c \restriction B \times B$ is constantly red or there is $B \subseteq Y$ such that $|B| = \theta$ and $c \restriction B \times B$ is constantly green. In the former case we get the conclusion of the claim. In the latter case we may assume $B \in N_0$, hence $B \subseteq N_0$, and let $k \leq n$ be maximal such that

$$B' = \{\xi \in B : \bigwedge_{l<k} c(\delta_l, \xi) = \text{red}\}$$

has cardinality θ. If $k = n$, any member of B' is as required in (*). So assume $k < n$. Now $B' \in N_k$, since $B \in N_0 \prec N_k$ and $\{N_l, A_l\} \in N_k$ and $\delta_l \in N_k$ for $l < k$. Also

$$\{\xi \in B' : c(\delta_k, \xi) = \text{red}\}$$

is a subset of B' of cardinality $< \theta$ by the choice of k. So for some $B'' \in N_0$, $c \restriction \{\delta_k\} \times (B' \setminus B'')$ is constantly green (e.g., as $B' \subseteq N_0$, and $N_0^{<\theta} \subseteq N_0$). Let

$$Z = \{\delta \in A_k : c \restriction \{\delta\} \times (B' \setminus B'') \text{ is constantly green}\}$$

and

$$Z' = \{\delta \in Z : (\forall \alpha \in B' \setminus B'')(\delta < \alpha \Leftrightarrow \delta_k < \alpha)\}.$$

So $Z \subseteq A_k$, $Z \in N_k$, $\delta_k \in Z$ and therefore $\text{otp}(Z) = \text{otp}(A_k) = (\lambda^+)^r$. Note that $k \neq m \Rightarrow Z' = Z$ and $k = m \Rightarrow Z' = Z \setminus \sup(B' \setminus B'')$, so Z' has the same properties. Now we apply the induction hypothesis; one of the following holds (note that we can interchange the colours): (a) There is $Z'' \subseteq Z'$, $\text{otp}(Z'') = \theta \times n$, $c \restriction Z'' \times Z''$ is constantly red, w.l.o.g. $Z'' \in N_k$, or (b) there is $Z'' \subseteq Z'$, $\text{otp}(Z'') = \theta \times (r-1)$, $c \restriction Z'' \times Z''$ green and w.l.o.g. $Z'' \in N_k$. If (a), we are done; if (b), $Z'' \cup (B' \setminus B'')$ is as required. □

Remark 4A. So $(\lambda^+)^{n+1} \to (\theta \times n)^2$ for $\lambda = \lambda^{<\theta}$, $\theta = \text{cf}(\theta) > \aleph_0$ (e.g., $\lambda = 2^{<\theta}$).

Remark 4B. Suppose $\lambda = \lambda^{<\theta}$, $\theta > \aleph_0$. If c is a 2-colouring of $(\lambda^{+r})^s \times n$ by k colours and every subset of it of order type $(\lambda^{+(r-1)})^s \times n$ has a monochromatic subset of order type θ for each of the colours, one of the colours being red, then by the last proof we get

(a) There is a monochromatic subset of order type $\theta \times n$ and of colour red *or*
(b) There is a colour d and a set Z of order type $(\lambda^{+r})^s$ and a set B of order type θ such that $B < Z$ or $Z < B$ and

$$\{(\alpha, \beta) : \alpha \in B,\ \beta \in Z \text{ or } \alpha \neq \beta \in B\}$$

are all coloured with d.

So we can prove that for 2-colourings by k colours c

$$(\lambda^{+r})^s \times n \to (\theta \times n_1, \ldots, \theta \times n_k)^2$$

when r, s, n are sufficiently large (e.g., $n \geq \min\{n_l : l = 1, \ldots, k,\ s \geq \sum_{l=1}^k n_l\}$) by induction on $\sum_{l=1}^k n_l$.

Note that if c is a 2-colouring of λ^{+2k}, then for some $l < k$ and $A \subseteq \lambda^{+2k}$ of order type $\lambda^{+(2l+2)}$ we have

(*) If $A' \subseteq A$, $\text{otp}(A') = \lambda^{+2l}$, and d is a colour which appears in A, then there is $B \subseteq A'$ of order type θ such that B is monochromatic of colour d.

We can conclude $\lambda^{+2k} \to (\theta \times n)_k^2$.

REFERENCES

[BH] J. BAUMGARTNER and A. HAJNAL, in preparation.

[EH] P. ERDŐS and A. HAJNAL, *Unsolved problems in set theory*, ***Axiomatic set theory***, Proceedings of symposia in pure mathematics, vol. XIII part I, American Mathematical Society, Providence 1971, pp. 17–48.

Institute of Mathematics
The Hebrew University
Jerusalem, Israel

Department of Mathematics
Rutgers University
New Brunswick, NJ, USA

ON THE STRUCTURE OF GAMMA DEGREES[1]

ALAN P. SILVER

In [4], Ladner, Lynch and Selman defined a non-deterministic polynomial time bounded version of many-one reducibility, denoted by $\leq_{m}^{\mathcal{NP}}$, as follows:

> For any sets A and B, we say $A \leq_{m}^{\mathcal{NP}} B$ if and only if there is a non-deterministic Turing transducer, M, that runs in polynomial time such that, $x \in A$ just in case there is a y, computed by M on input x, with $y \in B$.

However, their definition does not completely capture the essence of a many-one reducibility due to the fact that, given some $x \in A$, there only needs to be *some* y output by M such that $y \in B$. It may be the case that all computation branches of M halt on input x, but only one of the output values is actually in B. Seen in this light, their reducibility is obviously a candidate for a polynomial time bounded singleton reducibility rather than a many-one reducibility. This intuitive idea is borne out by the fact that, if we define $\leq_{s}^{\mathcal{P}}$, a polynomial time bounded version of singleton reducibility (see [7] for details), then $\leq_{m}^{\mathcal{NP}} \equiv \leq_{s}^{\mathcal{P}}$. We claim that gamma reducibility is the correct notion of a non-deterministic polynomial time bounded many-one reducibility and study its properties.

Gamma reducibility, $\leq_{\gamma}$, was first defined by Adleman and Manders [1] in 1977, and was later studied by Long in [5] and [6]. Its relevance to the world of time bounded computations comes, in part, from the fact that $\mathcal{P} = \mathcal{NP} \Rightarrow [\leq_{\gamma} \equiv \leq_{m}^{\mathcal{P}}] \Rightarrow \mathcal{P} = \mathcal{NP} \cap \text{co-}\mathcal{NP}$ and that if A is an $\mathcal{NP}$-complete set (with respect to $\leq_{\gamma}$), then $A \in \mathcal{NP} \cap \text{co-}\mathcal{NP} \Leftrightarrow \mathcal{NP} = \text{co-}\mathcal{NP}$.

DEFINITION 1 (Adelman and Manders [1]). $A \leq_{\gamma} B$ if there is a non-deterministic polynomial time bounded transducer, M, such that if

$$G(M) = \{\langle x, y\rangle \mid \text{On input } x,\ M \text{ outputs } y \text{ on some computation branch}\}$$

then:

(I) $\quad (\forall x)(\exists y)[\langle x, y\rangle \in G(M)]$

(II) $\quad (\forall x)(\forall y)[\langle x, y\rangle \in G(M) \Rightarrow [x \in A \Leftrightarrow y \in B]]$

Trivially, $A \leq_{\gamma} B \Leftrightarrow \overline{A} \leq_{\gamma} \overline{B}$.

---

[1] Research undertaken as part of the author's doctoral thesis [7], supported by the UK Science and Engineering Research Council, award number 88002274. The author wishes to thank his supervisor, Dr. Barry Cooper, and also Prof. Klaus Ambos-Spies for much useful advice in the preparation of this work.

We note without proof that $\langle \mathcal{D}_\gamma, \leq_\gamma \rangle$ is an upper semilattice with $A \oplus B$ as the least upper bound of A and B, that the $\mathcal{NP}$ γ-degrees form an ideal of $\langle \mathcal{D}_\gamma, \leq_\gamma \rangle$ and that the zero degree, $\mathbf{0}_\gamma$, is exactly $\mathcal{NP} \cap \text{co-}\mathcal{NP}$.

DEFINITION 2.

i) For any two sets, A and B, we say that A and B *differ finitely*, $A =^* B$, if and only if $A - B \cup B - A$ is finite.

ii) A class of sets $\mathcal{C}$ is *closed under finite variations* (c.f.v.) if, for $A \in \mathcal{C}$ and $B =^* A$, we have $B \in \mathcal{C}$.

iii) A class of recursive sets, $\mathcal{C}$, is *recursively presentable* (r.p.) if $\mathcal{C}$ is empty or there exists a recursive set U, called the *universal set for* $\mathcal{C}$, such that

$$\mathcal{C} = \{U^e \mid e \in \mathbb{N}\} \quad \text{where } U^e = \{x \mid \langle e, x \rangle \in U\}.$$

Note that if $\mathcal{C}$ is any recursively presentable class of sets, say $\mathcal{C} = \{A_i\}_{i \in \mathbb{N}}$, then for each i, $A_i \leq^p_m U$ via $\lambda x[\langle i, x \rangle]$, where U is the universal set for $\mathcal{C}$. Thus $A_i \leq_\gamma U$.

It is simple to show that if A is a recursive set, then $deg_\gamma(A)$ contains only recursive sets, and that the recursive γ-degrees form an ideal of the γ-degrees and are c.f.v. and r.p.

Let f be a strictly increasing function where $f(0) \neq 0$, and define f^n by $f^0(m) = m$ and $f^{n+1}(m) = f(f^n(m))$. The $(n+1)th$ f*-interval* is $I^f_n = \{x \mid f^n(0) \leq |x| < f^{n+1}(0)\}$, and we note that $\{I^f_n\}_{n \in \mathbb{N}}$ partitions $\{0,1\}^*$. Furthermore, if A is recursive, then let $I^f_A = \bigcup\{I^f_n \mid n \in A\}$.

Ambos-Spies defines a function as *polynomially honest* if f is recursive and there is some polynomial, p, such that for all x, we can compute $f(x)$ in less than $p(f(x))$ steps. This is clearly *not* equivalent to Homer's original definition of polynomial honesty. To avoid confusion, we will refer to a function satisfying Ambos-Spies's definition as *polynomially honest II*. Clearly, if f is polynomially honest II and $A \in \mathcal{P}$ (so A is recursive), then $I^f_A \in \mathcal{P}$. A function, g, *dominates* f if $(\forall n)[f(n) < g(n)]$. Note that if f is recursive, then there is some strictly increasing polynomially honest II function, g, that dominates f (see [2]).

A recursive set $A \notin \mathcal{P}$ is *super sparse II* if there is a strictly increasing polynomially honest II function f such that $A \subseteq Z_f = \{0^{f(n)} \mid n \in \mathbb{N}\}$ and "$0^{f(n)} \in A$?" can be answered in less than $f(n+1)$ steps. As above, we call this super sparse II to distinguish it from Hartmanis's well-known notion of super sparseness.

Finally, let

$$\begin{aligned} A^{(n)} &= \{x \mid \langle n, x \rangle \in A\} \\ A^{(\leq n)} &= \{\langle m, x \rangle \mid m \leq n \ \& \ \langle m, x \rangle \in A\} \\ kA &= \{kx \mid x \in A\} \\ kA + i &= \{kx + i \mid x \in A\} \end{aligned}$$

LEMMA 3 (Ambos-Spies [2]). *There exists a super sparse II set in EXP-TIME.*

PROOF. Let $f(0) = 1$ and $f(n+1) = 2^{f(n)}$. We will construct A as a subset of $\{0^{f(2n+1)} \mid n \in \mathbb{N}\}$.

Let $\{R_i\}_{i\in\mathbf{N}}$ be an enumeration of the sets in $\mathcal{P}$. Given some n, perform 2^n steps of the following algorithm:

i) Find the largest $m \leq n$ such that $f(2m+1) \leq n$.
ii) If $f(2m+1) < n$ then $0^n \notin A$. Otherwise;
iii) Set $0^n \in A \Leftrightarrow 0^n \notin R_k$ where $m = \langle k, l\rangle$.

If the computation does not terminate in 2^n steps, then $0^n \notin A$.

It is clear that $A \in \text{EXPTIME}$. Suppose, if possible, that $A \in \mathcal{P}$, then there is some $k \in \mathbf{N}$ such that $A = R_k$.

Now, consider n such that there is some m where $f(2m+1) = n$. Clearly, since there is a polynomial, p, such that $f(n+1)$ can be computed in less than $p(f(n))$ steps, then $f(2m+1)$ can be computed in less than $p(f(2m))$ steps. Thus, for sufficiently large n, $f(2m+1)$ can be computed in less than 2^n steps.

So, for large n we will have that $0^n \in A \Leftrightarrow 0^n \notin R_k$ where $m = \langle k, l\rangle$ and $f(2m+1) = n$.

Therefore, given k we will have, for sufficiently large l:

$$0^{f(2\langle k,l\rangle+1)} \in A \quad \Leftrightarrow \quad 0^{f(2\langle k,l\rangle+1)} \notin R_k$$

Thus $A \neq R_k$ and so $A \notin \mathcal{P}$.

Finally we note that $(\forall n)[\ 2^n \leq f(n+1)\ ]$, so "$0^{f(n)} \in A$?" can be answered in less than $f(n+1)$ steps.

Thus A is super sparse II. □

THE JOIN LEMMA (Ambos-Spies [2]). *Let C_0 and C_1 be any recursive sets and $\mathcal{C}_0$, $\mathcal{C}_1$ be r.p. and c.f.v. classes such that $C_0 \cup C_1 \notin \mathcal{C}_0$ and $C_1 \notin \mathcal{C}_1$. Then there is a recursive function, g_0, such that if g is a strictly increasing function that dominates g_0, and A is an infinite coinfinite recursive set, then $(C_0 \cap I_A^g) \cup C_1 \notin \mathcal{C}_0 \cup \mathcal{C}_1$.*

THE DENSITY THEOREM. *Let* **a** *and* **b** *be the γ-degrees of recursive sets such that* **b** < **a**. *Then there is a recursive γ-degree,* **d** *such that* **b** < **d** < **a**.

PROOF. Fix sets $A = \{2x \mid x \in A_0\}$ and $B = \{2x+1 \mid x \in B_0\}$ for some $A_0 \in \mathbf{a}$ and $B_0 \in \mathbf{b}$. Clearly A and B are recursive and γ-equivalent to A_0 and B_0 respectively.

Now suppose, if possible, that $A \cup B \in \mathbf{b}$, i.e., $A \cup B \leq_\gamma B$. Then there is some non-deterministic polynomial time bounded transducer, M, such that:

(I) $(\forall x)(\exists y)[\,\langle x, y\rangle \in G(M)]$

(II) $(\forall x)(\forall y)[\,\langle x, y\rangle \in G(M) \Rightarrow [x \in A \cup B \Leftrightarrow y \in B]]$

Note that $x \in A \cup B \Leftrightarrow (x = 2m \;\&\; x \in A) \vee (x = 2m+1 \;\&\; x \in B)$.

Define a non-deterministic machine, N, such that on input x, N checks if $x = 2m+1$. If so then N outputs some fixed $b \in \overline{B}$, otherwise N simulates M on input x and outputs $M(x)$.

Clearly, as M is polynomial time bounded, then N is too, and we have $(\forall x)(\exists y)[\langle x, y\rangle \in G(N)]$.

Now, suppose $\langle x, y\rangle \in G(N)$, then

i) $x = 2m+1$. Then $y = b$ and $x \notin A$ as $A \subseteq 2\mathbf{N}$.

Thus $x \notin A$ & $y \notin B$.

ii) $x = 2m$. Then if $x \in A$ we must have $y \in B$, and if $x \notin A$ then as $x \notin B \subseteq 2\mathsf{N}+1$, we must have $y \notin B$.

Thus $x \in A \Leftrightarrow y \in B$, i.e., $A \leq_\gamma B$ via N. This contradicts the fact that $A <_\gamma B$, so we must have $A \cup B \notin \mathbf{b}$.

Also, since $B \notin \mathbf{a}$, then we can apply the Join Lemma with $C_0 = A$, $C_1 = B$, $\mathcal{C}_0 = \mathbf{b}$ and $\mathcal{C}_1 = \mathbf{a}$. This will give the function g_0, so consider some strictly increasing polynomially honest II function, g, that dominates g_0. Clearly $2\mathsf{N}$ is infinite, coinfinite and recursive, so the Join Lemma gives that:

$$(A \cap I^g_{2\mathsf{N}}) \cup B \;\notin\; \mathbf{b} \cup \mathbf{a}$$

Now, let $D = (A \cap I^g_{2\mathsf{N}}) \cup B$.

If $B \leq_\gamma D \leq_\gamma A$, then we have $B <_\gamma D <_\gamma A$ as the Join Lemma ensures that $D \notin \mathbf{a}, \mathbf{b}$. Thus we have $\mathbf{b} < \mathbf{d} < \mathbf{a}$.

Claim. $B \leq_\gamma (A \cap I^g_{2\mathsf{N}}) \cup B \leq_\gamma A$.

Proof. We show that $B \leq^p_m (A \cap I^g_{2\mathsf{N}}) \cup B \leq^p_m A$, which will prove the claim. Pick some $a \in 2\mathsf{N} \cap I^g_{2\mathsf{N}+1}$.

Now, define $f(x) = \begin{cases} x & \text{if } x = 2m+1 \text{ some } m \\ a & \text{if } x = 2m \text{ some } m \end{cases}$

Clearly f is polynomial time computable. We consider three cases:

i) $x \in B \Rightarrow x = 2m+1 \Rightarrow f(x) = x \Rightarrow f(x) \in B$

$$\Rightarrow f(x) \in (A \cap I^g_{2\mathsf{N}}) \cup B$$

ii) $x \notin B$ & $x = 2m+1 \Rightarrow f(x) = x = 2m+1$

$$\Rightarrow f(x) \notin B \;\&\; f(x) \in 2\mathsf{N}+1 \Rightarrow f(x) \notin B \;\&\; f(x) \notin A$$
$$\Rightarrow f(x) \notin (A \cap I^g_{2\mathsf{N}}) \cup B$$

iii) $x \notin B$ & $x = 2m \Rightarrow f(x) \in 2\mathsf{N}$ & $f(x) \in I^g_{2\mathsf{N}+1}$

$$\Rightarrow f(x) \notin B \;\&\; f(x) \notin I^g_{2\mathsf{N}} \Rightarrow f(x) \notin (A \cap I^g_{2\mathsf{N}}) \cup B$$

Thus, $B \leq^p_m (A \cap I^g_{2\mathsf{N}}) \cup B$ via f.

Further, pick some $b \notin A \oplus B$ and define

$$g(x) = \begin{cases} 2x & \text{if } x = 2m \;\&\; x \in I^g_{2\mathsf{N}} \\ b & \text{if } x = 2m \;\&\; x \in I^g_{2\mathsf{N}+1} \\ 2x+1 & \text{if } x = 2m+1 \end{cases}$$

We consider three cases:

i) $x \in 2\mathsf{N}+1$
So $g(x) = 2x+1$, and so $x \in B \Leftrightarrow g(x) \in A \oplus B$. Since $A \subseteq 2\mathsf{N}$ then $x \in B \Leftrightarrow x \in (A \cap I^g_{2\mathsf{N}}) \cup B$. Thus $x \in (A \cap I^g_{2\mathsf{N}}) \cup B \Leftrightarrow g(x) \in A \oplus B$.

ii) $x \in 2\mathsf{N} \cap I^g_{2\mathsf{N}}$
So $g(x) = 2x$ and thus $x \in A \Leftrightarrow g(x) \in A \oplus B$. Again, as $B \subseteq 2\mathsf{N}+1$, then for $x \in 2\mathsf{N} \cap I^g_{2\mathsf{N}}$, we have $x \in A \Leftrightarrow x \in (A \cap I^g_{2\mathsf{N}}) \cup B$.
So $x \in (A \cap I^g_{2\mathsf{N}}) \cup B \Leftrightarrow g(x) \in A \oplus B$.

iii) $x \in 2\mathbb{N}$ & $x \in I^g_{2\mathbb{N}+1}$

So $g(x) = b$. Clearly $x \notin (A \cap I^g_{2\mathbb{N}}) \cup B$ and $g(x) \notin A \oplus B$.

Thus, $x \in (A \cap I^g_{2\mathbb{N}}) \cup B \Leftrightarrow g(x) \in A \oplus B$ follows immediately.

Thus, $(A \cap I^g_{2\mathbb{N}}) \cup B \leq^{\mathcal{P}}_{m} A \oplus B$ via g. Since $\mathbf{b} < \mathbf{a}$, then $A \equiv^{\mathcal{P}}_{m} A \oplus B$.

Thus $(A \cap I^g_{2\mathbb{N}}) \cup B \leq_\gamma A$ and this completes the proof. □

We can extend the proof of the Density Theorem to show that between any two distinct comparable recursive γ-degrees there exist two incomparable γ-degrees whose supremum is the higher degree. This result can then be further extended to give infinitely many incomparable degrees whose supremum is the higher one. Together with repeated applications of the Density Theorem this shows that, given recursive γ-degrees $\mathbf{b} < \mathbf{a}$, there are infinitely many distinct comparable and incomparable γ-degrees between them.

COMBINED SPLITTING AND DENSITY THEOREM. *Given any two recursive γ-degrees $\mathbf{a}$ and $\mathbf{b}$ with $\mathbf{b} < \mathbf{a}$, there exist recursive γ-degrees $\mathbf{d}_0$ and $\mathbf{d}_1$ such that:*

i) $\mathbf{b} < \mathbf{d}_i < \mathbf{a}$ *for* $i = 0, 1$

ii) $\mathbf{a}$ *is the least upper bound of* $\mathbf{d}_0$ *and* $\mathbf{d}_1$ *i.e.,* $\mathbf{a} = \mathbf{d}_0 \oplus \mathbf{d}_1$.

PROOF. As in the Density Theorem, fix A and B to be in $\mathbf{a}$ and $\mathbf{b}$ respectively, where $A \subseteq 2\mathbb{N}$ and $B \subseteq 2\mathbb{N}+1$. As before, $A \cup B \notin \mathbf{b}$ and $B \notin \mathbf{a}$, so apply the Join Lemma with $C_0 = A$, $C_1 = B$, $\mathcal{C}_0 = \mathbf{b}$ and $\mathcal{C}_1 = \mathbf{a}$.

We will obtain a recursive function, g_0, so let g be any strictly increasing polynomially honest II function that dominates g_0.

Let $D_0 = (A \cap I^g_{2\mathbb{N}}) \cup B$ and $\mathbf{d}_0 = deg_\gamma(D_0)$

$D_1 = (A \cap I^g_{2\mathbb{N}+1}) \cup B$ and $\mathbf{d}_1 = deg_\gamma(D_1)$.

Proof of i). By a similar argument to the Density Theorem we have that $B \leq_\gamma D_i \leq_\gamma A$ for $i = 0, 1$. By the Join Lemma, $D_i \notin \mathbf{a}, \mathbf{b}$, so $\mathbf{b} < \mathbf{d}_0$, $\mathbf{d}_1 < \mathbf{a}$.

Proof of ii). By our choice of A and B, $A \cup B \equiv_\gamma A \oplus B$, so for any set D,

$$(A \cap D) \cup B \equiv_\gamma (A \cap D) \oplus B.$$

Thus we have that $D_0 \equiv_\gamma (A \cap I^g_{2\mathbb{N}}) \oplus B$ and $D_1 \equiv_\gamma (A \cap I^g_{2\mathbb{N}+1}) \oplus B$.

We need to show that $[(A \cap I^g_{2\mathbb{N}}) \oplus B] \oplus [(A \cap I^g_{2\mathbb{N}+1}) \oplus B] \equiv_\gamma A \cup B$.

Pick some $a \notin A \cup B$. Define

$$f(x) = \begin{cases} a & \text{if } x = 2m \ \&\ m = 2n \ \&\ n \in I^g_{2\mathbb{N}+1} \\ a & \text{if } x = 2m+1 \ \&\ m = 2n \ \&\ n \in I^g_{2\mathbb{N}} \\ \left[\frac{[\frac{x}{2}]}{2}\right] & \text{otherwise} \end{cases}$$

Clearly f is polynomial time computable as $I^g_{2\mathbb{N}}, I^g_{2\mathbb{N}+1} \in \mathcal{P}$, and by definition, $D_0 \oplus D_1 \leq^{\mathcal{P}}_{m} A \cup B$ via f. Similarly, we can find a function to witness the reverse reduction and thus $D_0 \oplus D_1 \equiv_\gamma A \cup B$. However, $A \cup B \equiv_\gamma A \oplus B$ by our choice of A and B and since $\mathbf{b} < \mathbf{a}$ then we have $A \oplus B \equiv_\gamma A$. Thus we have $D_0 \oplus D_1 \equiv_\gamma A$, i.e., $\mathbf{d}_0 \oplus \mathbf{d}_1 = \mathbf{a}$. □

COROLLARY 7. *Given any two recursive γ-degrees* **a** *and* **b** *where* $\mathbf{b} < \mathbf{a}$, *and some* $n \in \mathsf{N}$, *there exist* $n+1$ *pairwise incomparable recursive* γ*-degrees* $\mathbf{d}_i$ *(for* $0 \le i \le n$*) such that:*

i) $(\forall i \le n)[\ \mathbf{b} < \mathbf{d}_i < \mathbf{a}\ ]$,
ii) **a** *is the least upper bound of the* $\mathbf{d}_i$.

PROOF. Pick some $n \in \mathsf{N}$. Follow the proof of Theorem 6, and define for all $i < n$ the set $D_i = (A \cap I^g_{n\mathsf{N}+i}) \cup B$.

Now, pick $a \notin 2\mathsf{N} \cap I^g_{n\mathsf{N}+j}$ where $j = (i+1) \bmod n$. Define

$$f_i(x) = \begin{cases} x & \text{if } x = 2m+1 \text{ some } m \\ a & \text{if } x = 2m \text{ some } m \end{cases}$$

Clearly, f_i is polynomial time bounded and, by a similar proof to before, we can see that $(\forall x)[\ x \in B \Leftrightarrow f_i(x) \in (A \cap I^g_{n\mathsf{N}+i}) \cup B\ ]$. Thus $B \le_\gamma D_i$ via f_i. Now, pick $b \notin A \oplus B$. Further define

$$g_i(x) = \begin{cases} 2x+1 & \text{if } x = 2m+1 \text{ some } m \\ 2x & \text{if } x = 2m \text{ some } m \text{ and } x \in I^g_{n\mathsf{N}+i} \\ b & \text{if } x = 2m \text{ some } m \text{ and } x \notin I^g_{n\mathsf{N}+i} \end{cases}$$

Again, g_i is polynomial time bounded as $I^g_{n\mathsf{N}+i} \in \mathcal{P}$. Further, $(A \cap I^g_{n\mathsf{N}+i}) \cup B \le^{\mathcal{P}}_m A \oplus B$ via g_i. Thus, $B \le_\gamma D_i \le_\gamma A \oplus B$ so, as before, $B <_\gamma D_i <_\gamma A$. Now, by adapting the proof of Theorem 6 in the obvious way, we see that $A \equiv_\gamma D_0 \oplus D_1 \oplus \cdots \oplus D_n$.

So, given any $n \in \mathsf{N}$, there exist $n+1$ incomparable recursive γ-degrees satisfying Theorem 6. □

It is immediate from the Density Theorem that minimal γ-degrees do not exist, and so we turn our attention to the question of minimal pairs of γ-degrees. We first show that minimal pairs do exist, and then show that every non-zero γ-degree below a super sparse II set is half of a minimal pair of γ-degrees.

THEOREM 8. *If A is super sparse II, then the γ-degrees of A and $\overline{A}$ form a minimal pair of γ-degrees.*

PROOF. Let A be a super sparse II set via the function f, and suppose that $B \le_\gamma A$ via M_{e_0} and $B \le_\gamma \overline{A}$ via M_{e_1}. Thus:

$$(\forall x)(\forall y)(\forall z)[[\ \langle x,y\rangle \in G(M_{e_0})\ \&\ \langle x,z\rangle \in G(M_{e_1})\ ] \Rightarrow y \neq z\ ] \qquad [*]$$

To test membership of B, use the following algorithm:

i) On input x, compute y such that $\langle x,y\rangle \in G(M_{e_0})$.
ii) If $y \neq 0^{f(n)}$ for any $n \in \mathsf{N}$ then $y \notin A$ and so $x \notin B$. Halt.
Note that to test if "$y \notin Z_f$?", we know that M_{e_0} is non-deterministic polynomial time bounded, so if $\langle x,y\rangle \in G(M_{e_0})$, then $|y| \le p_{e_0}(|x|)$. Thus, $f(n) \le p_{e_0}(|x|)$ and, since f is strictly increasing we have $n \le f(n)$. So to check if $y = 0^{f(n)}$ requires calculating $f(k)$ for $0 \le k \le f(n) \le p_{e_0}(|x|)$.

Thus, we must make a maximum of $p_{e_0}(|x|)$ calculations. Now, f is polynomially honest II, so $f(k)$ can be calculated in less than $r(f(k))$ steps, for some polynomial r.
Since $f(k) \leq f(n) \leq p_{e_0}(|x|)$, then $f(k)$ can be calculated in less than $r(p_{e_0}(|x|))$ steps.
Thus, testing if "$y \notin Z_f$?" can be performed in less than $p_{e_0}(|x|)\cdot r(p_{e_0}(|x|))$ steps.

iii) Otherwise compute z such that $\langle x, z\rangle \in G(M_{e_1})$.

iv) If $z \neq 0^{f(m)}$ for any $m \in \mathbf{N}$ then $z \notin A$ and so $x \in B$. Halt. As in step ii), this can be performed in polynomial time.

v) So now we know that $y = 0^{f(n)}$ and $z = 0^{f(m)}$. By [*], we know that $n \neq m$.

a) *Case 1.* $n < m$.
Now $x \in B \Leftrightarrow y \in A \Leftrightarrow 0^{f(n)} \in A$. Since A is super sparse II then "$0^{f(n)} \in A$?" can be answered in less than $f(n+1)$ steps. f is strictly increasing and $n < m$, so this can be done in less than $f(m) = |z|$ steps. Now M_{e_1} is polynomial time bounded, so for $\langle x, z\rangle \in G(M_{e_1})$ we must have $|z| \leq p_{e_1}(|x|)$. Thus, "$x \in B$?" can be answered in less than $p_{e_1}(|x|)$ steps.

b) *Case 2.* $m < n$.
Now, $x \in B \Leftrightarrow z \in \overline{A} \Leftrightarrow 0^{f(m)} \in \overline{A}$. As above, since A is super sparse II and $m < n$, then "$0^{f(m)} \in A$?" can be answered in less than $f(n) = |y| \leq p_{e_0}(|x|)$ steps. Thus "$x \in \overline{B}$?" can be answered in less than $p_{e_0}(|x|)$ steps, so "$x \in B$?" can too.

Clearly, steps i) and iii) can be performed in non-deterministic polynomial time, and as shown, the other steps can be performed in deterministic polynomial time. Thus $B \in \mathcal{NP}$. Now $B \leq_\gamma A \Leftrightarrow \overline{B} \leq_\gamma \overline{A}$, so by following the above analysis, we see that $\overline{B} \in \mathcal{NP}$ too.

Thus $B \in \mathcal{NP} \cap$ co-$\mathcal{NP} = \mathbf{0}_\gamma$. □

COROLLARY 9. *If A and $\overline{A}$ form a minimal pair, then for any B such that $\mathbf{0}_\gamma <_\gamma B \leq_\gamma A$, we have that B and $\overline{B}$ form a minimal pair in the γ-degrees. Thus, if A is super sparse II and $B \leq_\gamma A$, then B and $\overline{B}$ form a minimal pair.*

PROOF. Let A and $\overline{A}$ be minimal and $B \leq_\gamma A$ for some $B \notin \mathcal{NP} \cap$ co-$\mathcal{NP}$.

Suppose $C \leq_\gamma B$ and $C \leq_\gamma \overline{B}$. Then as $\leq_\gamma$ is transitive, we have that $C \leq_\gamma A$ and $C \leq_\gamma \overline{A}$. Since A and $\overline{A}$ are minimal, then $C \in \mathbf{0}_\gamma$.

Thus $deg_\gamma(B)$ and $deg_\gamma(\overline{B})$ are a minimal pair. □

Finally, we show that $\mathcal{D}_\gamma$ is not a lattice. This is achieved by showing that exact pairs of γ-degrees exist. Corollaries to this result will also show that every non-zero γ-degree is one half of a minimal pair and one half of an exact pair.

THEOREM 10. *Let C be a recursively presentable class of recursive sets and let B be a recursive set. Then there is a recursive set A such that:*

i) $A \notin C$

ii) $(\forall n)[\ A^{(n)} =^* B^{(n)}\ ]$

iii) $(\forall D)[[\ D \leq_\gamma A\ \&\ D \leq_\gamma B\ ] \Rightarrow (\exists n)[\ D \leq_\gamma B^{(\leq n)}\ ]]$

PROOF. For any set X and number n, we define $X|n = \{x \mid |x| \leq n \ \& \ x \in X\}$. Furthermore, for the purposes of clarity, we will use $\mathbb{N}$ to denote $\{0,1\}^*$, so that we have:

$$\mathbb{N}^{(\leq n)} = \{0,1\}^{*(\leq n)} = \{\langle m, x\rangle \mid m \leq n\}$$

Given C and B, we construct A in stages so that, at stage $s+1$ of the construction, we only add strings of length s to A. Thus, by the end of stage $s+1$ we will have completely determined $A|s$. The construction will be effective, thus ensuring that A is recursive.

In order to satisfy clause i), we will attempt to satisfy, for all e, the requirements:

$$R_{2e} : A \neq U^{(e)}$$

where U is a universal set for $\mathcal{C}$. This will ensure that $A \notin \mathcal{C}$.

We assume a recursive enumeration, $\{M_i, N_i\}_i$, of pairs of non-deterministic polynomial time bounded Turing transducers where M_i and N_i are time bounded by the polynomial $p_i(n)$. Now, if we have some set C such that $C \leq_\gamma A$ via M_i and $C \leq_\gamma B$ via N_i then $C = M_i(A) = N_i(B)$. Thus, in order to satisfy clause iii), it is sufficient to ensure that we satisfy, for all e, the requirements:

$$R_{2e+1} : M_e(A) = N_e(B) \ \Rightarrow \ N_e(B) \leq_\gamma B^{(\leq e)}$$

We will ensure that the construction used to meet these two requirements also satisfies clause ii).

Strategy.

We say that requirement R_n has a *higher priority* than requirement R_m if $n < m$. The action taken at stage $s+1$ of the construction will be designed to satisfy the highest priority requirement, R_n for $n \leq 2s+1$, which has not yet been satisfied at some previous stage and which can be satisfied by appropriately determining membership of A for strings of length s.

We say that R_{2e} is satisfied at stage $s+1$ if $A|s \neq U^{(e)}|s$. Recall that $A|s$ is completely determined by the end of stage $s+1$. If R_{2e} is satisfied for every e, then clause i) will be satisfied.

If R_{2e} is not satisfied at stage $s+1$, then we can satisfy it by setting $A(x) = 1 - U^{(e)}(x)$ for some string x of length s. Clearly, if R_{2e} is satisfied at stage $s+1$, then it will remain so at all later stages.

Note that clause ii) requires that $A^{(n)} =^* B^{(n)}$, so we must ensure that our action to satisfy R_{2e} does not cause this to fail. We do this by insisting that any string x of length s that is added to A to satisfy R_{2e} does not come from $\mathbb{N}^{(\leq e)}$. Then only finitely many strings z can be added to $A^{(n)}$ to make $A^{(n)} \neq B^{(n)}$. This will ensure that clause ii) is satisfied.

We say that R_{2e} *requires attention at stage* $s+1$ if $A|s = U^{(e)}|s$ (so R_{2e} is not yet satisfied) and there is some string $x \notin \mathbb{N}^{(\leq e)}$ such that $|x| = s$. In this case we will be able to satisfy R_{2e} by setting $A(x) = 1 - U^{(e)}(x)$.

In order to satisfy R_{2e+1}, we will try to construct A such that $M_e(A) \neq N_e(B)$. We will show that if we fail to do this then $N_e(B) \leq_\gamma B^{(\leq e)}$ and so the requirement will be satisfied anyway.

If $(\exists x)[\ |M_e(x)| < s\ \&\ A(M_e(x)) \neq B(N_e(x))\ ]$ then R_{2e+1} is already satisfied. Otherwise we say that R_{2e+1} *requires attention via x at stage $s+1$* if $(\exists x)[\ |M_e(x)| = s\ \&\ M_e(x) \notin \mathsf{N}^{(\leq e)}\ ]$.

The problem with this is that to check the above statement requires performing an unbounded search for an appropriate string x. This will cause the construction to be ineffective making A non-recursive. We can avoid this problem by bounding the search by some time bound, $t(n)$, for A. In this case then R_{2e+1} will require attention via x at stage $s+1$ if

$$(\exists x)[\ |x| \leq t(s)\ \&\ |M_e(x)| = s\ \&\ M_e(x) \notin \mathsf{N}^{(\leq e)}\ ] \qquad [*]$$

If this holds then, for all strings y of length s, we can set:

$$A(y) = \begin{cases} B(y) & \text{if } y \neq M_e(x) \\ 1 - B(N_e(x)) & \text{if } y = M_e(x) \end{cases}$$

Then $A(M_e(x)) \neq B(N_e(x))$ so we have $M_e(A) \neq N_e(B)$. Thus R_{2e+1} is satisfied.

We consider two possible cases:

1) $M_e(A) \neq N_e(B)$.
In this case R_{2e+1} will eventually require attention as we will find some x to witness this difference. Since there are only finitely many requirements of a higher priority and we will satisfy one at every stage, then R_{2e+1} will eventually be satisfied. Clearly, once satisfied R_{2e+1} will never again require attention and will remain satisfied forever.
2) Now there are two possible reasons why R_{2e+1} may not require attention at stage $s+1$. Firstly, R_{2e+1} may already be satisfied, in which case the above holds, or secondly there is no x to act as witness. In this case the search will be unsuccessful at every stage, and so there is some stage s_0 beyond which R_{2e+1} will never require attention.

Thus $\quad (\forall s \geq s_0)\neg(\exists x)[\ |x| \leq t(s)\ \&\ |M_e(x)| = s\ \&\ M_e(x) \notin \mathsf{N}^{(\leq e)}\ ]$

i.e., $(\forall s \geq s_0)(\forall x)[\ |x| \leq t(s) \Rightarrow [\ |M_e(x)| \neq s\ \vee\ M_e(x) \in \mathsf{N}^{(\leq e)}\ ]]$.

So, if this is true, then we need to show that there is a non-deterministic polynomial time bounded Turing transducer T such that $M_e(A) \leq_\gamma B^{(\leq e)}$ via T. Pick some $y_0 \in B^{(\leq e)}$ and some $y_1 \notin B^{(\leq e)}$. We construct T as follows; on input x, compute $M_e(x)$. We consider three cases:

i) $|M_e(x)| < s_0$.
Now, $A|s_0$ is finite and fixed by this stage, so information about it can be stored on one of T's worktapes and can be queried in polynomial time. Thus $A(M_e(x))$ can be computed in polynomial time. If $M_e(x) \in A$ then T outputs y_0, and if $M_e(x) \notin A$ then T outputs y_1.
Then $x \in M_e(A) \Leftrightarrow T(x) = y_0$
$\qquad \Leftrightarrow T(x) \in B^{(\leq e)}$.

ii) $|M_e(x)| \geq s_0$ and $M_e(x) \in \mathsf{N}^{(\leq e)}$.
We are assuming that $(\forall n)[\ A^{(n)} =^* B^{(n)}\ ]$, so by careful choice of s_0 we can ensure that for strings of length $\geq s_0$ we have $A^{(\leq e)} = B^{(\leq e)}$.

Now, $A^{(\leq e)}(M_e(x)) = B^{(\leq e)}(M_e(x))$, so we will have $A(M_e(x)) = B^{(\leq e)}(M_e(x))$. Thus, in this case, T outputs $M_e(x)$.
Then $x \in M_e(A) \Leftrightarrow T(x) = M_e(x)$
$\Leftrightarrow T(x) \in B^{(\leq e)}$.

iii) $|M_e(x)| \geq s_0$ and $M_e(x) \notin \mathbb{N}^{(\leq e)}$.
Since R_{2e+1} does not require attention at stage $s = |M_e(x)|$ then we have from [*] that
$(\forall y)[\ |M_e(y)| = s \Rightarrow [\ M_e(y) \in \mathbb{N}^{(\leq e)} \ \vee\ |y| > t(s)\ ]]$
In particular, since $M_e(x) \notin \mathbb{N}^{(\leq e)}$ and $|M_e(x)| = s$ we have $|x| > t(s) = t(|M_e(x)|)$. Now, $M_e(x)$ can be computed in less than $p_e(|x|)$ steps, and $A(M_e(x))$ can be computed in less than $t(|M_e(x)|) = t(s) \leq |x|$ steps, so $A(M_e(x))$ can be computed in non-deterministic polynomial time. Now, if $M_e(x) \in A$ then T outputs y_0 and if $M_e(x) \notin A$ then T outputs y_1.
Then $x \in M_e(A) \Leftrightarrow T(x) = y_0$
$\Leftrightarrow T(x) \in B^{(\leq e)}$

It is clear that, on input x, T always outputs some string y, so:

$$(\forall x)(\exists y)[\ \langle x, y\rangle \in G(T)\ ]$$

Furthermore, we have shown that:

$$(\forall x)(\forall y)[\ \langle x, y\rangle \in G(T) \Rightarrow [\ x \in M_e(A) \Leftrightarrow y \in B^{(\leq e)}\ ]]$$

Finally, since T was constructed to run in non-deterministic polynomial time, then $N_e(B) = M_e(A) \leq_\gamma B^{(\leq e)}$ via T as required.

Thus R_{2e+1} is satisfied in both cases.

We will construct the time bounds for A as we proceed through the construction. At any given stage, let the variable *count* record the *total* number of steps made during the whole construction so far. The time bounds will be defined in terms of *count*, where $t_e(s)$ will bound R_{2e+1} in its search at stage s. Then t, the final time bound for A will be defined by $t(n) = \Sigma_{e=0}^{2n+1} t_e(n)$.

The Construction.
Do nothing at stage 0.
Stage $s+1$.

1) For all $e \geq s+1$ let $t_e(s) :=$ *count*.
2) **for** $n := s$ **downto** 0
begin
We need to ascertain whether or not R_n requires attention at this stage and if so, what action it requires. Note that we do not perform this action yet, we are merely checking which requirements require attention. In the step 3) we will go back and take action to satisfy the requirement of highest priority that was found in this step.
    i) $n = 2e$.
    If R_n is not already satisfied then we note that R_n requires attention at stage $s+1$ and that it wants us to set, for all strings y of length s:

$$A(y) = \begin{cases} B(y) & \text{if } y \in \mathbb{N}^{(\leq e)} \\ 1 - U^{(e)}(y) & \text{if } y \notin \mathbb{N}^{(\leq e)} \end{cases}$$

Let $t_n(s) :=$ *count.*

ii) $n = 2e + 1$.

Check if R_n is already satisfied, i.e.:

$(\exists x)[\ |x| \leq t_{2e+2}(s-1)\ \&\ |M_e(x)| < s\ \&\ A(M_e(x)) \neq B(N_e(x))\ ]$

if not then check if

$(\exists x)[\ |x| \leq t_{2e+2}(s)\ \&\ |M_e(x)| = s\ \&\ M_e(x) \notin \mathbb{N}^{(\leq e)}\ ]$

If so, then note that R_n requires attention at stage $s + 1$ and that it wants us to set, for the least string y that satisfies the above :-

$$A(y) = \begin{cases} B(y) & \text{if } y \neq M_e(x) \\ 1 - B(N_e(x)) & \text{if } y = M_e(x) \end{cases}$$

Let $t_n(s) :=$ *count.*

end

3) Act on the requirement of highest priority (assuming that there is one) as decided in step 2).

End of stage $s + 1$.

It is clear that if R_n is satisfied at stage s then it remains satisfied at all later stages. Also, at every stage we satisfy the requirement of highest priority that requires attention (assuming that there is one). Thus, every requirement that requires attention will eventually be satisfied as there are only a finite number of requirements of higher priority that need to be satisfied first. It is clear from the construction that A is built so as to meet R_n for all n. Thus the construction works. □

COROLLARY 11. *Let* **b** *and* **c** *be recursive* γ*-degrees such that* $\mathbf{c} < \mathbf{b}$. *Then there exists a recursive* γ*-degree* $\mathbf{a} > \mathbf{c}$ *such that* **c** *is the infimum of* **a** *and* **b**.

Thus, by setting $\mathbf{c} = \mathbf{0}_\gamma$ we have that every non-zero recursive γ-degree is half of a minimal pair.

PROOF. Pick some $\hat{B} \in \mathbf{b}$ and $C \in \mathbf{c}$. Define B by

$$B(\langle n, x \rangle) = \begin{cases} \hat{B}(x) & \text{if } |x| \leq n \\ C(x) & \text{if } |x| > n \end{cases}$$

Clearly B is recursive as "$|x| \leq n$" is a recursive test and $\hat{B}$ and C are recursive sets.

Trivially we have that $B \leq^p_m \hat{B} \oplus C$ via $\Theta(\langle n, x \rangle) = \begin{cases} 2x & \text{if } |x| \leq n \\ 2x+1 & \text{if } |x| > n \end{cases}$

and $\hat{B} \oplus C \leq^p_m B$ via $\Theta(x) = \begin{cases} \langle |m|, m \rangle & \text{if } x = 2m \\ \langle 0, m \rangle & \text{if } x = 2m+1 \end{cases}$

Thus $B \equiv_\gamma \hat{B} \oplus C$ and so $B \in \mathbf{b}$. Furthermore $(\forall n)[\ B^{(n)} =^* C\ ]$ and so we have $(\forall n)[\ B^{(\leq n)} \equiv_\gamma B^{(n)} \equiv_\gamma C\ ]$.

Now, set $\mathcal{C} = \{D \mid D \leq_\gamma C\}$ and apply Theorem 10 with this B. This gives a recursive set A. By the theorem $(\forall n)[\ A^{(n)} =^* B^{(n)}\ ]$ so $(\forall n)[\ C \equiv_\gamma B^{(n)} \equiv_\gamma A^{(n)} \leq_\gamma A\ ]$. Furthermore, $A \notin \mathcal{C}$ by the theorem so $C <_\gamma A$.

Finally let $D \leq_\gamma A$ and $D \leq_\gamma B$. By the theorem $(\exists n)[\ D \leq_\gamma B^{(\leq n)}\ ]$ so $D \leq_\gamma C$ as $(\forall n)[\ C \equiv_\gamma B^{(\leq n)}\ ]$. Thus $\mathbf{a} = deg_\gamma(A)$ has the desired properties. □

DEFINITION 12.

i) A sequence, $\{\mathbf{c}_i\}_{i \in \mathbf{N}}$, of γ-degrees is *ascending* if $(\forall n)[\ \mathbf{c}_n \leq \mathbf{c}_{n+1}\ ]$ and $(\forall n)(\exists m)[\ \mathbf{c}_n < \mathbf{c}_m\ ]$.

ii) A sequence, $\{\mathbf{c}_i\}_{i \in \mathbf{N}}$, of γ-degrees is *recursive* if there is a recursive set, C, such that $\mathbf{c}_n = deg_\gamma(C^{(n)})$.

iii) γ-degrees $\mathbf{a}$ and $\mathbf{b}$ are an *exact pair* for $\{\mathbf{c}_i\}_{i \in \mathbf{N}}$ if $(\forall n)[\ \mathbf{c}_n \leq \mathbf{b}, \mathbf{c}\ ]$ and $(\forall \mathbf{d})[\ \mathbf{d} \leq \mathbf{b}, \mathbf{c} \Rightarrow (\exists n)[\ \mathbf{d} \leq \mathbf{c}_n\ ]]$.

It is clear from the Density Theorem that recursive ascending sequences of γ-degrees exist between any two comparable recursive γ-degrees. The next lemma will enable us to show that any upper bound for a recursive ascending sequence of γ-degrees is half of an exact pair for that sequence.

LEMMA 13 (Ambos-Spies [2]). *Let* $\mathbf{a}$ *and* $\mathbf{b}$ *be recursive* γ*-degrees. Then the following are equivalent:*

i) $\mathbf{a}$ *and* $\mathbf{b}$ *have no infimum.*

ii) There is an ascending sequence of γ*-degrees for which* $\mathbf{a},\mathbf{b}$ *is an exact pair.*

iii) There is a recursive ascending sequence of γ*-degrees for which* $\mathbf{a},\mathbf{b}$ *is an exact pair.*

PROOF. Clearly iii)⇒ii)⇒i), so it remains to show that i)⇒iii).

Assume $\mathbf{a}$ and $\mathbf{b}$ are recursive γ-degrees without infimum and let $\mathcal{C}$ be defined as $\mathcal{C} = \{D \mid deg_\gamma(D) \leq \mathbf{a}\ \&\ deg_\gamma(D) \leq \mathbf{b}\}$. It is obvious that $\mathcal{C}$ is recursively presentable, so let U be a universal set for $\mathcal{C}$ and let $C_n = U^{(\leq n)}$ and $\mathbf{c}_n = deg_\gamma(C_n)$. The sequence $\{\mathbf{c}_i\}_{i \in \mathbf{N}}$ is obviously recursive and $(\forall n)[\ \mathbf{c}_n \leq \mathbf{c}_{n+1} \leq \mathbf{a}, \mathbf{b}\ ]$.

Furthermore, for any $\mathbf{d} \in C = \{\mathbf{c} \mid \mathbf{c} \leq \mathbf{a}\ \&\ \mathbf{c} \leq \mathbf{b}\}$ we have $\mathbf{d} \leq \mathbf{c}_n$ for some n. Thus $\mathbf{a}$ and $\mathbf{b}$ are an exact pair for $\{\mathbf{c}_i\}_{i \in \mathbf{N}}$, and since $\mathbf{a}$ and $\mathbf{b}$ have no infimum then C has no greatest element. Therefore, the sequence $\{\mathbf{c}_i\}_{i \in \mathbf{N}}$ is ascending. □

LEMMA 14. *Let* $\{\mathbf{c}_i\}_{i \in \mathbf{N}}$ *be a recursive ascending sequence of* γ*-degrees such that* $(\forall n)[\ \mathbf{c}_n \leq \mathbf{b}\ ]$ *for some recursive* γ*-degree* $\mathbf{b}$*. Then there exists a recursive* γ*-degree* $\mathbf{a}$ *such that* $\mathbf{a}$ *and* $\mathbf{b}$ *are an exact pair for* $\{\mathbf{c}_i\}_{i \in \mathbf{N}}$.

Thus, every non-zero recursive γ-degree is half of an exact pair.

PROOF. Pick recursive sets C and $\hat{B}$ such that $(\forall n)[\ C^{(n)} \in \mathbf{c}_n\ ]$ and $\hat{B} \in \mathbf{b}$. Define B by

$$B(\langle n, x\rangle) = \begin{cases} \hat{B}(x) & \text{if } |x| \leq n \\ C^{(n)}(x) & \text{if } |x| > n \end{cases}$$

Clearly B is recursive as "$|x| \leq n$" is a recursive test and $\hat{B}$ and C are recursive sets. Since, by definition, $B^{(\leq n)} =^* C_n$, we have $B^{(\leq n)} \in \mathbf{c}_n$. Furthermore, $\hat{B} \leq_m^{\mathcal{P}} B$ via $\Theta(x) = \langle |x|, x \rangle$, so we have that $\hat{B} \leq_\gamma B$.

Now apply Theorem 10 with $C = \emptyset$. This gives a recursive set A such that $(\forall n)[\ A^{(n)} =^* B^{(n)}\ ]$. Thus $A^{(\leq n)} =^* B^{(\leq n)}$ so $A^{(\leq n)} \in \mathbf{c}_n$ for all n. Also, $(\forall D)[[\ D \leq_\gamma A\ \&\ D \leq_\gamma B\ ] \Rightarrow (\exists n)[\ D \leq_\gamma B^{(\leq n)}\ ]]$, so for any recursive γ-degree $\mathbf{d} \leq \mathbf{a}, \mathbf{b}$ we have $\mathbf{d} \leq \mathbf{c}_n$ for some n.

Thus $\mathbf{a}$ and $deg_\gamma(B)$ are an exact pair for $\{\mathbf{c}_i\}_{i \in \mathbf{N}}$, and since $\mathbf{b} \leq deg_\gamma(B)$ then $\mathbf{a}$ and $\mathbf{b}$ are also an exact pair for $\{\mathbf{c}_i\}_{i \in \mathbf{N}}$. □

It is clear from the above that every recursive ascending sequence of γ-degrees possesses an exact pair and that, given recursive γ-degrees $\mathbf{c} < \mathbf{b}$, there is some recursive γ-degree $\mathbf{a} > \mathbf{c}$ such that $\mathbf{a}$ and $\mathbf{b}$ have no infimum. Thus $\langle \mathcal{D}_\gamma, \leq_\gamma \rangle$ is not a lattice.

Remarks. There has been much debate, in the unbounded case, about the question of $\leq_r$-degrees within some $\leq_R$-degree, where $\leq_r \Rightarrow \leq_R$. In particular, Zakharov [10] proved that any non-zero $\leq_e$-degree contains at least two $\leq_s$-degrees. He also showed that there are infinitely many $\leq_m$-degrees within any Σ_2 $\leq_e$-degree. This was followed by Watson [8] who showed that any $\leq_e$-degree that is Δ_2 or Σ_2-high contains an infinite number of $\leq_s$-degrees. Finally, we know from Copestake [3] that all 1-generic $\leq_e$-degrees contain infinitely many $\leq_s$-degrees. However, she also showed that there exists a Σ_2 1-generic $\leq_e$-degree that is not in Δ_2, so it is still an open question as to whether there exists a non-zero $\leq_e$-degree that only contains finitely many $\leq_s$-degrees.

It is clear from the Density Theorem that any recursive $\leq_c^{\mathcal{NP}}$-degree that contains at least two distinct $\leq_\gamma$-degrees actually contains infinitely many $\leq_\gamma$-degrees. Since $\leq_c^{\mathcal{NP}}$ is the polynomial time bounded version of $\leq_e$, then this result is clearly of some interest. However, none of the known proofs of Zakharov's theorem (e.g., [10] and [8,9]) carry over to the polynomial time bounded case. We next show that the analogue of this result is actually false and that there are non-zero $\leq_c^{\mathcal{NP}}$-degrees that contain exactly one γ-degree. Thus any $\leq_c^{\mathcal{NP}}$-degree contains either exactly one or infinitely many γ-degrees. This gives an interesting contrast to the unbounded case. It should be noted that the reason that this result works is due to the fact that we can construct sets such that any reduction to them requires only one oracle question. This depends upon the fact that we are working with time bounded reducibilities as this result would obviously contradict Zakharov's theorem if it carried over to the unbounded case. Finally, we note that since $\leq_\gamma \Rightarrow \leq_s^{\mathcal{P}}$ then there are non-zero $\leq_c^{\mathcal{NP}}$-degrees that contain exactly one $\leq_s^{\mathcal{P}}$-degree.

THEOREM 15. *$(\exists A \in EXPTIME)[deg_c^{\mathcal{NP}}(A)$ contains exactly one γ-degree.]*

PROOF. Let A be the super sparse II set constructed in Lemma 3. Recall that A is super sparse II via the function f defined by:

$$f(0) = 1 \quad \text{and} \quad f(n+1) = 2^{f(n)}$$

and that $A \in$ EXPTIME.

We claim (following [2]) that in any oracle reduction to A, there is only one relevant oracle question, all of the others being redundant.

Recall that $A \subseteq Z_f = \{0^{f(n)} \mid n \in \mathbb{N}\}$ and let $B \leq_r A$ for some (non-deterministic) polynomial time bounded reduction $\leq_r$. We assume that the reduction is bounded by the polynomial $p(n)$. Now, to compute $B(x)$ we will perform some (non-deterministic) computation on x and ask at most $p(|x|)$ oracle questions of A. Clearly, since the reduction is bounded by p then the largest oracle question that can be asked is of length $p(|x|)$, and since we can check in polynomial time whether or not some y is in Z_f then the largest *relevant* oracle question is "$0^{f(n)} \in A$?", where $n = \max\{m \mid f(m) \leq p(|x|)\}$.

Now, given n, to compute $A(0^{f(n)})$ takes less than $2^{|0^{f(n)}|}$ steps, ie less than $2^{f(n)} = f(n+1)$ steps. To compute $A(0^{f(m)})$ for any $m < n$ takes $\leq f(m+1) \leq f(n) \leq p(|x|)$ steps. Thus all oracle questions with the exception of "$0^{f(n)} \in A$?" can be answered in polynomial time.

In other words, the only relevant oracle question is "$0^{f(n)} \in A$?", and so $B \leq_R A$ where $\leq_R$ is a polynomial time bounded reduction procedure that only allows one oracle question.

Thus, given $B \leq_c^{\mathcal{NP}} A$, we can replace the non-deterministic polynomial time bounded tt-condition generator by some other generator, g, that outputs 1-tt conditions, and we can replace the conjunctive evaluator by one, e, defined on $\alpha c\{0,1\}^*$ such that $B \leq_c^{\mathcal{NP}} A$ via g and e.

Now to check "$x \in B$?", we compute $g(x) = \alpha cy$ and then we know that since we are dealing with a conjunctive reducibility, then:

$$\begin{aligned} x \in B &\Leftrightarrow e(\alpha c C_A(y)) = 1 \\ &\Leftrightarrow C_A(y) = 1 \\ &\Leftrightarrow y \in A \end{aligned}$$

Now, define a non-deterministic polynomial time bounded Turing transducer, M, such that on input x, M first computes $g(x) = \alpha cy$ and then outputs y. Immediately we have that:

$$(\forall x)(\exists y)[\ \langle x, y\rangle \in G(M)\ ]$$

and

$$(\forall x)(\forall y)[\ \langle x, y\rangle \in G(M) \Rightarrow [x \in B \Leftrightarrow y \in A]]$$

Thus $B \leq_\gamma A$ via M. We have shown that $(\exists A \in \text{EXPTIME})(\forall B)[\ B \leq_c^{\mathcal{NP}} A \ \Rightarrow\ B \leq_\gamma A\ ]$. We now show the complement to this result. Assume that $A \leq_c^{\mathcal{NP}} B$ via a reduction bounded by polynomial q and $B \leq_\gamma A$. This latter fact gives that $B \leq_{1\text{-}tt}^{\mathcal{NP}} A$, so using the well-known characteristic of $\leq_{1\text{-}tt}^{\mathcal{P}/\mathcal{NP}}$ (see [2] for example), we know that there exist non-deterministic polynomial time bounded functions $g : \Sigma^* \mapsto \Sigma^*$ and $h : \Sigma^* \times \{0,1\} \mapsto \{0,1\}$ and a polynomial, p, such that

$$(\forall x)[\ B(x) = h(x, A(g(x)))\ ] \quad \& \quad |g(x)| \leq p(|x|)$$

Define $mf(x) = max\{n \mid f(n) \leq p(q(|x|))\}$ and say that a string y (where $|y| \leq q(|x|)$) is *x-relevant* if $g(y) = 0^{f(mf(x))}$ and $h(y,0) \neq h(y,1)$.

If y is not x-relevant then "$y \in B$?" can be answered in polynomial time as follows:

i) If $h(y,0) = h(y,1)$ then $B(y) = h(y,0)$.

ii) If $g(y) \notin \{0^{f(0)}, \ldots, 0^{f(mf(x))}\}$ then $g(y) \notin A$ so $B(y) = h(y,0)$.

iii) If $g(y) \in \{0^{f(0)}, \ldots, 0^{f(mf(x)-1)}\}$ then $B(y) = h(y, A(g(y)))$ where $A(g(y))$ can be computed in $\leq p(q(|x|))$ steps. This fact follows since A is super sparse II, so $A(0^{f(m)})$ can be computed in $\leq f(m+1)$ steps. Thus $A(g(y))$ can be computed in $\leq f(mf(x))$ steps. By definition of $mf(x)$ this is $\leq p(q(|x|))$ steps.

So, assuming that we have strings y and y', both x-relevant oracle questions, we note that $B(y) = h(y, A(0^{f(mf(x))}))$, $h(y,0) \neq h(y,1)$ and $B(y') = h(y', A(0^{f(mf(x))}))$.

Clearly, if we are given $B(y)$, then we can compute

$$A(0^{f(mf(x))}) = \begin{cases} 0 & \text{if } h(y,0) = B(y) \\ 1 & \text{if } h(y,1) = B(y) \end{cases}$$

in non-deterministic polynomial time, and consequently we can compute $B(y') = h(y', A(0^{f(mf(x))}))$ in non-deterministic polynomial time.

Thus, given the answer to any one oracle question, we can efficiently compute the answer to all others. Thus $A \leq_{1\text{-}tt}^{\mathcal{NP}} B$ and so we can find non-deterministic polynomial time bounded functions g_1 and h_1 such that $A(x) = h_1(x, B(g_1(x)))$, and a machine M, bounded by polynomial p_1, such that $B \leq_\gamma A$ via M.

Thus, for all x, $B(x) = A(M(x))$, so $A(x) = h_1(x, A(M(g_1(x))))$. [*]

Define $mf'(x) = max\{n \mid f(n) \leq p_1(|x|)\}$ and fix $x_0 \notin B$ and $x_1 \in B$. Construct a non-deterministic machine as follows; on input x:

i) If $h_1(x,0) = h_1(x,1)$ then output $x_{h(x,0)}$ and halt.

ii) If $M(g_1(x)) \neq 0^{f(mf'(x))}$ then
  a) if $h_1(x,i) = i$ output $x_{A(M(g_1(x)))}$ and halt;
  b) if $h_1(x,i) = 1-i$ output $x_{1-A(M(g_1(x)))}$ and halt.

iii) If $M(g_1(x)) = 0^{f(mf'(x))}$ then
  a) if $x \neq M(g_1(x))$ then output $x_{A(x)}$ and halt;
  b) if $x = M(g_1(x))$ then output $g_1(x)$ and halt.

Note that in step iii) b), we have the case where $x = M(g_1(x))$, so by [*] we have that $h_1(x,i) = i$ and so $A(x) = B(g_1(x))$. The machine clearly runs in non-deterministic polynomial time as all steps are polynomially bounded (note that $A(M(g_1(x)))$ and $A(x)$ can be computed in $\leq p_1(|x|)$ steps as A is super sparse II), and from [*] it follows immediately that $A \leq_\gamma B$ via this machine.

Thus $(\forall B)[\ A \leq_c^{\mathcal{NP}} B \ \Rightarrow\ A \leq_\gamma B\ ]$ and so $deg_\gamma(A) = deg_c^{\mathcal{NP}}(A)$ as required. □

REFERENCES

[1] LEONARD ADLEMAN and K. MANDERS, *Reducibility, randomness and intractibility*, ***Proc. 9th Annual ACM Symposium on the Theory of Computing***, 1977, pp. 151–163.

[2] KLAUS AMBOS-SPIES, *On the structure of the polynomial time degrees of recursive sets*, Habilitationsschrift, Forschungsbericht Nr. 206, Universität Dortmund, 1985.

[3] KATE COPESTAKE, *Enumeration degrees of Σ_2 sets*, Ph.D. thesis, Leeds University, 1987.

[4] RICHARD LADNER, NANCY LYNCH and ALAN L. SELMAN, *A comparison of polynomial time reducibilities*, ***Theoretical Computer Science***, **1** (1975), pp. 103–123.

[5] TIMOTHY J. LONG, *On γ-reducibility versus polynomial time many-one reducibility*, ***Theoretical Computer Science***, **14** (1981), pp. 91–101.

[6] TIMOTHY J. LONG, *Strong non-deterministic polynomial-time reducibilities*, ***Theoretical Computer Science***, **21** (1982), pp. 1–25.

[7] ALAN P. SILVER, *On polynomial time bounded reducibilities*, Ph.D. thesis, Leeds University, May 1991.

[8] PHIL WATSON, *Concerning D-recursively enumerable sets and s-reducibility of Σ_2 Sets*, Ph.D. thesis, Leeds University, 1988.

[9] PHIL WATSON, *On restricted forms of enumeration reducibility*, ***Annals of Pure and Applied Logic***, **49** (1990), pp. 75–96.

[10] S. D. ZAKHAROV, *e- and s-degrees*, ***Algebra and Logic***, **23** (1984), pp. 273–281.

Printed in the United States
By Bookmasters